Advances in Natural and Technological Hazards Research

Volume 48

The book series entitled *Advances in Natural and Technological Hazards* is dedicated to serving the growing community of scholars, practitioners and policy makers concerned with the different scientific, socio-economic and political aspects of natural and technological hazards. The series aims to provide rapid, refereed publications of topical contributions about recent advances in natural and technological hazards research. Each volume is a thorough treatment of a specific topic of importance for proper management and mitigation practices and will shed light on the fundamental and applied aspects of natural and technological hazards. Comments or suggestions for future volumes are welcomed.

More information about this series at http://www.springer.com/series/6362

Hamid Reza Pourghasemi
Mauro Rossi
Editors

Natural Hazards GIS-Based Spatial Modeling Using Data Mining Techniques

Springer

Editors
Hamid Reza Pourghasemi
Natural Resources and Environmental Engineering
Shiraz University
Shiraz, Iran

Mauro Rossi
IRPI
National Research Council
Perugia, Italy

ISSN 1878-9897 ISSN 2213-6959 (electronic)
Advances in Natural and Technological Hazards Research
ISBN 978-3-319-73382-1 ISBN 978-3-319-73383-8 (eBook)
https://doi.org/10.1007/978-3-319-73383-8

Library of Congress Control Number: 2018957631

This Springer imprint is published by the registered company Springer Nature Switzerland AG
The registered company address is: Gewerbestrasse 11, 6330 Cham, Switzerland

Foreword by Leonardo Cascini

Nowadays, the society hardly accepts loss of lives and damages to properties caused by natural hazards and require a safer, reliable and resilient community much more than in the past. An effective way to face with this problem is a proper *risk management strategy* that is usually developed through several steps. The first one is *risk analysis* that calls for the evaluation of factors generating the risk, namely the hazards and related consequences. Generally speaking, *hazard analysis* requires the estimation of the spatial and temporal probability of an adverse event in a given area. The *consequence analysis* involves identifying and quantifying the elements at risk (i.e. people and properties) and estimating their vulnerability.

Focusing on the hazard analysis, several methods are currently available depending on the size of the study area, the available dataset, the time given to technicians to estimate the risk and other relevant factors (e.g. financial resources). The scientific literature describes steps and procedures for the applicability of these methods over either small or large areas, suggesting for all of them the proper individuation of the area affected by the adverse existing or potential event. To this aim, the availability of a good inventory of the events occurred in the past is a requirement of particular concern.

In this regard, difficulties in gathering information dramatically increase as the size of the area to be investigated increases or even when Authorities, in charge of the land management, prompt the technicians to provide inventory and hazard zoning maps in a short time and on the basis of the available dataset only. In such a case, the main questions for technicians are what are the chances to provide consistent answers to so complex and demanding questions? And more in general, which are the most suitable methods that, in case of data availability, may be used to provide quantitative and as objective as possible analysis over large areas?

The book *Natural Hazards GIS-Based Spatial Modeling Using Data Mining Techniques* provides consistent answers to these questions, and more in general, it applies as a reference point for the specialists involved in such challenging tasks. The well-known expertise of the authors allows the readers to go in-depth to all the available procedures and suggests them new approaches that will be diffusely adopted in a near future to real case studies, provided that GIS-based spatial models are properly calibrated and validated. Moreover, the book covers a number of hazards that usually are analysed separately, thus providing a comprehensive overview of potentialities and limits of the proposed approaches.

These represent real strong points of the text and a guarantee of quality for all people, directly or indirectly, involved in the risk management process since it allows verifying if the methods adopted by technicians are based on well-known, consistent and advanced procedures, independently from the specific field of application. Indeed, I wish all the best to the authors, the editors and the publisher having in mind that the entire new proposal to succeed must have many of the components of the book that I had the pleasure of presenting.

Fisciano, Italy

Leonardo Cascini
Director of LARAM School
Department of Civil Engineering
University of Salerno

Foreword by Manfred F. Buchroithner

Having grown up in the mountains of the Eastern Alps, I was exposed to natural hazards and disasters from my early childhood on. I witnessed snow avalanches, mud flows, floodings and inundations, landslides, heavy (gully) erosion, rockfalls, forest fires as well as heavy ice- and hailstorms which frequently lead to windthrow and deadfall. All these experiences influenced my later research activities and my consciousness of their importance.

Later, in the 1980s and 1990s, data mining slowly began to penetrate the geosciences, I was introduced to this promising field by colleagues from informatics, both from East and West. Thus, in my function as Vice Chairman of the European Association of Remote Sensing Laboratories (EARSeL), I tried to be instrumental in setting up conferences and workshops dealing with this topic. This intention was corroborated by my decades-long research activities in High Asia, in particular in Pamir, Hindu Kush, Tian Shan and the Nepalese Himalaya, as well as in the Andes, and the cumbersome search for data from these regions where I also experienced several earthquakes of noticeable intensity. Fieldwork in desert areas, but also in other regions like the Sahel, made me aware of droughts. By means of GIS-based spatial modelling using data mining techniques all the mentioned geohazards will become predictable with a higher degree of certainty.

Out of the approximately 70 Ph.D. students I had the pleasure to supervise and evaluate so far, some 25–30 dealt with natural disasters resp. hazards. All these doctoral researchers had to base their studies upon their own but also on existing data. Hence, data mining also played an important role for their work, not to mention remote sensing and GIS-based modelling. I would have been happy if I could have referred them to a book like the present one. Therefore, I can—in times of "big data" in the widest sense—only congratulate the two editors Hamid Reza Pourghasemi and Mauro Rossi for their decision and effort to publish a book about *Natural Hazards GIS-Based Spatial Modeling Using Data Mining Techniques*.

In this sense, and based on my firm personal conviction, I am sure that in times of an increasing number of natural events the present book is filling an important gap in the textbooks resp. compendia dealing with natural hazards. May it find its way into many libraries and private bookshelfs, be they analogue or digital.

Dresden, Germany
October 2017

Manfred F. Buchroithner
Professor Emeritus
Senior Professor
TU Dresden

Foreword by Nicola Casagli

The reduction of disaster risks, including those caused by natural hazards, is one of the main problems of the Planet, addressed by most governments and international organizations.

Population growth, deforestation and climate change open new challenges for the safety and the protection of citizens and communities, which require a strong co-operation between scientists and policy makers.

The UNISDR Sendai Framework for Disaster Risk Reduction 2015–2030 considers the "understanding" of risk as the first priority in the global effort for the mitigation of the consequences of disasters.

In particular, it is necessary to better understand the risk of natural hazards in all its components, such as probability of occurrence, magnitude and intensity, vulnerability and exposure of people and assets, and their changes in space and in time.

Such knowledge is the starting point for any action of risk prevention, mitigation, preparedness and response, in order to reduce the environmental impact, to improve the resilience of communities and the safety of the citizens.

This general objective requires international and interdisciplinary research to better understand the Earth system dynamics, the effects of climate change, the population variations, the past and future trends.

In this framework geographic information system and remote sensing play a crucial role, as they provide scientists and policy makers with powerful tools for the rapid acquisition of relevant data, for representing risk factors and their spatial distribution and properties, and for supporting forecasting models and scenario analyses.

Data mining has been strongly developed in recent years in many fields of application, due to the availability of large digital data collections, the increased capacity of big data storage and processing, and the new technology of data harvesting (remote sensing, sensor networks) and analysis (machine learning, pattern recognition).

This book provides an updated review of the scientific state of the art at international level on these topics, focusing on the three main categories of natural disasters: geophysical (landslide and earthquakes), hydrological (floods and erosion) and climatological (subsidence and wildfires).

Several advanced techniques of data mining are presented by the authors, showing how the scientific community has already developed effective tools for extracting relevant information from large and complex datasets.

Florence, Italy

Nicola Casagli
Professor at University of Florence

Foreword by Norman Kerle

Natural hazards have been a challenge throughout human history. People's attitude towards the multitude of threats has been highly variable, from fatalistically accepting hazards to trying to outsmart nature through advanced engineering. In recent decades, it has become increasingly clear that only a comprehensive understanding of the nature of hazards, when properly integrated in a risk assessment context, is the key to successful coexistence. It is clear that naturally occurring processes such as earthquakes, flooding or landsliding cannot be stopped or fundamentally altered in their mechanics, and that attempts to tame nature that became popular in the first half of the twentieth century rarely succeed. Instead, we have seen tremendous progress in understanding risk, resulting in better planning and management tools. Disasters still occur regularly and are as much a reminder of the prevalence of potentially damaging phenomena and processes, as of the fact that the number of human being and their accumulated wealth continues to grow, leading to more frequent and costly encounters with hazardous events. The latter development especially applies to dense urban area, with a rapidly rising number of megacities, of which many are located in highly exposed regions such as flood plains, or close to fault lines or low-lying coastal stretches. The number of disaster events has actually been declining in the last decade, which can be seen as a promising signal that risk management can yield fruits. Also, unusually severe events such as typhoons causing hundreds of thousands of fatalities that marked especially the 1970s have declined in number, being now more limited to less predictable earthquake or tsunami events.

Reducing risk and mitigating consequences can be done through avoiding exposure, i.e. relocating communities or establishing effective early warning and evacuation, or by reducing vulnerability, such as through people better understanding the threat and adequate counter measures, or though the enforcement of better building codes. However, all measures begin with a clear understanding of the hazard, which determines who is affected when, where, how often and with what severity. Like most natural processes, hazards are highly variable in their behaviour, defying easy statistical analysis and posing a challenge to decision makers (a "100-year flood" is a typical example). In addition, the often massive changes to natural systems due to anthropogenic activities have fundamentally altered the nature of many hazards: floods occur more often or more quickly, while deforested mountain slopes fail more readily. Many of these changes have accelerated in the last few years or decades, leaving too little time to understand their effect on nature and the frequency of events. Since all those changes lead to a reconfiguration of older exposure scenarios (through hazard zones now having different shapes), changes in vulnerability (since it is a function of both hazard type and magnitude), risk is again becoming an increasingly uncertain property. Even where complex risk assessments have been done, often based on historical data, hazard events and their associated consequences often substantially depart from expectations.

How then do we best deal with those changing hazard scenarios? Geospatial data and methods have become the best tool chest in disaster risk management. Numerous methods have been developed to observe the environment with a huge array of remote sensing and many other data gathering techniques, while the collected information is processed with sophisticated statistical or modelling tools. With the exception of seismic activity, every natural hazard is well understood in its genesis and relevant parameters, allowing a detailed assessment of a given threat. The most commonly used input in hazard assessment remains the data from past events. Landslide inventories, or the extent of earlier flood events, are used in empirical modelling. There are limits though. With extensive remote sensing databases only stretching back to the 1970, and base data such as on geology or water table depth being coarse or patchy in many parts of the work, the modelling basis is often thin and incomplete. Especially, to assess the susceptibility to a given hazard over larger areas, next to empirical analysis modelling based on physical parameters known to be relevant remains desirable. However, detailed physical modelling quickly reaches its limits, when parameters vary spatially. This natural complexity quickly leads to significant simplifications, with, for example, only standard parameters such as slope, aspect, land cover and rainfall records being used to assess the susceptibility of a slope to failure. More detailed studies that consider all system properties and information from diverse sensor and databases are desirable, but yield new challenges, not least due to vast amounts of data and parameters.

With the growing amounts of geospatial information, data mining has been gaining in relevance, allowing complex multivariate data to be processed, the most explanatory environmental parameters to be identified, and hidden patterns and trends to be found. With growing computing power and better models, this has morphed into advanced machine learning, where sophisticated models can determine the nature also of complex environmental systems based on limited input data. Numerous approaches have been developed, from basis decision trees to complex but less transparent artificial neural networks (ANN), to more advanced approaches based on random forests or convolutional neural networks that can learn complex processes also based on training data from other locations. Despite all this sophistication, however, the number of challenges remains high, precluding a ready use of those recent methods. Problems as diverse as dealing with a small number of training sites (e.g. when landslide inventories are small or events are very infrequent), how to sample from a larger pool or extensive area, how to deal with fuzzy or poorly defined boundaries of relevant natural features, or how to work with point data of features with a polygon shape, still pose problems.

For those reasons, a book that assesses the diverse use of spatial modelling with machine learning in hazard assessment is timely and useful. The present book spans a wide area, addressing hazards as diverse as the susceptibility of landslides or wild fires, erosion and land subsidence, floorplan analysis, as well as earthquake and rainfall prediction. Numerous methods are demonstrated and evaluated in a range of case studies set in eight countries in Europe and Asia. The chapters compare many comparative analysis studies where different machine learning tools are evaluated, but also show how multiple hazards can be assessed in a common spatial modelling framework. Finally, one chapter also shows how an analysis based on spatial segments derived with object-based image analysis (OBIA) can lead to more realistic scenarios. Problems with more opaque methods are critically assessed, such as the black box nature of ANN, or the frequent overfitting resulting from models that is also well known from OBIA. Hence, the book can help researchers to understand the advantages and disadvantages of different machine learning-based spatial modelling techniques. By illuminating problems and showing solutions, it can also be valuable for decision makers who need to identify suitable operational tools to aid in their hazard assessment work.

Enschede, The Netherlands
October 2017

Prof. Dr. Norman Kerle
Professor of Remote Sensing and
Disaster Risk Management
University of Twente

Foreword by Saro Lee

Prof. Dr. Saro Lee is from Korea Institute of Geosciences and Mineral Resources (KIGAM) and Korea University of Science and Technology (UST), Daejeon, Korea. He studied many years on GIS-based spatial modelling in different geological fields. We applied many data mining/machine learning models in diverse natural hazards cases such as landslides, flood, and ground subsidence in these years.

In general, spatial modelling in GIS is known as an important tool in the modern digital world. When the mentioned tools combine to data mining/machine learning techniques, it could serve as a good source of information to widespread sciences community such as students, researchers and academic staffs.

The book *Natural Hazards GIS-Based Spatial Modeling Using Data Mining Techniques* introduces to readers as an ensemble of GIS and RS tools by data mining techniques for spatial modelling on geological, hydrological, and climatological disasters. It will be able to solve limitation of traditional and statistical models applied in the mentioned fields. These algorithms cause both increasing accuracy in dealing with complex and uncertain problems and developing new application in different research areas.

The proposed book is a collection of essays with fourteen chapters that written by many famous researchers of different countries. In general, chapters consisted of different cases such as gully erosion modelling, landslides mapping, land subsidence cases, multi-hazard assessment, flood susceptibility and hazard modelling, earthquake events modelling, and fire susceptibility mapping.

The editors (Dr. Hamid Reza Pourghasemi and Dr. Mauro Rossi) state in the preface that this book will become the reference of choice for researchers in different fields including land surveying, remote sensing, cartography, GIS, geophysics, geology, natural resources, and geography. I know two editors, they have many peer review publications according to Google Scholar; so, I am especially pleased to introduce this book to readers by different multi-disciplinary experts.

Daejeon, Korea

Prof. Dr. Saro Lee
Korea Institute of Geosciences and Mineral Resources (KIGAM)
Korea University of Science and Technology (UST)

Preface

Natural hazards such as landslides, floods, earthquakes, forest fires, droughts and erosion processes impact severely every year structures, infrastructures and population producing financial damages and human casualties. Based on the Centre for Research on the Epidemiology of Disasters (CRED) database, 22.3 million people were killed by natural disasters between 1900 and 2006, an average of about 208,000 people per year. A proper evaluation of the susceptibility, hazard and risk posed by these natural phenomena is fundamental for planners, managers and decision makers in developed and developing countries. In this context, geographic information system (GIS) and remote sensing (RS) tools can be effectively used in order to assess and manage the hazard and risk before, during and after the occurrence of these natural events. A large variety of expert knowledge, statistical and analytical methods and models were applied worldwide, according to data availability and accessibility. However, choosing the best and efficient method or model remains one of the main concerns in the scientific literature.

Traditional methods for modelling natural hazards rely upon the use of deterministic conceptual descriptions linking the spatial and temporal occurrence of the natural phenomena and the geo-environmental settings in which they occur. Errors and uncertainties in using such models are inevitable, mainly due to the limited knowledge (i.e. lack of accurate spatial and temporal information) of the geo-environmental factors, but also to the simplified (i.e. inappropriate) modelling schema adopted in such deterministic description. Another important limitation is the absence of precise borders/classes for some conditioning factors commonly used in the modelling and represented as categorical (i.e. classified) variables, such as soil, land use and lithology. In addition, the determination of natural border for continuous factors such as elevation, slope, distance from linear elements, topographic indices and density elements is very difficult. In these conditions, the use of deterministic modelling tools is not straightforward and may lead to biased estimates of hazard and risk.

These issues have led to the use of data mining techniques to model geological, hydrological, soil erosion and other geo-environmental processes. These algorithms may increase the accuracy in dealing with complex and uncertain problems, and they have been largely applied in other scientific fields with positive outcomes. Data mining techniques proved to be effective in assessing the susceptibility, hazard and risk posed by natural disasters, often leading to highly accurate predictions, even where limited information on these phenomena are available.

In this book, we give and overview of the application of data mining algorithms for the spatial modelling of natural hazards in different study areas. The book is a collection of essays written by expert researchers from different countries. We believed that the book could be a useful guide for researchers, students, organizations and decision makers in different fields including land surveying, remote sensing, cartography, GIS, geophysics, geology, natural resources and geography that in their work are facing problems related to the hazard management and more generally to the land use planning.

The book contains the following 12 chapters:

1. Gully Erosion Modeling Using GIS-Based Data Mining Techniques in Northern Iran; A Comparison Between Boosted Regression Tree and Multivariate Adaptive Regression Spline;
2. Concepts for Improving Machine Learning Based Landslide Assessment;
3. Assessment of the Contribution of Geo-environmental Factors to Flood Inundation in a Semi-arid Region of SW Iran: Comparison of Different Advanced Modelling Approaches;
4. Land Subsidence Modelling Using Data Mining Techniques. The Case Study of Western Thessaly, Greece;
5. Application of Fuzzy Analytical Network Process Model for Analyzing the Gully Erosion Susceptibility;
6. Landslide Susceptibility Prediction Maps: From Blind-Testing to Uncertainty of Class Membership: A Review of Past and Present Developments;
7. Earthquake Events Modelling Using Multi-criteria Decision Analysis in Iran;
8. Prediction of Rainfall as One of the Main Variables in Several Natural Disasters;
9. Landslide Inventory, Sampling and Effect of Sampling Strategies on Landslide Susceptibility/Hazard Modelling at a Glance;
10. GIS-Based Landslide Susceptibility Evaluation Using Certainty Factor and Index of Entropy Ensembled with Alternating Decision Tree Models;

11. Evaluation of Sentinel-2 MSI and Pleiades 1B Imagery in Forest Fire Susceptibility Assessment in Temperate Regions of Central and Eastern Europe. A Case Study of Romania;
12. Monitoring and Management of Land Subsidence Induced by Over-exploitation of Groundwater.

Shiraz, Iran — Dr. Hamid Reza Pourghasemi
Perugia, Italy — Dr. Mauro Rossi

Contents

Gully Erosion Modeling Using GIS-Based Data Mining Techniques in Northern Iran: A Comparison Between Boosted Regression Tree and Multivariate Adaptive Regression Spline

Mohsen Zabihi, Hamid Reza Pourghasemi, Alireza Motevalli and Mohamad Ali Zakeri

Abstract Land degradation occurs in the form of soil erosion in many regions of the world. Among the different type of soil erosion, high sediment yield volume in the watersheds is allocated to gully erosion. So, the purpose of this research is to map the susceptibility of the Valasht Watershed in northern Iran (Mazandaran Province) to gully erosion. For this purpose, spatial distribution of gullies was digitized by sampling and field monitoring. Identified gullies were divided into a training (two-thirds) and validating (one-third) datasets. In the second step, eleven effective factors including topographic (elevation, aspect, slope degree, TWI, plan curvature, and profile curvature), hydrologic (distance from river and drainage density), man-made (land use, distance from roads), and lithology factors were considered for spatial modeling of gully erosion. Then, Boosted Regression Tree (BRT) and Multivariate Adaptive Regression Spline (MARS) algorithms were implemented to model gully erosion susceptibility. Finally, Receiver Operating Characteristic (ROC) used for the assessment of prepared models. Based on the findings, BRT model (AUC = 0.894) had better efficiency than MARS model) AUC = 0.841) for gully erosion modeling and located in very good class of accuracy. In addition, altitude, aspect, slope degree, and land use were selected as the most conditioning agents on the gully erosion occurrence. The results of this research can be used for the prioritization of critical areas and better decision making in the soil and water management in the Valasht Watershed. In addition, the used models are recommended for spatial modeling in other regions of the worlds.

M. Zabihi · A. Motevalli · M. A. Zakeri
Department of Watershed Management Engineering,
Faculty of Natural Resources, Tarbiat Modares University, Tehran, Iran

H. R. Pourghasemi (✉)
Department of Natural Resources and Environmental Engineering,
College of Agriculture, Shiraz University, Shiraz, Iran
e-mail: hr.pourghasemi@shirazu.ac.ir; hamidreza.pourghasemi@yahoo.com

H. R. Pourghasemi and M. Rossi (eds.), *Natural Hazards GIS-based Spatial Modeling Using Data Mining Techniques*, Advances in Natural and Technological Hazards Research 48, https://doi.org/10.1007/978-3-319-73383-8_1

Keywords Gully erosion · Boosted regression tree
Multivariate adaptive regression spline · Coupling GIS and R · Valasht watershed

1 Introduction

1.1 Soil Erosion by Water and Its Types

One of the main causes of land degradation in the entire world is soil erosion by water which caused extensive changes in the earth's surface (Pimentel 2006). The main cause of water erosion is runoff that caused by rainfall. Water erosion is due to the movement of organic and mineral particles of soil by water and the accumulation of degraded materials in the downstream places (Franzluebbers 2010). Substances that are destroyed by water erosion reduce water quality, reduce water capacity of dams, threaten aquatic life, and increase the risk of flood and other harmful environmental issues (Robertson et al. 2004; Sadeghi and Zakeri 2015).

Soil erosion can be considered as a major change; because the rate of soil degradation is 10 to 40 times higher than the rate of soil formation by innate proceedings and several kilometers of agricultural land are lost every year due to soil degradation (Luffman et al. 2015). There are various forms of water erosion, including splash, sheet, rill, gully and tillage erosion, landslides and river or stream bank erosion (Osman 2014). This article studies the gully erosion, axially. In fact, the effects of various factors, such as land use, rainfall, soil, lithology and topography lead to erosion called Gully (Dotterweich et al. 2013; Conoscenti et al. 2014; Superson et al. 2014; Luffman et al. 2015). One of the important sedimentation factors in different parts of the earth is Gully erosion (Vanwalleghem et al. 2005; Bouchnak et al. 2009). The extent of soil loss caused by the gully destruction from 10 to more than 90% of the total sediment produced by various types of destruction of water, that is variable and significant amount of erosion (Poesen et al. 2003). This type of erosion is a serious problem in many parts of the world (Martínez-Casasnovas 2003). It can be a main factor in road destruction (Jungerius et al. 2002) and can be affected on water pollution or threaten the aquatic Creatures (Wantzen 2006).

1.2 GIS Techniques for Gully Erosion Modeling

In order to understand the various processes governing soil erosion, the need for modeling this phenomenon is essential. By modeling, it is possible to estimate runoff and sediment in order to maintain and control measures. To reduce the effects of water and wind erosion, issues such as understanding the effective factors of erosion, assessing the both internal and external effects of erosion, identifying strategies and assessing the performance of protective operations are important, which is the modeling of the first step to accomplish these (Franzluebbers 2010).

Today, the use of Geographic Information System (GIS) as a tool for modeling is common. This system can be used to preserve process and analyze geospatial factors such as soil, land use, topography, etc. In general, the use of GIS can be very useful and accelerated in hydrologic modeling (Jain et al. 2001; Jungerius et al. 2002). Recently, GIS and data mining techniques have been increased for modeling of natural hazards and different types of land degradation. In this regards, gully erosion as an effective factor on soil and water resources degradation have been conducted using different algorithms by many researches (Dube et al. 2014; Monsieurs et al. 2015; Shruthi et al. 2015; Bergonse and Reis 2016; Goodwin et al. 2017). In this Regards, Different methods have been conducted for spatial modeling of gully erosion in last years; these methods are Logistic Regression (Akgun and Turk 2011; Conoscenti et al. 2014), Conditional Analysis (Conoscenti et al. 2013), Classification and Regression Trees (Geissen et al. 2007), Weights of Evidence (Rahmati et al. 2016), Frequency Ratio (Rahmati et al. 2016), and Random Forest (Kuhnert et al. 2010). However, the low number of studies used BRT and MARS methods for gully erosion susceptibility mapping. Gutiérrez et al. (2009) in order to model gully as an independent variable against an independent variable used two methods, including Classification and Regression Trees (CART) and MARS. They founded a better efficiency of MARS for gully predicting with the area under the curves of 0.98 and 0.97 for the training and validation datasets, while CART presented values of 0.96 and 0.66, respectively. Gutiérrez et al. (2011) used the MARS model to predict gully creation locations. The results showed that this model is a good performance in geomorphic research. In addition, Gutiérrez et al. (2015) used MARS algorithm for gully erosion susceptibility mapping in two basins in Italy and Spain using topographical properties. Based on the findings of this study, the use of topographic properties as an independent factor in the prediction of gully erosion has been acceptable in both regions. So, the aim of this study is gully erosion modeling based on BRT and MARS data mining techniques and their comparison in Iran. Moreover, the BRT method is not used for gully erosion modeling so far. This can be considered as distinguishing aspects of this study comparing with previous researches.

2 Materials and Methods

2.1 Study Area

The Valasht Watershed is located in 30 km of southwest of the Chalus City in the Mazandaran Province. This area is belonging to the Chalus River Basin. The Valasht Watershed between latitudes of 36° 32′ 19″ to 36° 34′ 39″ north and longitudes of 51° 15′ 00″ to 51° 19′ 26″ east, with an area of 1544 ha. The altitude variation of Valasht Watershed is from 1005 to 1839 m. The Valasht Watershed as an isolate topographical almost is circular (like bowl) and small. The lithology of study area was delineated using Chalus Sheet at 1:100,000-scale and is presented in

Table 1 Lithology of the Valasht Watershed

Code	Lithology	Age	Era
Q_{AL}	Recent loose alluvium in the river channels	Quaternary	Cenozoic
Q_2	Young alluvial fans and terraces, river terraces, and mainly cultivated	Quaternary	Cenozoic
k_2L_1	Globotruncana limestone, marly limestone	Cretaceous	Mesozoic
k_2LM_2	Globotruncana limestone, marl, marly limestone	Cretaceous	Mesozoic
K_2M	Marl, marly limestone, limestone	Cretaceous	Mesozoic
K_2VT	Undivided Upper Cretaceous volcanites	Cretaceous	Mesozoic

Table 1. The land use is divided into 8 categories, including: dense forestland with an area of 245 ha (18.9%), thin forest with area of 304.5 ha (19.7%), dry farming with area of 776.5 ha (50.3%), irrigated farming with area of 5.84 ha (0.4%), orchard with area of 49.8 ha (3.2%), rangeland with area of 97.1 ha (6.3%), residential with area of 41.2 ha (2.66%), and lake with area of 23.73 ha (1.53%), respectively. The location of the study area is shown in Fig. 1.

2.2 Methodology

This study consists of several main steps, including (i) Gully erosion inventory mapping, (ii) preparation of gully erosion effective factors, (iii) gully erosion susceptibility spatial modeling using two data mining techniques, (iv) assessment of the variables importance applied on gully erosion, and (v) accuracy assessment of gully erosion susceptibility models. The details of this research as flowchart are given in Fig. 2.

2.3 Gully Erosion Inventory Mapping

For modeling of gully erosion, spatial distribution of gullies was digitized by Global Positioning System (GPS) and extensive field survey. It can be stated that the majority of gullies type was linear with the mechanism of shear stress Then, two-thirds of the samples (76) were selected as training and one-third of locations (32) were used for validation purposes (Stumpf and Kerle 2011; Pourghasemi et al. 2013; Pourghasemi and Kerle 2016). Also, the existence (1) or absence (0) of Gully was defined in relation to the factors influencing gullies occurrence and gullies (Rahmati et al. 2016).

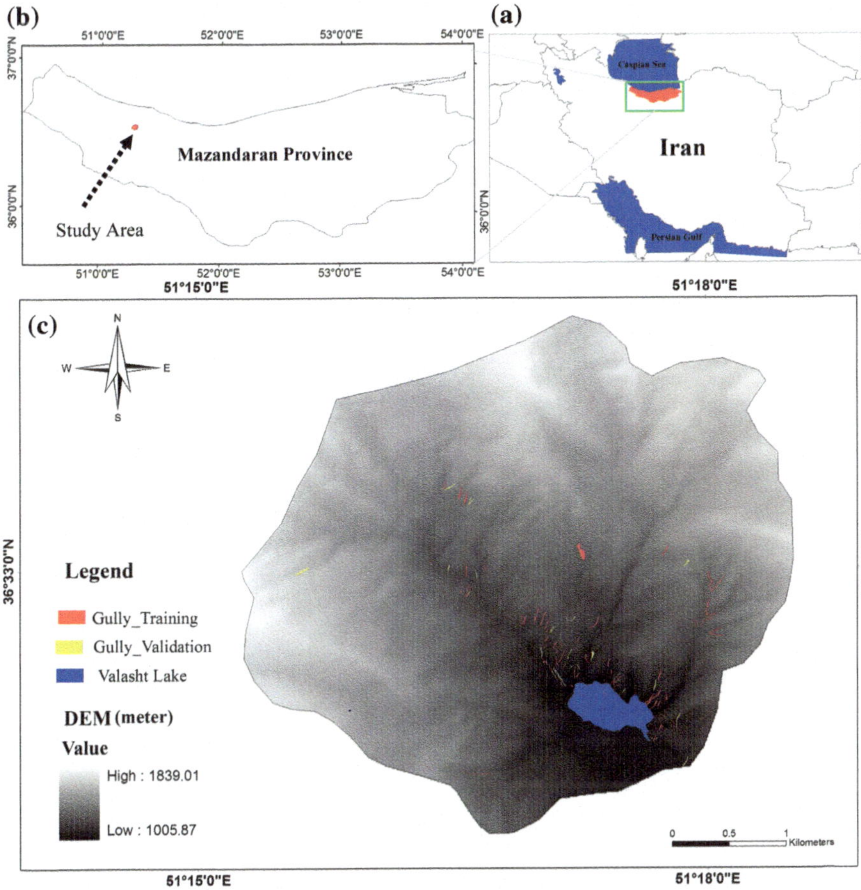

Fig. 1 Location of the Valasht watershed in Iran

2.4 *Preparation of Gully Erosion Effective Factors*

Gully occurrence is a result of several factors that it is important to identify and conceptualize these factors. According to the mechanism of formation of gully, several factors control the development of gullies (Li et al. 2017). It is worthwhile to identify the factors associated with the creation and development of gully erosion. In fact, the preparation of the different layers in order to consider the gully erosion controlling factors is essential. Thus, effective factors in this study are defined as: slope aspect, elevation, drainage density, land use, lithology, plan curvature, profile curvature, distance from river, distance from road, slope degree, and Topographic Wetness Index (TWI).

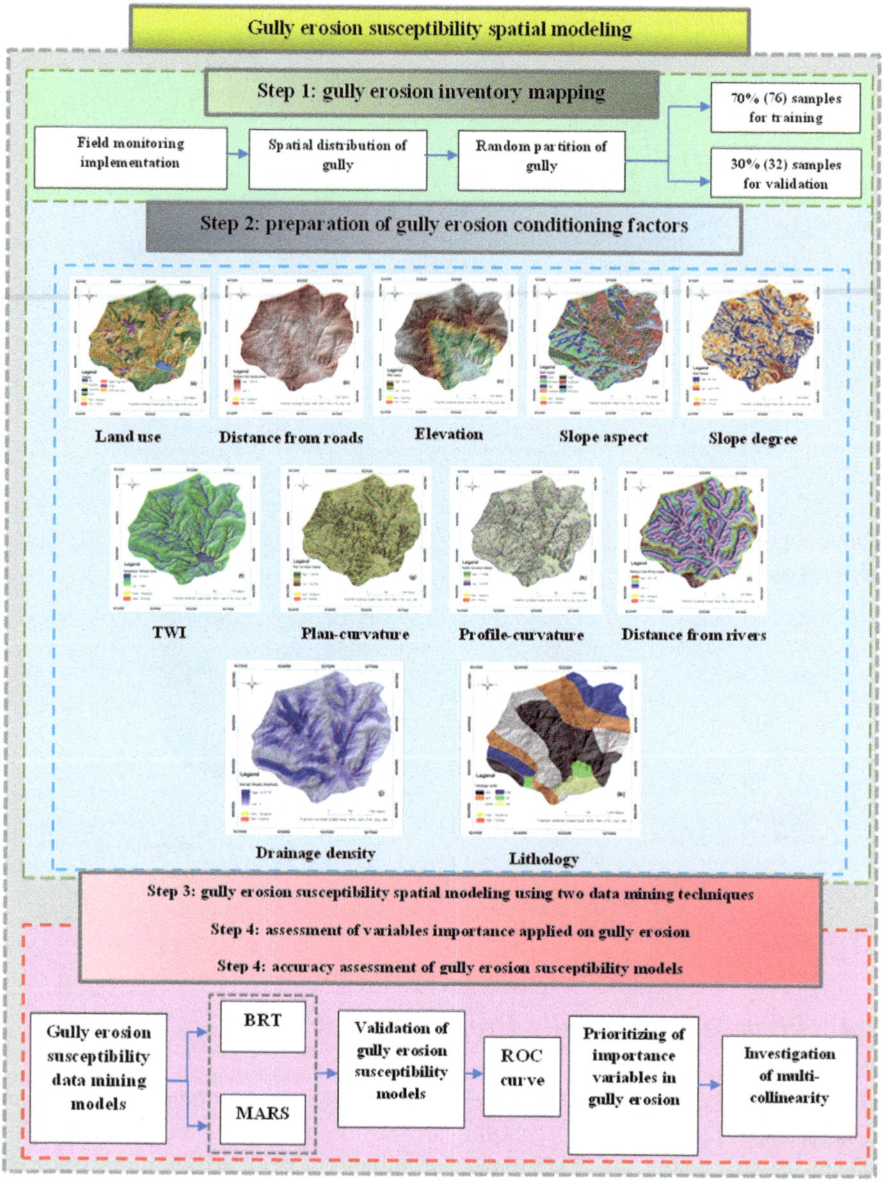

Fig. 2 The flowchart of gully erosion susceptibility spatial modeling

2.5 Man-Made Factors

Many factors can affect the type and amount of gully erosion in a watershed. One of these factors is the improper land use (Dube et al. 2014; Rahmati et al. 2016). Changes in forest and turning it to agricultural land, commercial and residential, provide gully erosion potential creation and its development. Therefore, identifying and studying the relationship between changes in different land uses is possible in order to identify and management of gully erosion (Sadeghi et al. 2007; Desta and Adunga 2012; Dube et al. 2014; Dymond et al. 2016). The land use map was extracted using supervised classification of Landsat 7/ETM + images (year 2014) and the maximum likelihood method. The land use map reclassified to eight types including forestland, thin forest, dry farming, irrigated farming, orchard, rangeland, thin forest, and, irrigated farming, residential and lake (Fig. 3a). In addition to improper of land use, road construction with poor drainage due to changes along the route on vegetation, hydrology, and soil will disrupt the natural balance of areas, will increase surface runoff on the road and thus, causes the initiation and development of gully erosion (Desta and Adunga 2012; Dymond et al. 2016). Naturally, susceptibility of areas near the road due to incorrect drainage and excess runoff is greater than other regions in a watershed (Desta and Adunga 2012; Bergonse and Reis 2016; Dymond et al. 2016; Li et al. 2017). The distance from road map for the Valasht Watershed was prepared based on general directorate of roads and urban development (Fig. 3b).

2.6 Topographic Factors

The impact of topographic features on the hydrological response of a basin is an undeniable in connection with the excess runoff and its focus on the formation of gully erosion and development (Conforti et al. 2011; Desta and Adunga 2012; Luffman et al. 2015; Barnes et al. 2016; Bergonse and Reis 2016). Landform characteristic (elevation, slope aspect, slope degree, TWI, plan curvature, and profile curvature) were prepared from Digital Elevation Model (DEM) with a pixel size of 10 m × 10 m. Elevation levels play a significant role in climate indices. According to Li et al. (2017) climatic conditions vary by changing elevation and the potential of gully erosion occurrence will be different (Fig. 3c). Slope aspect plays an important role in obtaining the required moisture to creating runoff and occurrence of gully erosion (Conforti et al. 2011; Barnes et al. 2016; Bergonse and Reis 2016). This layer was prepared from DEM and classified into nine categories (Fig. 3d). Gully erosion is created in hilly region or mountainous areas with steep slopes. In fact, the slope is a key factor for the critical drainage of a region (Valentin et al. 2005). Steep sloping areas have high-velocity runoff and have high gully potential initial conditions; although, climate and soil conditions are the same (Valentin et al. 2005; Desta and Adunga 2012) (Fig. 3e). The TWI indicated that geomorphic pattern of topography to rainfall or wet condition (Beven et al. 1984). The TWI is calculated according to Eq. (1).

$$TWI = ln\left(\frac{A_s}{\tan \beta}\right) \quad (1)$$

where, A_s is the specific catchment area (meter) which is determined by the up-slope area via generic spot, and per unit contour length; β is the local slope (degree) (Fig. 3f). Plan curvature has a major role in triggering of the gully. The impact of plan curvature on gully erosion in association with convergence or divergence water flow and its focus is to water fall (Valentin et al. 2005; Conforti et al. 2011; Desta and Adunga 2012) (Fig. 3g). Profile curvature morphometric parameters that have an important impact on start of gully process, So that concave topography (depression region), are more sensitive to initial of gully appearance (Desta and Adunga 2012) (Fig. 3h).

2.7 Hydrological Factors

Hydrological controlling factors are described often, in relation to the amount of surface runoff and its density (Tebebu et al. 2010; Desta and Adunga 2012; Ollobarren et al. 2016). Naturally, distance from rivers and drainage density factors in a region increase the potential of triggering gully erosion (Shellberg et al. 2016). So, the distance from river for the Valasht Watershed was prepared according to topographical map (Fig. 3i). Drainage density was extracted from stream network; the sum of the drainage lengths in the total cells of watershed was divided on total area of the watershed cells (Montgomery and Dietrich 1989) (Fig. 3j).

2.8 Lithology Factor

Geology is an intrinsic factor in relation to shear stress and the hydraulic conductivity of the water to start the process of gully (Dai and Lee 2002; Rahmati et al. 2017). Also, formation types in a rock unit (Marl, silt and etc.) are very important at the initiation of gully erosion (Rahmati et al. 2017). The lithology of the Valasht Watershed was delineated using Chalus Geological Sheet at 1:100,000-scale (Fig. 3k). Based on geological survey of Iran (GSI 1997), most gullies (Fig. 4) are located in units of K_2M and K_2LM_2 including: marl, marly limestone, limestone and Globotruncana limestone, marl, and marly limestone, respectively (Table 1).

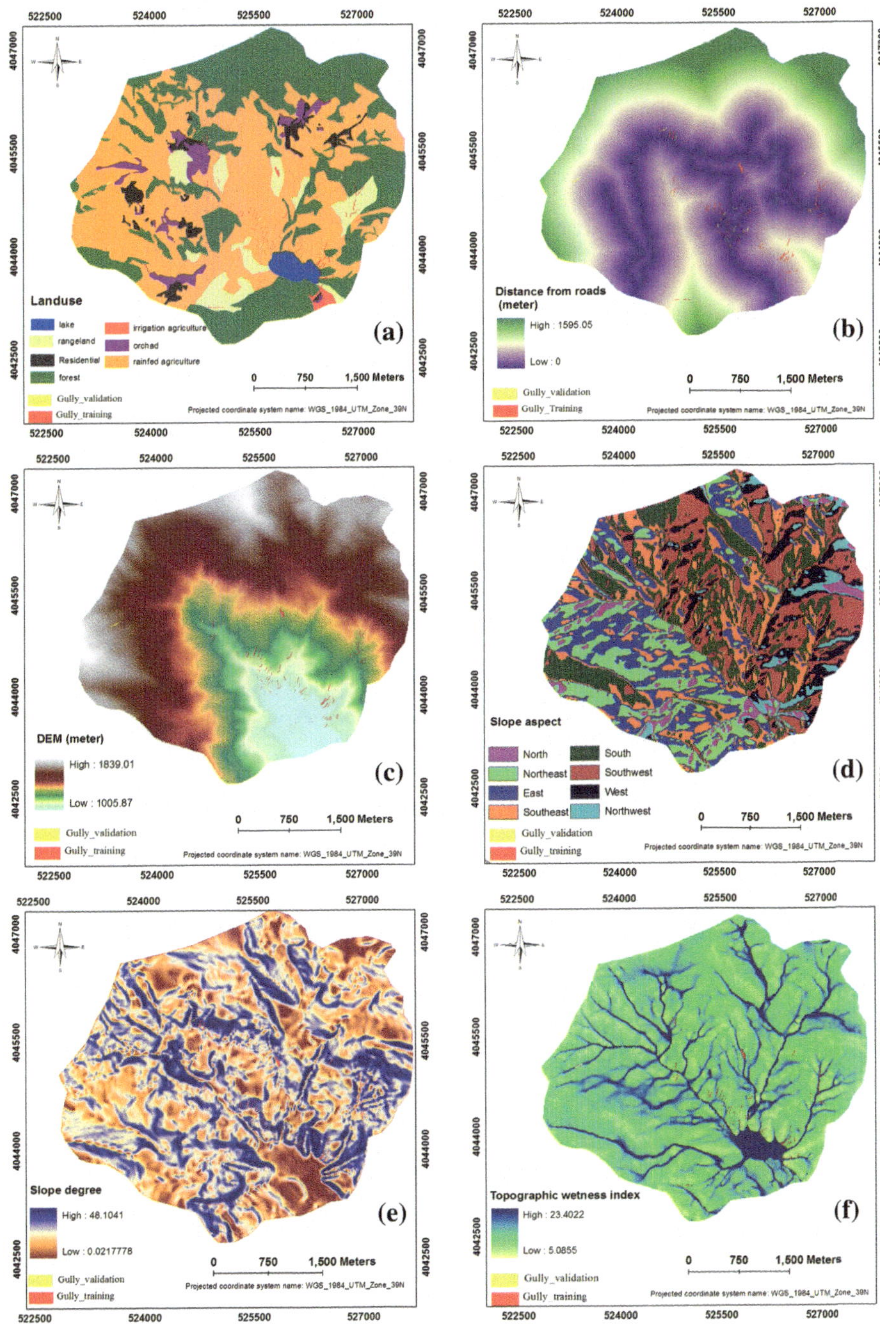

Fig. 3 Effective factors in spatial modeling of gully erosion in the Valasht Watershed

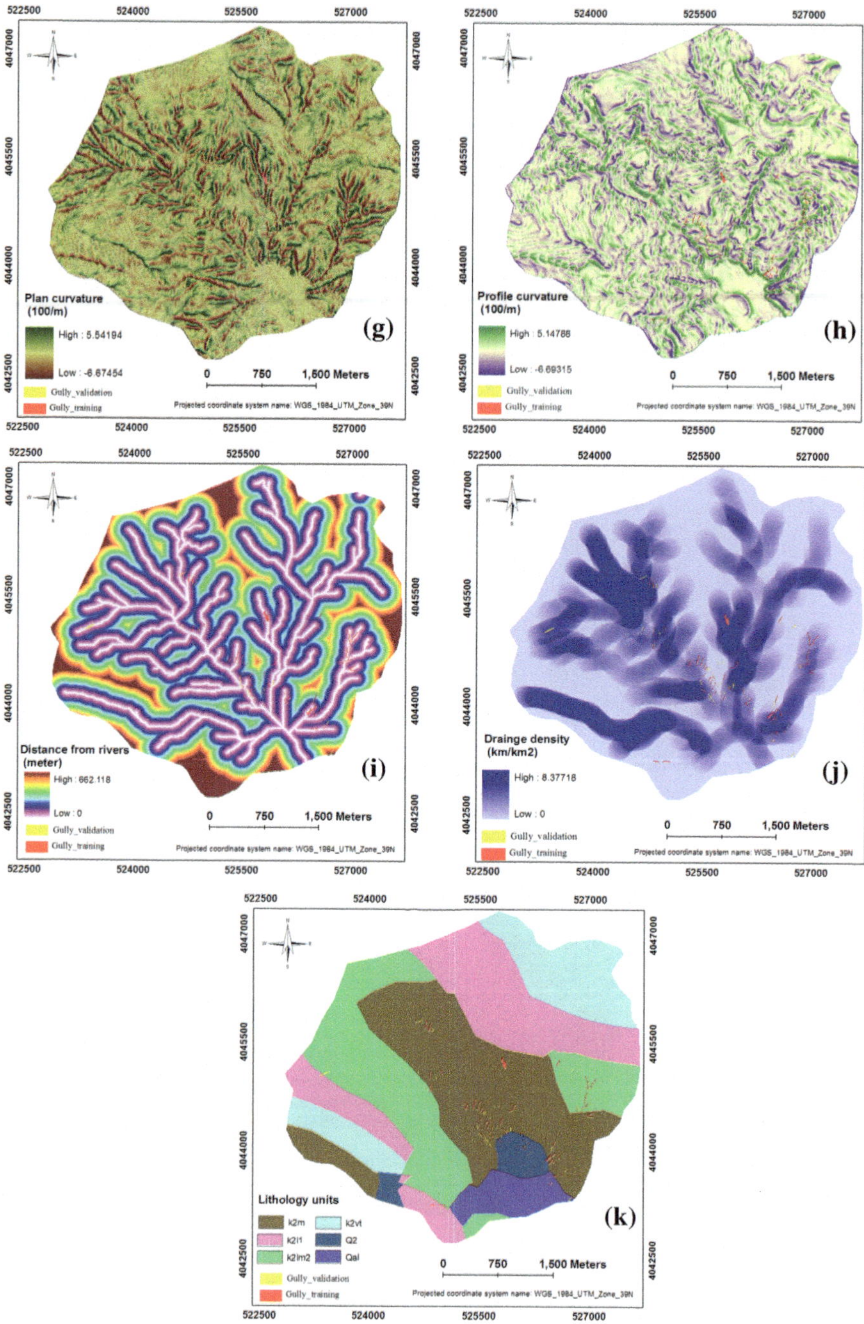

Fig. 3 (continued)

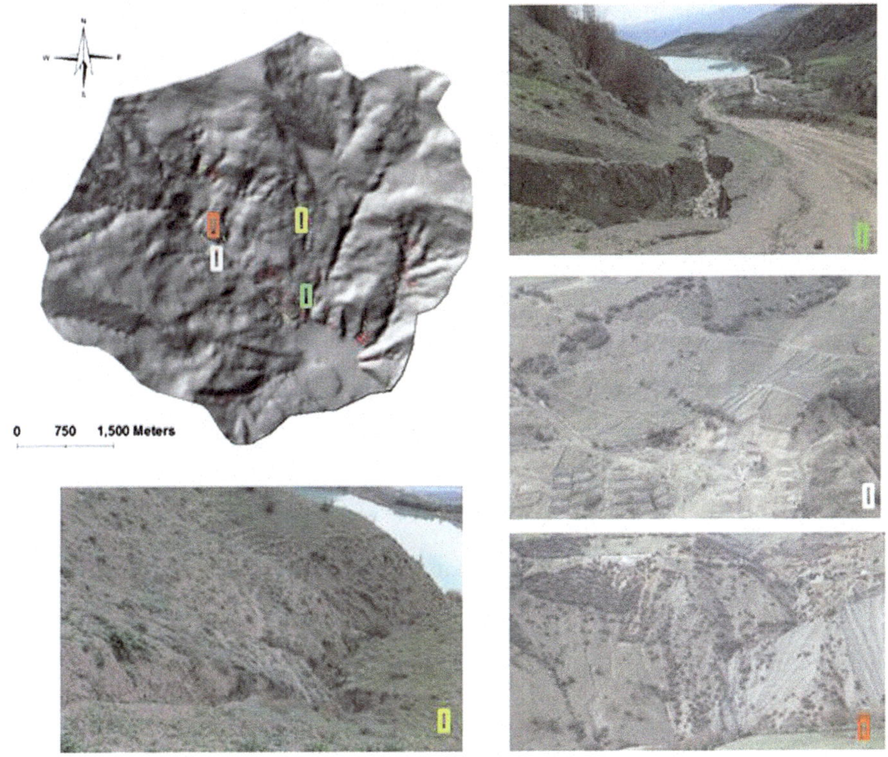

Fig. 4 Some of identified gully erosion in the Valasht Watershed

2.9 Gully Erosion Susceptibility Spatial Modeling Using Data Mining Techniques

2.9.1 Boosted Regression Tree (BRT)

BRT method has been used by different researchers in several studies (Schapire 2003; Leathwick et al. 2006; Elith et al. 2008; Youssef et al. 2015; Liu et al. 2016; Naghibi et al. 2016; Salazar et al. 2016). This method combines the techniques of statistical and machine learning algorithms (Breiman et al. 1984). BRT method is defined with two algorithms: a series of models can be fitted with an average of decision trees and output model can be combined to calculate the overall prediction using the boosting (Friedman 2001). BRT is a comparative method to combine many simple models for providing acquisition of proper functioning (Elith et al. 2008; Schapire 2003). High speed in large data analysis and less sensitivity to over-fitting are the advantages of BRT (Liu et al. 2016; Salazar et al. 2016). Performance of this algorithm depends on the setting of boosted trees and pruning trees (Leathwick et al. 2006; Elith et al. 2008). BRT fit many decision trees to increase the accuracy of the model. The BRT algorithm is presented in Eq. (1) (Schonlau 2005; Naghibi et al. 2016):

$$MVC : sign\left[\sum_{m}^{M} a_m \cdot c_m(x)\right] \tag{2}$$

where, MCV: Majority Vote Classification, a_m : log $((1-r_m)/r_m)$ which r_m: compute the (weighted) misclassification rate, Fit classifier c_m to the weighted data. In this equation, recalculate weights $w_i = w_i$ exp(m I(yi $\neq$ Cm)) that initialize weights equal to wi = 1/n For m = 1 to the next category of c_m (Schonlau 2005).

2.9.2 Multivariate Adaptive Regression Splines (MARS)

Alike BRT, recently from the MARS models used for spatial modeling in environmental sciences (Felicísimo et al. 2013; Conoscenti et al. 2016; Pourghasemi and Rossi 2016). The MARS method is implemented using non-parametric modeling techniques (Friedman 1991). This method can be implemented regardless of the link between the dependent and independent factor (Friedman 1991; Zabihi et al. 2016). The MARS method is based on basis functions for each explanatory variable and is defined as follows:

$$\text{Max}(0, x-k) \text{ or } \text{Max}(0, k-x) \tag{3}$$

where, k is a knots and observations is one of the explanatory variables and x is an independent variable (Friedman 1991; Zabihi et al. 2016). Thus, MARS model is described as follows:

$$\hat{Y} = \gamma + \sum_{m}^{M} \beta_m H_m(x) \tag{4}$$

where, y is the dependent variable predicted by the function, γ is a constant, M is the number of terms, and (x) is the explanatory variables. H_m is basis functions and β_m, coefficients that are determined by minimizing the sum of squared residuals (Friedman 1991; Zabihi et al. 2016). In MARS model, the best model is selected based on the minimization of the Generalized Cross validation (GCV) (Friedman 1991; Golub et al. 1979). The determination of GCV is based on Eq. (5).

$$\text{GCV} = \frac{\frac{1}{N}\sum_{i=1}^{N}\left[y_i - \hat{f}(x_i)\right]^2}{\left[1 - \frac{C(H)}{N}\right]^2} \tag{5}$$

where, N is the number of data and *C (H)* a dependent variable that increases with the number of basis function (BF) in the model and is calculated based on the following equation:

$$C(H) = (H + 1) + dH \tag{6}$$

where, *d* is retribution for each basis function is considered in the model and *H* is number of basic functions in Eq. 4 (Friedman 1991; Zabihi et al. 2016). The MARS model doesn't consider assumptions about the relationship between the response variable and the conditioning factors. Because the use of a similar iterative approach, MARS method is similar to machine learning algorithms.

2.10 Assessment of Variables Importance Applied to Gully Erosion

To better understanding of Gully behavior and its progression, the effect of multiple factors need to be measured simultaneously in the process of gully erosion (Rahmati et al. 2017). So, identifying the most important of effective factors on gully erosion and prioritization of involved factors should be provided. This is kind of sensitivity analysis of factors affecting on modeling of gully erosion. In the present study, BRT model was implemented for determination of effective factors on gully erosion, too. In addition, multicollinearity examination was performed to determine the effective factors. Multicollinearity analysis is one of the methods that indicate non-independent gully erosion conditioning factors which can be observed in the dataset because of their high correlation (Dormann et al. 2013). In this regards, Variance Inflation Factor (VIF) and tolerance are two commonly indices (Mousavi et al. 2017) that applied in the present study. In the case of VIF, values greater than 5 or 10 for each conditioning factor is not acceptable and it should be removed for further analysis in the process of modeling (O'Brien 2007). Also, values less than 0.2 or 0.1 of tolerance shows multicollinearity problem (O'Brien 2007; Mousavi et al. 2017).

2.11 Accuracy Assessment of Gully Erosion Susceptibility Models

Receiver Operational Curve (ROC) is a suitable method for verifying and comparing model predictions. (Swets 1988; Pourghasemi et al. 2012; Hong et al. 2016; Motevalli and Vafakhah 2016; Pham et al. 2016). The ROC curve is a graph of sensitivity (y-axis) versus 1—specificity (x-axis) (Beguería 2006). Five classes of capability are defined based on the AUC: 50–60% (low accuracy), 60–70%

(medium accuracy), 70–80% (well accuracy), 80–90% (very well accuracy), and 90–100% (high accuracy) (Yesilnacar 2005; Rahmati et al. 2017).

3 Results and Discussion

3.1 Examination of Multicollinearity

The results of examination of multicollinearity are presented in Fig. 5 (tolerance) and Fig. 6 (VIF). As is shown in Fig. 5, it can be seen that aspect (0.88), profile curvature (0.86), lithology (0.80), land use (0.79), altitude (0.61), and distance from roads (0.57) have the highest values of tolerance in the study area, respectively. In addition, TWI has been devoted to the last rank with the minimum value of tolerance (0.30). By the way, based on VIF calculation results provided in Fig. 6, all of independent factors don't have any multicollinearity problem for gully erosion susceptibility modeling. In this case, TWI, distance from river, plan curvature, drainage density, slope, and distance from the roads with the values of 3.31, 2.69, 2.10, 2.09, 1.79, and 1.74 is devoted in rank 1 to 6 of VIF. Other affecting factors, including altitude, land use, lithology, profile curvature, and aspect are located in next ranks. Based on these descriptions, there is no problem in term of multicollinearity of factors in the current study. In other words, all of selected conditioning factors on gully erosion susceptibility can be used for further analyses and investigations.

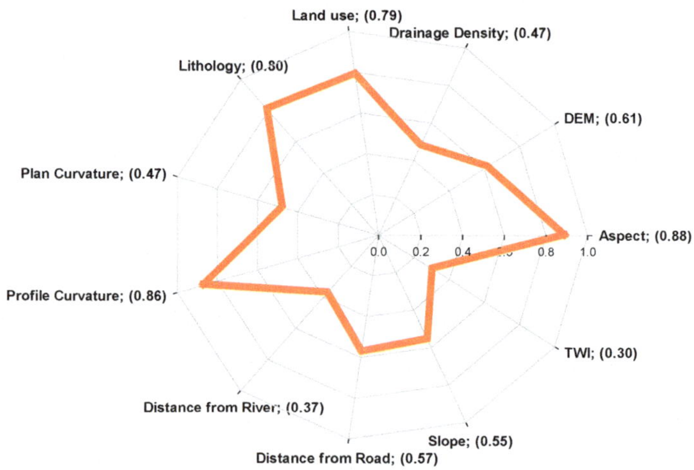

Fig. 5 Tolerance values of gully erosion effective factors for multicollinearity analysis

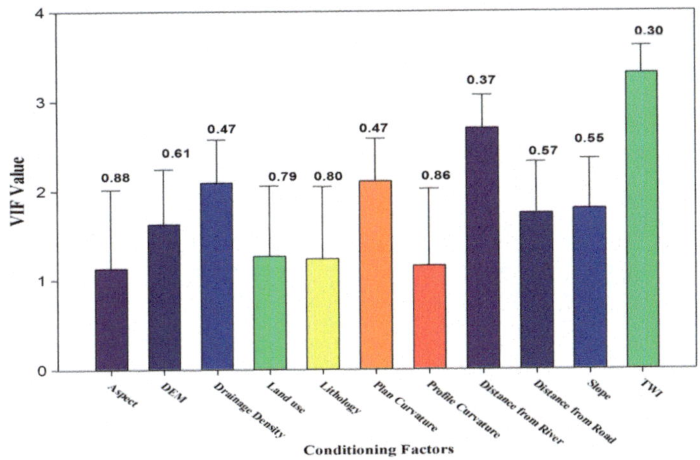

Fig. 6 Variance Inflation Factor (VIF) values of effective factors on gully erosion after multicollinearity analysis

3.2 Application of BRT Model

BRT as an advanced data mining technique was implemented for gully erosion susceptibility modeling in the Valasht Watershed, northern Iran. At first, the importance of conditioning factors on gully erosion susceptibility is specified using BRT model which is presented in Fig. 7. Based on our findings, altitude (DEM) is the most important factor (with the value of 43.03%) for gully erosion susceptibility in the current study. This result indicated that the gully erosion susceptibility depends on altitude. Other factors including aspect (13.64%), slope degree (7.21%), land use (7.12%), distance from road (5.34%), distance from river (5.33%), drainage density (5.20%), plan curvature (3.79), TWI (3.52%), profile curvature (3.51%), and lithology (2.26%) are located in next ranks. It is obvious that the topographic factors have a significant relationship with gully erosion susceptibility. Based on the results of Le Roux and Sumner (2012), the most important topographic factor in the gully erosion occurrence was slope in the Eastern Cape Province of South Africa. Based on their results, the upslope area in the topographic wetness index has a significant effect on gully erosion. Although, prioritization of condition factors in Le Roux and Sumner (2012) study were not exactly consistent with the current research finding, but both researches have been acknowledged the significant role of topographic factors in the occurrence of gully erosion. These differences may be attributed by different study areas.

According to gully erosion susceptibility map produced by the BRT model (Fig. 8) and corresponding calculation results (Fig. 9), the highest area of the Valasht Watershed with the value of 42.85% or 661.64 ha is covered by moderate class of gully erosion susceptibility. The highest class of susceptibility is devoted to rank 2 (24.84%). Also, 343.36 and 155.44 ha of the study area are located in very

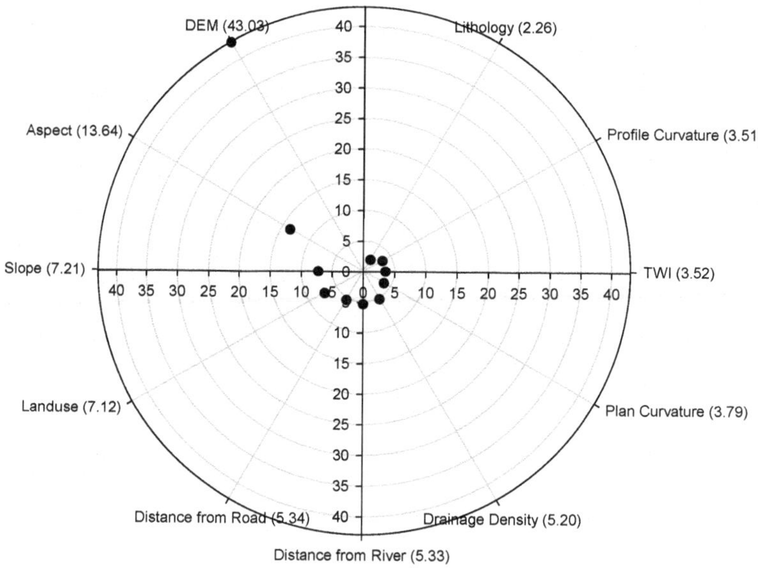

Fig. 7 The percent of importance of each conditioning factor based on BRT model

high and low classes of gully erosion susceptibility, respectively (Fig. 9). As is shown in Fig. 8, in the point of spatial distribution of gully erosion susceptibility, high and very high classes are covered the middle part of the watershed; while low class of susceptibility envelope the around of the mentioned watershed.

3.3 *Application of MARS Model*

The next machine learning (data mining) model used for gully erosion susceptibility mapping was MARS. The final map of gully erosion susceptibility produced by MARS model is provided in Fig. 10. The mentioned map created by Eq. 7. The final map of gully erosion susceptibility was divided to four classes based on the natural break method containing low, moderate, high, and very high classes the same with BRT algorithm (Basofi et al. 2015; Colkesen et al. 2016). According to MARS model results (Fig. 10), the classes of gully erosion susceptibility including low, moderate, high, and very high have covered 13.12, 32.86, 27.57, and 26.45% of area (Fig. 9) in the Valasht Watershed. In the point of spatial distribution view, the low class of susceptibility has an irregular spatial distribution which has a low area in the Valasht Watershed, too. The high and very high classes of susceptibility are located in the middle of study area toward outlet of watershed (Fig. 10). Because of high volume of eroded soil and sediment in gully erosion process, it is very important and necessary to prevent from entering of this sediment to the

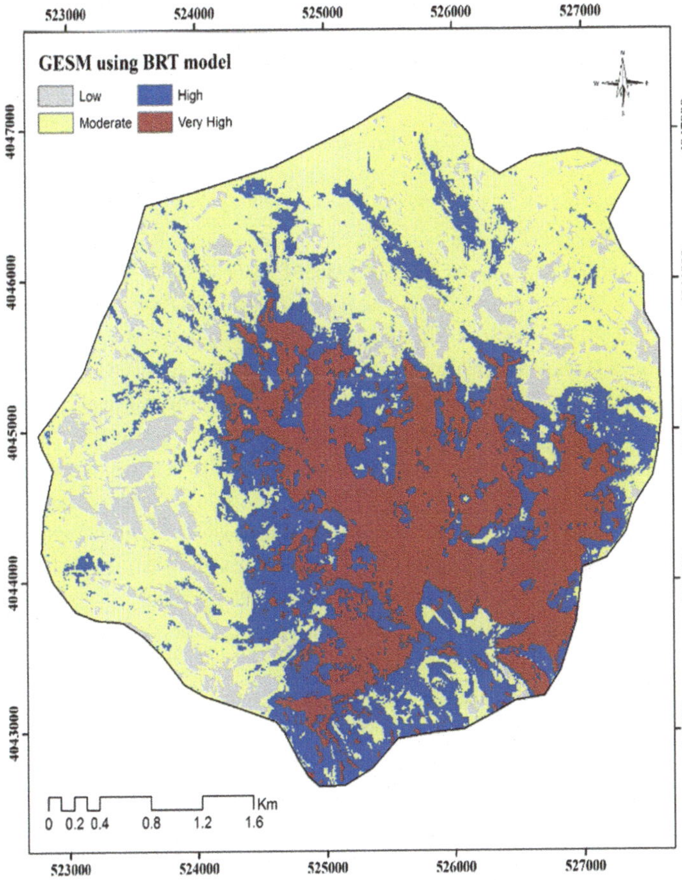

Fig. 8 Gully erosion susceptibility map produced by BRT model

Fig. 9 Gully erosion susceptibility classes according to BRT and MARS models

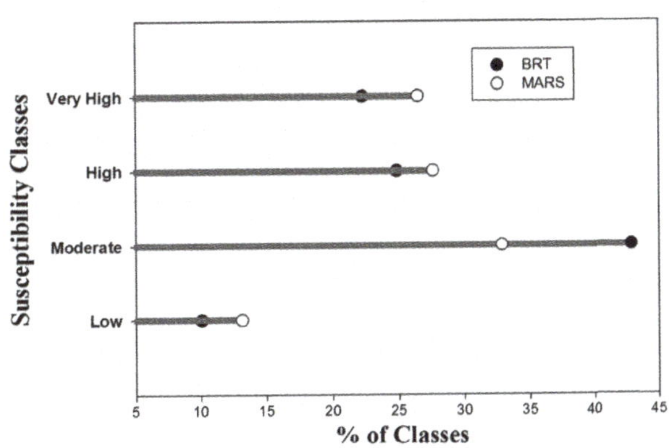

Valasht Lake in the output of mentioned watershed. This subject can be considered by managers and decision makers for selecting of appropriate solution of the mentioned problem.

$$
\begin{aligned}
\mathrm{GESM_{MARS}} =\;& 0.1874707 + 0.5021102(\mathrm{South}) + 0.1453134(\max(0, 1.438846 - \mathrm{PlanC}))\\
& + 0.3666469(\mathrm{South})(\mathrm{Forest}) + 0.4623038(\mathrm{Southwest})(\mathrm{Qal}) - 0.6258331\\
& (\mathrm{Rangeland})(\mathrm{K2l2}) - 3.886562(\mathrm{East})(\max(0, \mathrm{ProfC} - 2.044398))\\
& - 0.001312126(\mathrm{South})(\max(0, \mathrm{DEM} - 1170.622)) - 0.004548146(\mathrm{South})\\
& (\max(0, 1170.622 - \mathrm{DEM})) - 0.8485331(\mathrm{South})(\max(0, \mathrm{Drainage\ Density}\\
& - 4.288856)) + 0.6355912(\mathrm{South})(\max(0, \mathrm{Drainage\ Density} - 3.590517)\\
& - 0.001046979(\mathrm{South})(\max(0, 317.6476 - \mathrm{DisRoad})) + 0.09513996(\mathrm{Southwest})\\
& (\max(0, \mathrm{Drainage\ Density} - 2.57988)) + 0.006357844(\mathrm{Southwest})\\
& (\max(0, 40 - \mathrm{DisRiver})) + 0.05349974(\mathrm{Southwest})(\max(0, \mathrm{Slope} - 65.36047))\\
& - 0.1220924(\mathrm{Southwest})(\max(0, \mathrm{Slope} - 78.49104)) - 0.01029525\\
& (\max(0, \mathrm{DEM} - 1349.472))(\mathrm{Rangeland}) + 0.002747312(\max(0, 1292.027 - \mathrm{DEM}))\\
& (\mathrm{Rainfed\ Agriculture}) - 0.2422385(\max(0, 2.199255 - \mathrm{Drainage\ Density}))\\
& (\mathrm{Rangeland}) - 0.2070098(\max(0, \mathrm{Drainage\ Density} - 2.199255))(\mathrm{Rangeland})\\
& + 1.42692(\mathrm{Rangeland})(\max(0, \mathrm{PlanC} - 0.6826829)) + 0.01334593(\mathrm{Rangeland})\\
& (\max(0, \mathrm{DisRiver} - 219.54)) + 0.002243662(\mathrm{Rangeland})\\
& (\max(0, 219.54 - \mathrm{DisRiver})) + 0.001496906(\mathrm{Rangeland})(\max(0, \mathrm{DisRoad}-\\
& 416.173)) + 0.0006178245(\mathrm{Rangeland})(\max(0, 416.173 - \mathrm{DisRoad}))\\
& + 0.05900507(\mathrm{K2m})(\max(0, 1.438846 - \mathrm{PlanC})) - 0.0004876583\\
& (\max(0, 1391.002 - \mathrm{DEM}))(\max(0, 1.438846 - \mathrm{PlanC})) - 0.002044421\\
& (\max(0, \mathrm{DEM} - 1391.002))(\max(0, 1.438846 - \mathrm{PlanC})) + 0.002543124\\
& (\max(0, \mathrm{DEM} - 1483.44))(\max(0, 1.438846 - \mathrm{PlanC})) + 1.770632e - 05\\
& (\max(0, 1292.027 - \mathrm{DEM}))(\max(0, 130.384 - \mathrm{Road})) - 3.821526e - 05\\
& (\max(0, \mathrm{DEM} - 1514.782))(\max(0, \mathrm{Slope} - 23.52937) - 0.06118585(\max(0,\\
& 1.320009 - \mathrm{DrainageDensity}))(\max(0, 1.438846 - \mathrm{PlanC})\\
& - 0.03516613(\max(0, \mathrm{Drainage\ Density} - 1.320009))(\max(0, 1.438846 - \mathrm{PlanC}))\\
& + 0.006461668(\max(0, 1.438846 - \mathrm{PlanC}))(\max(0, 29.24419 - \mathrm{Slope}))\\
& - 0.005925642(\max(0, 2.044398 - \mathrm{ProfC}))(\max(0, 41.59864 - \mathrm{Slope}))
\end{aligned} \tag{7}
$$

Based on the obtained results, the rank of conditioning factors was different in BRT and MARS models. This is due to the nature of the applied models. Although, BRT model consider the reaction between each conditioning factor, but MARS model doesn't take into account it.

3.4 Assessment of BRT and MARS Models

Assessment is one of the most important phases in the modeling process (Chung-Jo and Fabbri 2003; Zabihi et al. 2016). For this purpose, an independent dataset of

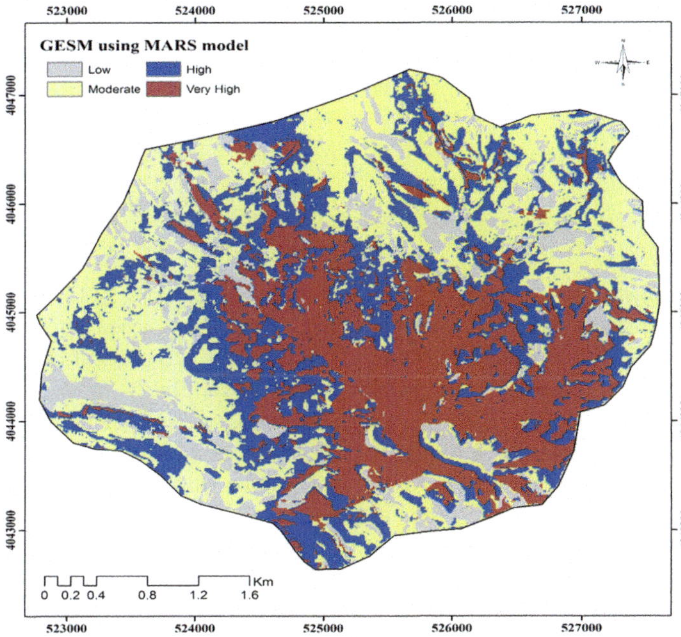

Fig. 10 Gully erosion susceptibility map produced by MARS model

Fig. 11 The ROC of BRT and MARS models for spatial modeling of gully erosion susceptibility

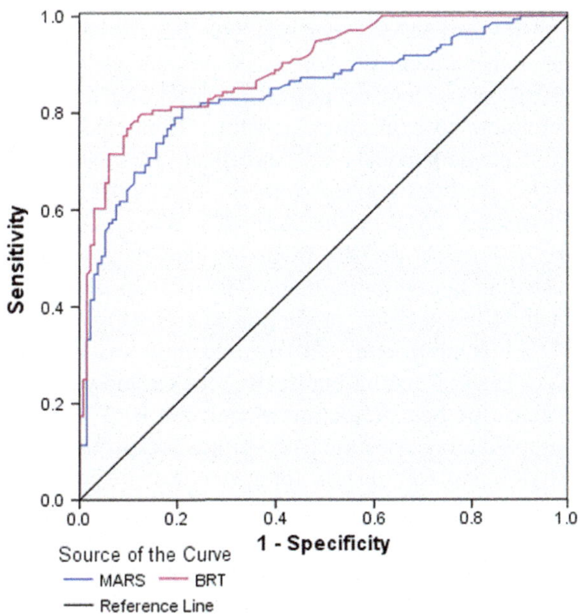

Table 2 Detailed information of ROC computation in BRT and MARS models

Models	Area (AUC)	Standard error	Asymptotic significant	Asymptotic 95% confidence interval	
				Lower bound	Upper bound
MARS	**0.841**	0.025	0.000	0.792	0.889
BRT	**0.894**	0.019	0.000	0.856	0.931

gully erosion implemented for validation of the used models. In this regards, from 109 observed gully erosion in field surveys, 33 locations (30%) applied for validation step that not considered in training phase. Also, 70 percent of locations (76 cases) used for constructing of gully erosion susceptibility model in the study watershed. To assess the accuracy of implemented models (BRT and MARS), ROC was employed by reconciling two gully erosion susceptibility maps (Figs. 8 and 9) and existing locations of gully erosion (locations for validation step). The results of ROC calculation are provided in Fig. 11 and Table 2. Based on the findings of this research, BRT model has the better performance versus MARS model. The value of AUC in BRT model was equal to 0.894 and prediction accuracy obtained 89.4%, while, AUC and prediction accuracy of MARS model are equal to 0.841 and 84.1%, respectively. Accordingly, the accuracy of two used models is located in very good class, but BRT model had the best performance in the current study. So, the results of BRT model can be used for decision making and soil erosion management. Detailed information about ROC calculation is presented in Fig. 11 and Table 2.

BRT and MARS methods have been used in some studies in the past to model phenomena related to soil and water management. In this regards, Conoscenti et al. (2015) using logistic regression (LR) and MARS techniques tried to assess susceptibility to earth-flow landslide. Their finding showed that Overall accuracy of implemented models was excellent. However, AUC values of MARS proved a higher predictive power of mentioned model (AUC = 0.881–0.912) with respect to LR models (AUC = 0.823–0.870). Wang et al. (2015) applied three mathematical methods such as LR, bivariate statistical analysis (BS), and MARS to create landslide susceptibility maps. The findings of their research showed that the MARS method has a better prediction rate (79%) when compared to LR (75%) and BS (77%). Naghibi et al. (2016) used three machine learning models including BRT, CART, and Random Forest (RF) for groundwater potential mapping. Based on their results, the best prediction model was BRT while CART and RF models selected for predicting locations of springs, respectively. Zabihi et al. (2016) compared MARS and RF models for groundwater spring potential mapping in Iran. They indicated MARS and RF as good estimators for groundwater spring potential in the Bojnourd Township, northeast of Iran. The advantage of BRT model based on Naghibi et al. (2016) study about groundwater potential mapping is modeling of different types of effective factors on considered events and overcoming in the lost data situation. In addition, assessment of BRT in modeling of different fields including groundwater spring potential mapping, landslide susceptibility mapping, and ecology are confirmed (Abeare 2009; Aertsen et al. 2011; Naghibi and Pourghasemi 2015; Youssef et al. 2015).

The data mining algorithms are as a new method that implemented for modeling by researchers in recent years. On the opposite side, statistics and statistical modeling as old methods are the traditional fields. In this field, quantification, collection, analysis, interpretation, and drawing are performed by data. But, data mining algorithm trying to investigates large existing databases in order to discover patterns and relationships in the data (Benjamini and Leshno 2005). The size of data is one of differences between new and old methods of modeling. In addition, the use of data mining methods in recent years proved the capability of it's against old models (Conoscenti et al. 2015). With these explanations and because of dynamics of the natural systems and its uncertainty, the application of data mining techniques can lead to understanding the natural systems and finally the management of mentioned system.

4 Conclusion

Land degradation is one of the most concerns that managers, decision makers, and researchers have always faced in recent years. Among different types of land degradation, soil erosion is most important of its. Nevertheless, gully erosion because of high contribution in the rate of output sediment should be seriously considered. In this regards, determination of susceptible zones to gully erosion in a watershed is the first step for soil erosion control and management. So, the present study is planned to map the gully erosion susceptibility in northern Iran (Mazandaran Province). For achieving this aim, eleven effective factors on gully erosion including altitude, aspect, slope degree, distance from road, distance from river, drainage density, plan curvature, profile curvature, TWI, land use, and lithology are determined based on literature review, availability, and accessibility to information. There are no multicollinearity problems based on tolerance and VIF indices. Therefore, all of considered effective factors are implemented in investigations and analyses. Application of data mining BRT and MARS models for spatial modeling of gully erosion susceptibility was the next step of this research. In the used models validation step, although both models have been located in very good class based on prediction accuracy value, but BRT model with the AUC value of 0.894 is selected and preferred for identifying susceptible zones to gully erosion. Results of BRT model revealed among studied conditioning factors, altitude (DEM), Aspect, and slope with the value of 43.03, 13.64, and 7.21% are assigned to rank 1 to 3 of important factors, respectively. It should be said that the BRT model unify the important advantages of tree-based models. This model doesn't need for initial data transformation or deletion of outliers. In addition, fitting complex nonlinear relationships is one of other advantages of BRT method (Elith et al. 2008). Also, feature selection and pruning are the benefits of the tree-based methods. Feature selection is a process which the most important and effective factors are selected for modeling and making decision. With applying these cases, the results are more accurate and acceptable. High speed in large data analysis and less sensitivity to over-fitting are the advantages of BRT (Liu et al. 2016; Salazar et al. 2016). Running with at least two factors is one of the disadvantages of aforementioned model. Working well with a large number of predictor variables, detecting and identifying interactions between variables, efficient and fast algorithm, despite its complexity, and Robust to outliers are the advantages of the MARS model. Difficulty to understand, and susceptibility to overfitting are the dis advantages of MARS algorithm. These results of current study can be applied for appropriate management of soil and water resources in the Valasht Watershed. Also, the results of the current study can be used as an important key of management in correct decision making including selection of critical and high sensitive to gully erosion for its control by policy makers, planners, and managers. In other words, the prioritization of different areas of gully erosion control in a watershed is the most important practical application of these results. Finally, more researches is recommended in order to comparison of findings and the final conclusions in the study area and other regions.

References

Abeare SM (2009) Comparisons of boosted regression tree, GLM and GAM performance in the standardization of yellowfin tuna catch-rate data from the Gulf of Mexico Longline Fishery. PhD thesis, University of Pretoria

Aertsen W, Kint V, Van Orshoven J, Muys B (2011) Evaluation of modelling techniques for forest site productivity prediction in contrasting ecoregions using stochastic multicriteria acceptability analysis (SMAA). Environ Model Softw 26(7):929–937

Akgun A, Turk N (2011) Mapping erosion susceptibility by a multivariate statistical method: a case study from the Ayvalık region, NW Turkey. Comput Geosci 37:1515–1524

Barnes N, Luffman I, Nandi A (2016) Gully erosion and freeze-thaw processes in clay-rich soils, northeast Tennessee, USA. Geo Res J 9:67–76

Basofi A, Fariza A, Ahsan AS, Kamal IM (2015) A comparison between natural and Head/tail breaks in LSI (Landslide Susceptibility Index) classification for landslide susceptibility mapping: A case study in Ponorogo, East Java, Indonesia. In: IEEE, 2015 International Conference on Science in Information Technology (ICSITech), Yogyakarta, 27–28 October, pp 337–342

Beguería S (2006) Validation and evaluation of predictive models in hazard assessment and risk management. Nat Hazards 37(3):315–329

Benjamini Y, Leshno M (2005) Statistical methods for data mining. Data mining and knowledge discovery handbook. Springer, US, pp 565–587

Bergonse R, Reis E (2016) Controlling factors of the size and location of large gully systems: A regression-based exploration using reconstructed pre-erosion topography. CATENA 147:621–631

Beven KJ, Kirkby MJ, Schofield N, Tagg AF (1984) Testing a physically-based flood forecasting model (TOPMODEL) for three U.K. Catchments. J Hydrol 69:119–143

Bouchnak H, Felfoul MS, Boussema MR, Snane MH (2009) Slope and rainfall effects on the volume of sediment yield by gully erosion in the Souar lithologic formation (Tunisia). CATENA 78(2):170–177
Breiman L, Friedman JH, Olshen RA, Stone CJ (1984) Classification and regression trees. Wadsworth & Brooks, Monterey, CA
Chung-Jo F, Fabbri AG (2003) Validation of spatial prediction models for landslide hazard mapping. Nat Hazards 30:451–472
Colkesen I, Sahin EK, Kavzoglu T (2016) Susceptibility mapping of shallow landslides using kernel-based Gaussian process, support vector machines and logistic regression. J Afr Earth Sci 118:53–64
Conforti M, Aucelli PPC, Robustelli G, Scarciglia F (2011) Geomorphology and GIS analysis for mapping gully erosion susceptibility in the Turbolo stream catchment (Northern Calabria, Italy). Nat Hazards 56(3):881–898
Conoscenti C, Agnesi V, Angileri S, Cappadonia C, Rotigliano E, Märker M (2013) A GIS-based approach for gully erosion susceptibility modelling: a test in Sicily, Italy. Environ Earth Sci 70 (3):1179–1195
Conoscenti C, Angileri S, Cappadonia C, Rotigliano E, Agnesi V, Märker M (2014) Gully erosion susceptibility assessment by means of GIS-based logistic regression: a case of Sicily (Italy). Geomorphology 204:399–411
Conoscenti C, Ciaccio M, Caraballo-Arias NA, Gómez-Gutiérrez Á, Rotigliano E, Agnesi V (2015) Assessment of susceptibility to earth-flow landslide using logistic regression and multivariate adaptive regression splines: a case of the Belice River basin (western Sicily, Italy). Geomorphology 242:49–64
Conoscenti C, Rotigliano E, Cama M, Caraballo-Arias NA, Lombardo L, Agnesi V (2016) Exploring the effect of absence selection on landslide susceptibility models: a case study in Sicily, Italy. Geomorphology 261:222–235
Dai FC, Lee CF (2002) Landslide characteristics and slope instability modeling using GIS, Lantau Island, Hong Kong. Geomorphology 42(3–4):213–228
Desta L, Adugna B (2012) A field guide on gully prevention and control. Nile Basin Initiative Eastern Nile Subsidiary Action Program (ENSAP), Addis Ababa, Ethiopia, p 67
Dormann CF, Elith J, Bacher S, Buchmann C, Carl G, Carré G, …, Münkemüller T (2013) Collinearity: a review of methods to deal with it and a simulation study evaluating their performance. Ecography 36(1):27–46
Dotterweich M, Stankoviansky M, Minár J, Koco Š, Papčo P (2013) Human induced soil erosion and gully system development in the Late Holocene and future perspectives on landscape evolution: The Myjava Hill Land, Slovakia. Geomorphology 201:227–245
Dube F, Nhapi I, Murwira A, Gumindoga W, Goldin J, Mashauri DA (2014) Potential of weight of evidence modelling for gully erosion hazard assessment in Mbire District-Zimbabwe. Phys Chem Earth 67:145–152
Dymond JR, Herzig A, Basher L, Betts HD, Marden M, Phillips CJ, Roygard J (2016) Development of a New Zealand SedNet model for assessment of catchment-wide soil-conservation works. Geomorphology 257:85–93
Elith J, Leathwick JR, Hastie T (2008) A working guide to boosted regression trees. J Anim Ecol 77(4):802–813
Felicísimo ÁM, Cuartero A, Remondo J, Quirós E (2013) Mapping landslide susceptibility with logistic regression, multiple adaptive regression splines, classification and regression trees, and maximum entropy methods: a comparative study. Landslides 10:175–189
Franzluebbers AJ (2010) Principles of Soil Conservation and Management. Vadose Zone J 9 (1):199–2001
Friedman JH (1991) Multivariate adaptive regression splines. Ann Stat 1–67
Friedman JH (2001) Greedy function approximation: a gradient boosting machine. Ann Stat 1189–1232

Geissen V, Kampichler C, López-de Llergo-Juárez JJ, Galindo-Acántara A (2007) Superficial and subterranean soil erosion in Tabasco, tropical Mexico: development of a decision tree modeling approach. Geoderma 139:277–287
Geology Survey of Iran (GSI) (1997) Geology map of the Mazandaran Province. http://www.gsi.ir
Golub GH, Heath M, Wahba G (1979) Generalized cross-validation as a method for choosing a good ridge parameter. Technometrics 21(2):215–223
Gómez-Gutiérrez Á, Conoscenti C, Angileri SE, Rotigliano E, Schnabel S (2015) Using topographical attributes to evaluate gully erosion proneness (susceptibility) in two mediterranean basins: advantages and limitations. Nat Hazards 79(1):291–314
Goodwin NR, Armston JD, Muir J, Stiller I (2017) Monitoring gully change: A comparison of airborne and terrestrial laser scanning using a case study from Aratula, Queensland. Geomorphology 282:195–208
Gutiérrez ÁG, Contador FL, Schnabel S (2011) Modeling soil properties at a regional scale using GIS and multivariate adaptive regression Splines. Geomorphometry 2011:53–56
Gutiérrez ÁG, Schnabel S, Contador JFL (2009) Using and comparing two nonparametric methods (CART and MARS) to model the potential distribution of gullies. Ecol Modell 220 (24):3630–3637
Hong H, Pourghasemi HR, Pourtaghi ZS (2016) Landslide susceptibility assessment in Lianhua County (China): A comparison between a random forest data mining technique and bivariate and multivariate statistical models. Geomorphology 259:105–118
Jain SK, Kumar S, Varghese J (2001) Estimation of soil erosion for a Himalayan watershed using GIS technique. Water Resour Manage 15(1):41–54
Jungerius PD, Matundura J, Van De Ancker JAM (2002) Road construction and gully erosion in West Pokot, Kenya. Earth Surf Proc Land 27(11):1237–1247
Kuhnert PM, Henderson AK, Bartley R, Herr A (2010) Incorporating uncertainty in gully erosion calculations using the random forests modelling approach. Environmetrics 21:493–509
Le Roux JJ, Sumner PD (2012) Factors controlling gully development: comparing continuous and discontinuous gullies. Land Degrad Dev 23(5):440–449
Leathwick JR, Elith J, Francis MP, Hastie T, Taylor P (2006) Variation in demersal fish species richness in the oceans surrounding New Zealand: An analysis using boosted regression trees. Mar Ecol Prog Ser 321:267–281
Li Z, Zhang Y, Zhu Q, Yang S, Li H, Ma H (2017) A gully erosion assessment model for the Chinese Loess Plateau based on changes in gully length and area. CATENA 148:195–203
Liu J, Sui C, Deng D, Wang J, Feng B, Liu W, Wu C (2016) Representing conditional preference by boosted regression trees for recommendation. Inf Sci 327:1–20
Luffman IE, Nandi A, Spiegel T (2015) Gully morphology, hillslope erosion, and precipitation characteristics in the Appalachian Valley and Ridge province, southeastern USA. CATENA 133:221–232
Martinez-Casasnovas JA (2003) A spatial information technology approach for the mapping and quantification of gully erosion. Catena 50(2-4):293–308
Monsieurs E, Poesen J, Dessie M, Adgo E, Verhoest NE, Deckers J, Nyssen J (2015) Effects of drainage ditches and stone bunds on topographical thresholds for gully head development in North Ethiopia. Geomorphology 234:193–203
Montgomery D, Dietrich WE (1989) Source areas, drainage density, and channel initiation. Water Resour Res 25(8):1907–1918
Motevalli A, Vafakhah M (2016) Flood hazard mapping using synthesis hydraulic and geomorphic properties at watershed scale. Stochast Environ Res Risk Assess 30(7):1889–1900
Mousavi SM, Golkarian A, Naghibi SA, Kalantar B, Pradhan B (2017) GIS-based groundwater spring potential mapping using data mining boosted regression tree and probabilistic frequency ratio models in Iran. AIMS Geosc 3(1):91–115
Naghibi SA, Pourghasemi HR (2015) A comparative assessment between three machine learning models and their performance comparison by bivariate and multivariate statistical methods in groundwater potential mapping. Water Resour Manage 29(14):5217–5236

Naghibi SA, Pourghasemi HR, Dixon B (2016) GIS-based groundwater potential mapping using boosted regression tree, classification and regression tree, and random forest machine learning models in Iran. Environ Monit Assess 188(1):44

O'brien RM (2007) A caution regarding rules of thumb for variance inflation factors. Qual Quant 41(5):673–690

Ollobarren P, Capra A, Gelsomino A, La Spada C (2016) Effects of ephemeral gully erosion on soil degradation in a cultivated area in Sicily (Italy). CATENA 145:334–345

Osman KT (2014) Soil erosion by water. In: Soil degradation, conservation and remediation. Springer, Netherlands, pp 69–101

Pham BT, Pradhan B, Tien Bui D, Prakash I, Dholakia MB (2016) A comparative study of different machine learning methods for landslide susceptibility assessment: A case study of Uttarakhand area (India). Environ Model Softw 84:240–250

Pimentel D (2006) Soil erosion: a food and environmental threat. Environ Dev Sustain 8(1):119–137

Poesen J, Nachtergaele J, Verstraeten G, Valentin C (2003) Gully erosion and environmental change: importance and research needs. CATENA 50(2):91–133

Pourghasemi HR, Kerle N (2016) Random forests and evidential belief function-based landslide susceptibility assessment in Western Mazandaran Province, Iran. Environ Earth Sci 75(3):1–17

Pourghasemi HR, Mohammady M, Pradhan B (2012) Landslide susceptibility mapping using index of entropy and conditional probability models in GIS: Safarood Basin, Iran. Catena 97:71–84

Pourghasemi HR, Moradi HR, Fatemi Aghda SM (2013) Landslide susceptibility mapping by binary logistic regression, analytical hierarchy process, and statistical index models and assessment of their performances. Nat Hazards 69(1):749–779

Pourghasemi HR, Rossi M (2016) Landslide susceptibility modeling in a landslide prone area in Mazandarn Province, north of Iran: a comparison between GLM, GAM, MARS, and M-AHP methods. Theor Appl Climatol, 1–25

Rahmati O, Haghizadeh A, Pourghasemi HR, Noormohamadi F (2016) Gully erosion susceptibility mapping: the role of GIS-based bivariate statistical models and their comparison. Nat Hazards 82(2):1231–1258

Rahmati O, Tahmasebipour N, Haghizadeh A, Pourghasemi HR, Feizizadeh B (2017) Evaluating the influence of geo-environmental factors on gully erosion in a semi-arid region of Iran: an integrated framework. Sci Total Environ 579:913–927

Robertson GP, Broome JC, Chornesky EA, Frankenberger JR, Johnson P, Lipson M, …, Thrupp LA (2004) Rethinking the vision for environmental research in US agriculture. Bioscience 54(1):61–65

Sadeghi SH, Zakeri MA (2015) Partitioning and analyzing temporal variability of wash and bed material loads in a forest watershed in Iran. Earth Syst Sci 124(7):1503–1515

Sadeghi SHR, Rangavar AS, Bashari M, Abbasi AA (2007) Waterfall erosion as a main factor in ephemeral gully initiation in a part of northeastern Iran. In: 2007 International Symposium on gully erosion: Pamplona, 17–19 September, pp 114–115

Salazar F, Toledo MÁ, Oñate E, Suárez B (2016) Interpretation of dam deformation and leakage with boosted regression trees. Eng Struct 119:230–251

Schapire RE (2003) The boosting approach to machine learning: an overview. Nonlinear Estimation Classif 171:149–171

Schonlau M (2005) Boosted regression (boosting): an introductory tutorial and a Stata plugin. Stata 5(3):330–354

Shellberg JG, Spencer J, Brooks AP, Pietsch TJ (2016) Degradation of the Mitchell River fluvial megafan by alluvial gully erosion increased by post-European land use change, Queensland, Australia. Geomorphology 266:105–120

Shruthi RB, Kerle N, Jetten V, Abdellah L, Machmach I (2015) Quantifying temporal changes in gully erosion areas with object oriented analysis. CATENA 128:262–277

Stumpf A, Kerle N (2011) Object-oriented mapping of landslides using Random Forests. Remote Sens Environ 115(10):2564–2577

Superson J, Rodzik J, Reder J (2014) Natural and human influence on loess gully catchment evolution: a case study from Lublin Upland, E Poland. Geomorphology 212:28–40
Swets JA (1988) Measuring the accuracy of diagnostic systems. Science 240(4857):1285–1293
Tebebu TY, Abiy AZ, Zegeye AD, Dahlke HE, Easton ZM, Tilahun SA, …, Steenhuis TS (2010) Surface and subsurface flow effect on permanent gully formation and upland erosion near Lake Tana in the northern highlands of Ethiopia. Hydrol Earth Syst Sci 14(11):2207–2217
Valentin C, Poesen J, Li Y (2005) Gully erosion: impacts, factors and control. Catena 63(2–3):132–153
Vanwalleghem T, Bork HR, Poesen J, Schmidtchen G, Dotterweich M, Nachtergaele J, …, De Bie M (2005) Rapid development and infilling of a buried gully under cropland, central Belgium. Catena 63(2):221–243
Wang LJ, Guo M, Sawada K, Lin J, Zhang J (2015) Landslide susceptibility mapping in Mizunami City, Japan: a comparison between logistic regression, bivariate statistical analysis and multivariate adaptive regression spline models. Catena 135:271–282
Wantzen KM (2006) Physical pollution: effects of gully erosion on benthic invertebrates in a tropical clear-water stream. Aquat Conserv Mar Freshwater Ecosyst 16(7):733–749
Yesilnacar E, Topal T (2005) Landslide susceptibility mapping: a comparison of logistic regression and neural networks methods in a medium scale study, Hendek region (Turkey). Eng Geol 79(3):251–266
Youssef AM, Pourghasemi HR, Pourtaghi ZS, Al-Katheeri MM (2015) Erratum to: landslide susceptibility mapping using random forest, boosted regression tree, classification and regression tree, and general linear models and comparison of their performance at Wadi Tayyah Basin, Asir Region, Saudi Arabia. Landslides 13(5):839–856
Zabihi M, Pourghasemi HR, Pourtaghi ZS, Behzadfar M (2016) GIS-based multivariate adaptive regression spline and random forest models for groundwater potential mapping in Iran. Environ Earth Sci 75:1–19

Concepts for Improving Machine Learning Based Landslide Assessment

Miloš Marjanović, Mileva Samardžić-Petrović, Biljana Abolmasov and Uroš Đurić

Abstract The main idea of this chapter is to address some of the key issues that were recognized in Machine Learning (ML) based Landslide Assessment Modeling (LAM). Through the experience of the authors, elaborated in several case studies, including the City of Belgrade in Serbia, the City of Tuzla in Bosnia and Herzegovina, Ljubovija Municipality in Serbia, and Halenkovice area in Czech Republic, eight key issues were identified, and appropriate options, solutions, and some new concepts for overcoming them were introduced. The following issues were addressed: Landslide inventory enhancements (overcoming small number of landslide instances), Choice of attributes (which attributes are appropriate and pros and cons on attribute selection/extraction), Classification versus regression (which type of task is more appropriate in particular cases), Choice of ML technique (discussion of most popular ML techniques), Sampling strategy (overcoming the overfit by choosing training instances wisely), Cross-scaling (a new concept for improving the algorithm's learning capacity), Quasi-hazard concept (introducing artificial temporal base for upgrading from susceptibility to hazard assessment), and Objective model evaluation (the best practice for validating resulting models against the existing inventory). All of them are followed by appropriate practical examples from one of abovementioned case studies. The ultimate objective is to provide guidance and inspire LAM community for a more innovative approach in modeling.

Keywords Landslide inventory · Susceptibility · Hazard · Machine learning Sampling · Validation · Cross-scaling

---

M. Marjanović (✉) · B. Abolmasov
Faculty of Mining and Geology, University of Belgrade, Belgrade, Serbia
e-mail: milos.marjanovic@rgf.bg.ac.rs

M. Samardžić-Petrović · U. Đurić
Faculty of Civil Engineering, University of Belgrade, Belgrade, Serbia

H. R. Pourghasemi and M. Rossi (eds.), *Natural Hazards GIS-based Spatial Modeling Using Data Mining Techniques*, Advances in Natural and Technological Hazards Research 48, https://doi.org/10.1007/978-3-319-73383-8_2

1 Introduction

Machine Learning (ML) techniques overtook the primacy in the field of Landslide Assessment Modeling (LAM) in the past couple of decades. Heuristic, statistical, and deterministic approaches gradually gave way to ML. This is evident from the publishing trends in the main thematic journals and conferences that involve LAM. Numerous comparative LAM studies show this increasing trend, mostly regarding comparisons between heuristic versus statistic (Shahabi et al. 2014; Yalcin et al. 2011), statistic versus deterministic (Ciurleo et al. 2017), but most importantly, statistic and/or heuristic versus ML (Steger et al. 2016; Choi et al. 2012; Erener et al. 2016; Yilmaz 2009). There are also numerous comparative studies that followed, and regarded comparisons between different ML techniques (Pradhan 2013; Pham et al. 2016; Youssef et al. 2016; Chen et al. 2017). Some of the above will be briefly discussed hereinafter to depict the general state of the matter and illustrate the trend of transition to ML techniques in LAM.

In their work, Yao et al. (2008) performed landslide susceptibility mapping based on support vector machines (SVM) for Hong Kong case study using one class and two class SVM methods and compared them to logistic regression models. They have concluded that two class SVM outputs better results compared to logistic regression and they recommended cross-validation usage for avoiding overestimation of landslide class.

Gallus and Abecker (2008) compared logistic regression, Gaussian Process models and the SVM for LAM for Voralberg case study (Austria), and they also showed that kernel-based methods outperformed logistic regression in their models.

Goetz et al. (2015) evaluated six different methods for LAM: generalized linear model logistic regression (GLM or LR), generalized additive models (GAM), weights of evidence (WofE), SVM, random forest classification (RF), and bootstrap aggregate classification trees with penalized discriminant analysis (BPLDA) for multiple areas of Lower Austria. Their study showed small differences of prediction performances between statistical and ML approach for LAM. RF and BDPA showed the best prediction performance, but visually the GAM and GLM were with best interpretation. Their general conclusion is that SVM, RF and BPLDA are useful for high-dimensional prediction problems, with large number of predictors.

Kavzoglu et al. (2014) compared multi-criteria decision analysis (MCDA), SVM and support vector regression (SVR) to predict the shallow landslide susceptibility of Trabzon province (NE Turkey) using 10 different predictors. Their results showed that the MCDA and SVR gave satisfactory results and they outperformed logistic regression method by 8% (of overall accuracy), but in their case study MCDA method produced slightly better results than the SVM method.

The study of Park et al. (2013) compared the abilities of frequency ratio (FR), analytic hierarchy process (AHP), logistic regression (LR), and artificial neural network (ANN) for LAM for the Inje area (Korea). Their results indicate that FR had the best performance, followed by AHP, LR, and ANN. They also found that all results were acceptable, but it should be highlighted that they did not include geological factors as predictors. Their results should be considered with caution regarding LAM, due to lower accuracy compared to similar studies in the available literature (Yilmaz 2009; Marjanović 2013; Youssef et al. 2016; Ciurleo et al. 2017).

Yilmaz (2009) compared Conditional Probability (CP), LR, ANN, and SVM for producing landslide susceptibility maps of Koyulhisar area in Turkey. His results showed that maps obtained by SVM and ANN have a slightly better accuracy compared to conventional statistical methods, but he also stated that all models were with similar accuracies.

Marjanović (2013) compared five modelling methods: stability index, AHP, Fuzzy sets, CP, and SVM for landslide susceptibility of Fruška Gora (Serbia). The study showed that SVM outperformed other methods, wherein stability index had the lowest performance (lowest area under the curve AUC, referring to Random Operating Characteristic or ROC curve).

Tsangaratos and Ilia (2017) in their review paper compared results from three different studies that have used SVM, Naïve Bayes (NB), AHP and LR for LAM, and they also concluded that SVM, NB and LR (all of them had >80% accuracy) outperformed AHP (with 69% accuracy).

Although Ciurleo et al. (2017) did not deal with ML technique, it is safe to infer some conclusions from their work. They suggested that statistical (simple bivariate) approach based on the informative value method, tends to reach similar performance as a deterministic TRIGRS model (based on Richards infiltration model coupled with infinite slope model). At the expense of somewhat higher overestimation of high landslide susceptibility class, they emphasize that a simple statistical approach was sufficient even for a large-scale study. The simplicity of the bivariate model is contrasted with the complexity of the TRIGRS model, which requires numerous physical and mechanical properties (limiting the application only to a well investigated sites). Given such circumstances, it is possible to assume that ML would perform equally well as simple bivariate statistics, which is probably the first next thing this research will attempt to compare.

Steger et al. (2016) challenged ML techniques, Random Forest and SVM in particular, with LR and general additive models. The conclusion was again that the ML techniques considerably outperform the others, but they also emphasized that model validation cannot rely only on performance metrics, but also include qualitative and spatial context of output maps, thereby drawing attention to biases that each model inherits from the input data.

There are also reports of better performance of non-ML technics in comparison to ML techniques (Yalcin et al. 2011; Choi et al. 2012). However, the validation metrics in these was questionable, as they both gave way to non-ML techniques, disregarding their large number of false positives (overestimations).

The current ML practice in LAM is becoming oversaturated with case studies that merely reapply familiar techniques or several of them simultaneously. Very slight differences between most popular ML techniques are compared, but even such comparative studies lead to "yet another Landslide Assessment map", with little or no innovation in the process. Sudden heaps in numbers of research papers are taking place only when entirely new, or some useful but forgotten ML technique comes into play, but even then, mere re-application is usually taking place.

On the other hand, LAM researchers, especially the inexperienced ones, are agonized with several common issues, and driven to cope with the same problems, usually making same mistakes. Herein, the most common of these issues were identified and will be addressed in the respective sub-chapters. The objective is to try to guide potential novice LAM community throughout, and inspire new ideas to build upon. The following issues are included: landslide inventory enhancements, choice of attributes and attribute selection, choice of ML task—classification versus regression, choice of ML technique, sampling strategy, cross-scaling, Quasi-Hazard —artificially introducing temporal base, and objective model evaluation. Each of these will be elaborated in separate sub-chapters, followed by a brief overall conclusion. All addressed issues for consideration are complemented with practical examples from our experiences, in various case studies.

1.1 Case Studies

Practical examples addressing target issues will be demonstrated on completed and ongoing case studies. Hereinafter, landsliding history, setting and other characteristics of these case study areas will be briefly described.

- The City of Belgrade in Serbia, which has over 1.6 million inhabitants within the area of 3222 km^2, was assessed for landslide susceptibility in several occasions (Božović et al. 1981; Gojgić et al. 1995), including detailed landslide mapping during 2006–2010 (Lokin et al. 2012), but there are also ongoing aspects of this study area. It is especially in focus after the 2014 disastrous flooding and landsliding events in Serbia, wherein 1.6 million people were affected across Serbia, with 51 casualties, around 32,000 evacuated, total damage up to 1.53 billion USD or 5% of Gross Domestic Product (GDP). The area includes central and eastern suburban parts of the City, since northern outskirts are flat plains with no landsliding potential (Fig. 1a), the geological setting is relatively complex, with Cretaceous flysch and limestone base, overlain by Neogene clays, marls, which together with Quaternary deluvial deposits, host the most of the landslides. Deep-seated composite slides are dominant, many with active or temporarily suspended kinematic status (270 of around 1000 landslides), and the largest ones are about 850,000 m^2, while average size is around 90,000 m^2.
- The City of Tuzla in Bosnia and Herzegovina was a subject of "Detailed flood and landslide risk assessment for the urban areas of Tuzla and Doboj" in 2015–16. It followed a general landslide risk assessment for the entire Bosnia and Herzegovina in 2015, as a detailed example of the proposed methodology with ML implementation (Marjanović and Đurić 2016). The city is located in the north-eastern part of the Tuzla Canton in Federation of Bosnia and Herzegovina (Fig. 1b). The territory of the inner case study area covers around 113 km^2, with a total population of 120,441. It was previously known as unstable terrain, with steep slopes and pronounced relative height differences over short distances, with an average slope of 11° and maximum slope of 56°. The historical landslide

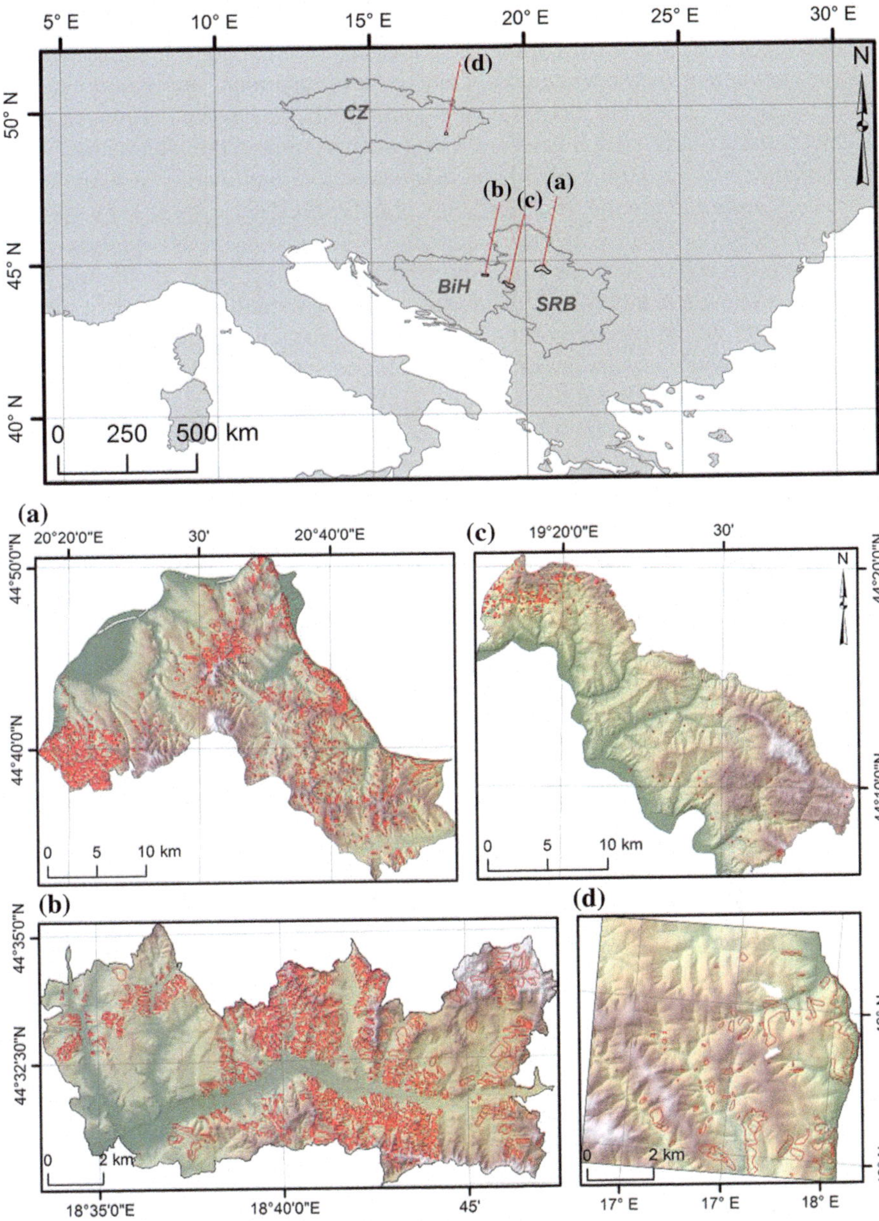

Fig. 1 Case studies setting: **a** The City of Belgrade, **b** The City of Tuzla, **c** Ljubovija Municipality, **d** Halenkovice Area (registered landslides are given as red contours)

inventory contains 941 records, with largest landslides of over 400,000 m^2, and an average size of 17,900 m^2. These are primarily hosted in high-plasticity soft clay and incoherent to moderately coherent sedimentary rock, and mainly shallow translational slides, but there is a great number of flows as well. The area was in focus, particularly after Bosnia and Herzegovina suffered the aftermath of floods and landslides in 2014, with 90,000 evacuated, 25 killed, while the total damage was estimated at around 2 billion USD, or 15% of GDP.

- Ljubovija Municipality in Serbia was also in focus after the 2014 events and is being approached from several aspects in an ongoing study, including deterministic, heuristic and ML landslide susceptibility modeling methods. Ljubovija is located in the westernmost part of Serbia, occupying an area of around 331 km^2, with population of 3929 (2011 census) (Fig. 1c). The terrain is dominated by smaller but steep valleys of the Drina River tributaries, with a maximum slope of 65°. Geologically, the most of the area is covered by Paleozoic schist formations, which host landslides in their thick weathering crusts, generally well known for instabilities. Shallow slides and debris flows were the most common types in the May 2014 event, which is unusual (in western Serbia, geological and geotechnical engineering practice indicates that slides are more abundant than flows due to geological setting and climate). Therefore, the attention in this study area was directed to flows. There were 271 landslide occurrences recorded in 2014, mostly active or suspended, with an average size of about 2,000 m^2, while the largest one was over 90,000 m^2.
- Halenkovice area (60 km^2) is located in the southeastern Czech Republic, close to Brno (Fig. 1d). It has been investigated for shallow movements (Marjanović 2014; Kircher et al. 2000; Bíl and Müller 2008) which are typically hosted in a deluvial mantle, developing translational shallow earth-sides. Due to a high clayey and loamy proportions in the mantle, the instabilities are pervasive even on relatively gentle slopes (maximal just over 30°, average around 9°). As of 2013, there were 24 active and 113 dormant landslides, with an average size of about 42,000 m^2, although in many cases multiple instances were merged together as a single landslide polygon. The area is sparsely populated so these landslides do not pose imminent threats to human lives or property.

2 Landslide Inventory Enhancements

Landslide Assessment is usually practiced as landslide susceptibility, i.e. the spatial probability of landslide occurrence (Varnes 1984), whereas landslide hazard is its temporal extension that is usually difficult to finalize since it requires historical data on frequency and/or magnitude of registered landslides. The latter is rarely archived for any meaningful return periods, e.g. 50-year periods or longer, therefore landslide susceptibility is largely preferred to hazard assessment. In any case, the analysis first requires an inventory of registered landslides. In hazard case, the inventory is multi-temporal, whereas for susceptibility studies the inventory usually contains landslides originating from a single massive landsliding event, or more often, all historical landslides ever-recorded in the target area. Landslide inventory standards usually require further details about landslide type, activity status and other useful pieces of information, depending on the level of acquisition detail (e.g. landslides recorded via remote sensing will not contain as many details as those mapped in the field), and they are usually implying that the instances are mapped as polygon features. For detailed scales, features are even further split into source and accumulation parts (Guzzetti et al. 2012). Furthermore, in all ML implementations for LAM the original polygonal geometry of the inventory is converted to raster format of desired resolution, as landslide localities or so-called positives, which are used for sampling, learning, and validating protocols. This also led to sampling non-landslide localities, so-called negatives. The final raster is usually called class label comprising landslide and non-landslide binary classes.

The key problem with any landslide inventory is usually the scarcity of landslide instances in comparison to the size of the target area (Malamud et al. 2004). In western Serbia, it is quite often that landslides cover less than 10% of the target area (Đurić et al. 2017), which entails further difficulties. Optimal operative resolution is defined by the smallest distinguishable size of recorded landslides, which means that the smallest ones might be represented by a single pixel. Thus, there would generally be too few landslide and too many non-landslide instances to train and validate the models on, especially in detailed studies where only source areas are considered, so even fewer landslide instances are available. In addition, validation requires about 10–20% of landslide instances and equal number of non-landslide instances (Marjanović et al. 2011; Lombardo et al. 2014; Cama et al. 2016), although there are some other views that support against balanced, and suggest splitting that is proportional to the original class distribution (Oommen et al. 2011), which is still less practical in the LAS due to the landslide samples scarcity. It is understood that validation instances never took part in the analysis, but serve only for validation purposes, which introduces another deduction of the training instances count. For instance, in the case of Ljubovija municipality the initial number of all landslide pixels was 427 against 386,780 non-landslide pixels (30 m resolution), or 427 available landslide instances for training and validating over an area of about 355 km^2 (Fig. 2a). One way of dealing with such shortcoming was to increase the resolution of landslide pixels. One-fold resolution increase resulted in 1708 instances (Fig. 2b), which then made it easier to sample for training (e.g. randomly, uniformly distributed or by some other sampling strategies (see Section "Sampling strategy") and spare 10–20% for validation. Ideally (Hastie et al. 2009), there should be up to 25% for validation, 25% for testing and 50% for training, but this is a very general rule of thumb, and does not apply in the cases with strong class size disproportion (herein non-landslides $\gg$ landslides). Testing and validating sets are usually separated (training set for training the model, validating set tuning the model, testing set to challenge it against unseen instances, as in real-world applications), but it is applicable in data-rich situation (Hastie et al. 2009), which is usually not the case in LAM.

Another way is to account for acquisition precision and precision of the conversion of vector polygon to raster, thereby using one-half pixel dimension

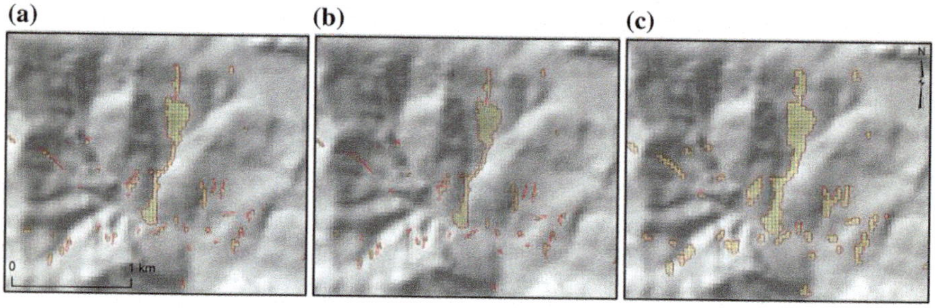

Fig. 2 Rasterization of landslide polygons from the inventory—a detail of Ljubovija Municipality case study: **a** Direct rasterization at 30 m resolution, **b** Direct rasterization at 15 m resolution, **c** Rasterization at 30 m resolution with 15 m tolerance buffer around polygons

(e.g. 15 m) as a tolerance buffer. In the case of Ljubovija, 1475 landslide instances were generated (Fig. 2c), providing similarly improved sampling choice as with the previous technique with resolution increase. Non-landslide instances usually do not require any intervention, and it would not be justified to use the same artificial boost of resolution, unless the input data support it (i.e. inputs resolution should dictate the pixel reduction).

3 Choice of Attributes

Beside class label (landslide/non-landslide), which is (as previously described) obtained from the landslide inventory, ML learning protocol training/testing requires attributes (also called predictors, independent variables, features, dimensions - in the context of input space dimensions), that are appended to each training instance. Set of attributes must be the same in training and testing/validating samples. These are usually separated into conditioning and triggering factors, depending on their temporal stationarity, as will be explained in greater detail later on. Conditioning factors are usually in the focus of susceptibility assessment, while hazard assessment can benefit from using triggering factors (see Section "Quasi-hazard"). There is more-or-less common ground among researchers on which conditioning factor attributes are relevant for LAM. These are chiefly geological (Yilmaz 2009; Pradhan 2013; Dumlao and Victor 2015), morphometric (Foumelis et al. 2004; Sharma et al. 2012; Kukemilks and Saks 2013), and environmental attributes (Heymann et al. 1994; Jones 2002; Shahabi et al. 2014), acquired from various sources, including corresponding thematic maps, elevation data, satellite imagery and linked products. Developing of GIS and Geostatistics allowed even further synthesizing of additional attributes from existing inputs, such as XY-lat/long coordinates, various spatial buffers or interpolations, higher-order derivatives, etc. (Fig. 3). Using XY coordinates works only if there is a high spatial autocorrelation of landslide distribution, which is not always the case (unlike other phenomena such as temperature vs. latitude for instance).

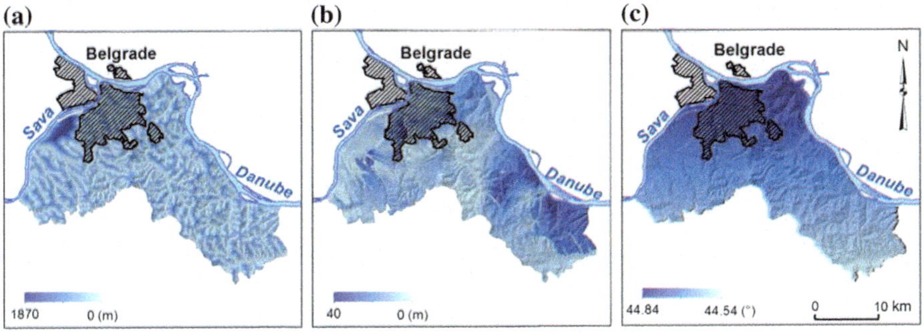

Fig. 3 Examples of synthetic attributes for The City of Belgrade case study: **a** Distance to stream, **b** Kriged depths to groundwater, **c** Latitude

Attributes can be either numeric or discrete (such as lithological or land cover maps). Discrete attributes usually require some additional processing, i.e. quantification of discrete values (e.g. in Weka software for ML, only class labels can be discrete in the classification task). One way is to assign an expert-driven weights for each class of discrete attribute, thereby transforming it into a numerical input. Another choice is to generate binary (dummy) attributes, e.g. 5-class discrete attribute will be split into 5 new binary attributes, wherein each class (1) will be represented against all remaining 4 classes (0) put together (one vs. all, Fig. 4).

Given all these possibilities to generate higher-order morphometric, synthetic and dummy attributes, the operative number of attributes can grow considerably, thereby increasing the dimensionality of ML model. However, high-dimensionality is not always in the best interest of the model, as it can entail redundancy and poorer generalization. It is therefore advisable to optimize it by introducing some attribute selection or extraction technique.

When processing operative attributes, one can basically turn to ranking-and-selecting, or correlating-and-extracting approaches (Hall and Holmes 2003). The former rank the attributes by their importance and leave to the user to decide whether to reject any or possibly use the ranking in the weighting process for simple modeling approaches. The latter implies cross-correlating attributes against the class label and making a reduced subset of attributes.

Several ranking techniques are very popular in LAM: gain ratio, information gain (IG), principal components, etc. All these are based on entropy values, which determines how informative (useful) an attribute is for revealing the classification/regression rule within the class label. In fact, all tree-based ML algorithms work by using these rankers for shaping the classification/regression rule (Mitchell 1997), so it is safe to say that attribute selection is somewhat redundant for tree-based ML implementation, as it is performed during the learning process. Anyhow, attributes are ranked from the most important to the least important, which then leaves the user to define a threshold which will be used as a criterion for excluding unimportant attributes (reducing feature space dimensionality). One way of defining it is using leave-last-out approach (Marjanović 2014), wherein the modeling is

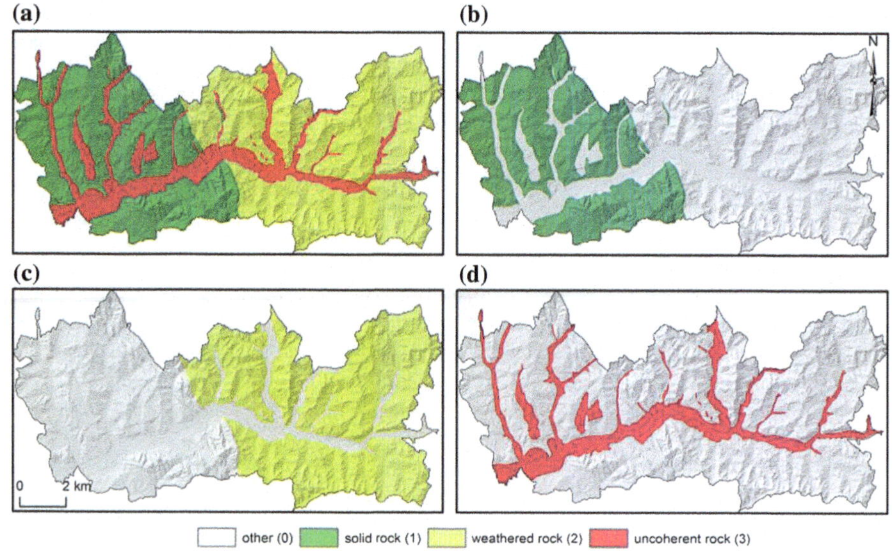

Fig. 4 Example of generating dummy attributes from discrete rasters: **a** Engineering geological units, **b** Unit of solid rock against other, **c** Unit of weathered rock against other, **d** Unit of uncoherent rock against other

performed repeatedly by using smaller and smaller attribute sets. In every successive run, the lowest ranked attribute is left out of the attribute set. The threshold is reached when performance parameters are starting to drop significantly, e.g. if it drops more than 2–3% of the given evaluation parameter. In the example of Halenkovice study, the threshold can be drawn at attribute 21, meaning that least ranked attributes between 21 and 26 could be easily excluded without affecting the prediction performance (Fig. 5). Naturally, there is no need to test all ranks from 26 to 1, but it is simply verifiable via removing several least-ranked attributes to see if anything drastic happens. In most cases, these rankers create similar ranking lists, but some differences may occur.

Correlation Feature Selection (CFS) is one of the widely-used examples for feature extraction (Witten et al. 2011). It creates reduced attributes subsets that contain only attributes correlated with the class label, but uncorrelated with each other. It rarely increases performance (Table 1), but if the performance remains similar, it is justified to practice this technique for saving time and hardware capacity. Given that at the current state of software/hardware development, time consumptions for processing model variants with complete and reduced attribute sets do not differ much, the attribute extraction is not entirely justified.

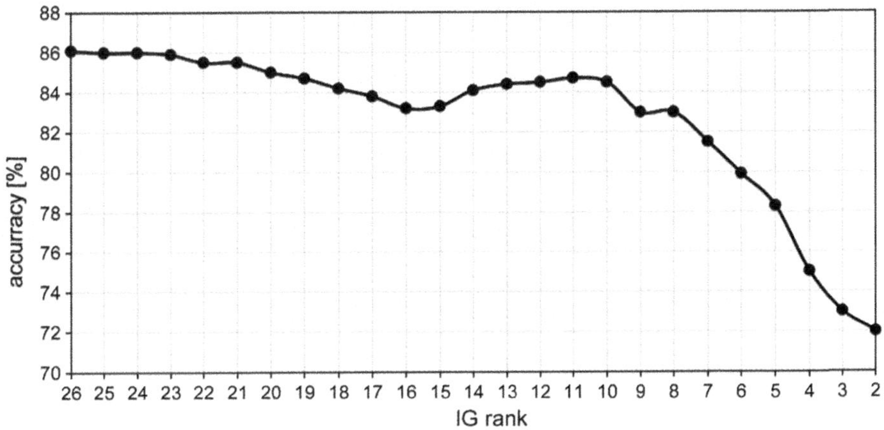

Fig. 5 Accuracy versus IG rank of attributes in leave-last-out concept for Halenkovice case study, using the SVM algorithm in a typical classification task (after Marjanović 2014)

4 Classification Versus Regression

Another fundamental problem that is commonly taken for granted is the choice of the ML task. Namely, the task can be to map and predict landslide class, or more commonly, to map the spatial probability (susceptibility) of landslides. ML implementation offers two different task definitions to adjust to these two approaches. The latter, i.e. probabilistic mapping, implies the regression ML task. The chosen ML algorithm is therein used to learn a regression function from supplied training instances. Training instances have a numeric character, i.e. they represent landslide probabilities in a range [0–1]. It is convenient that there are extreme examples with 0 or 1 probability, but also intermediate examples with probabilities e.g. 0.2, 0.5, 0.7 etc., which helps the algorithm to learn about such inexplicit cases and predict better. This is sometimes difficult to determine if there is no reliable evidence (temporal/historical record of failure probability over the specified return period, which is common for hazard analysis, or a static factor of safety as a reference for the probability of failure). For instance, if there are active landslides, it is advisable to assign them probability 1, while dormant landslides (by definition) are inactive within two years, which would mean that their annual probability should be 0.5 or less. It is possible to use some deterministic models, which outputs factor of safety, as learning support for assigning probabilities, but in regional case studies deterministic models are not very applicable for large scales. After the training is completed, all unused instances are projected along regression function learned by a particular ML algorithm. Direct output is a susceptibility map with a 0–1 span of probability (Fig. 6a).

The other option, i.e. mapping landslide class across the area, implies the classification ML task, wherein the algorithm learns from the training set of

Table 1 AUC comparison before and after attribute selection for The City of Belgrade case study, using SVM and RF algorithms and their CFS and IG ranker (first 10 best-ranked attributes) variants

Model	ROC area			
	Class 0	Class 1	Class 2	Average AUC
SVM	0.68	0.56	0.82	0.69
SVM_CFS	0.62	0.55	0.53	0.57
SVM_IG	0.69	0.56	0.83	0.69
RF	0.70	0.58	0.78	0.68
RF_CFS	0.63	0.57	0.60	0.60
RF_IG	0.69	0.58	0.78	0.68

Class 0 = non-landslide, class 1 = dormant landslide, class 2 = active landslide

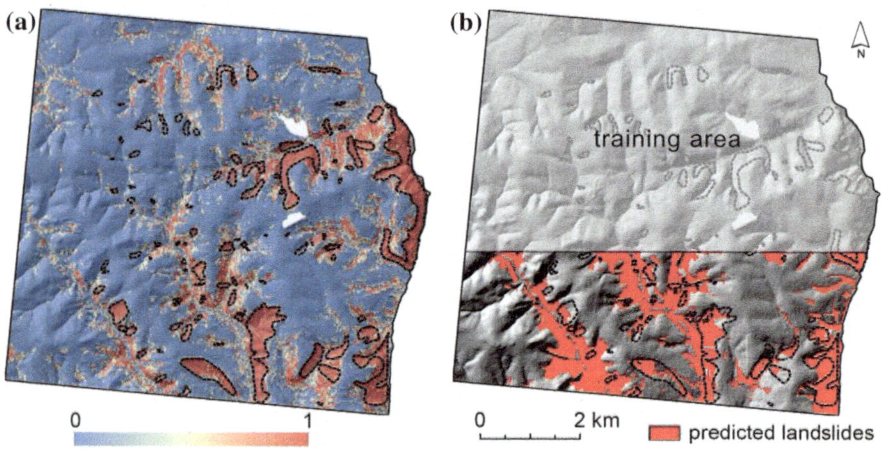

Fig. 6 Landslide susceptibility (**a**), and a classified landslide prediction map (**b**), from Halenkovice case study, using SVM algorithm (after Marjanović 2014). Note how classification task can result in either probabilistic or class-specific predictive maps

landslide and non-landslide instances how to map specific class. This case is more in use when practitioners have difficulties assigning probabilities to their inventoried landslides. Instead, they would rather keep the original discrete class labels, e.g. by activity status: active = 1, suspended = 2, dormant = 3, marginally stable slopes = 4 and non-landslides = 5 (any number can be assigned instead of 1, 2, 3, 4, 5 because these numbers represent just the class name). Learned classification function is then used to separate each unseen instance into either of these landslide classes. Resulting map is discrete, usually binary (landslide vs. non-landslide) or it can involve multi-class case, e.g. 1–5, depending on the number of classes introduced in the training. Even though discrete, these maps can be transformed into probabilistic susceptibility maps if the model is iteratively run and the final result is obtained by averaging all intermediate models (Fig. 6). Iterations are usually implying different sets of training samples, so that each intermediate model has a different layout of training instances. Many ML algorithms, especially ensemble multi/meta-techniques, such as Random Forest, allocate probability of class prediction per each predicted instance, which (similarly as above mentioned averaging) can be used to produce a 0–1 span of values to transform predictive maps to probabilistic.

5 Choice of ML Technique

Choosing the most appropriate ML technique to perform desired classification/regression task is fundamental. Users are usually practicing techniques they are most familiar with, but which one generalizes better in particular cases cannot be determined beforehand. It is therefore advisable to use several techniques simultaneously and observe which one best suits a particular case study. There is no general preference, as all the techniques have their pros and cons (Table 2). One could go for quick-yielding techniques that do not require deeper understanding of their black-box concepts, such as Extreme Learning (EL) or similar ANN, or even ensemble learning techniques such as Random Forests (RF). Others might prefer simple, but effective techniques, such as Decision Trees (DT), especially if they would like some insight in the function-making process (grey-box). DTs are therein unique, and drift from typical black-box, because they allow user to observe the aggregated rules that lead to a specific classification/regression function. In any case, it is quite important to optimize the chosen algorithm and to use at least a couple of them, and cross-compare results. Optimization usually requires to define the best pairs or triplets of relevant algorithm parameters, e.g. misclassification penalty C and kernel dimension γ in SVM, number of hidden layers and iterations in ANN, number of trees and number of randomly sampled attributes in RF. Some reports (Caruana and Niculescu-Mizil 2006) are suggesting that best-yielding techniques in general are ensemble learning algorithms, such as RF, closely followed by SVM, EL, ANN, therein outperforming DT and LR. Similar findings to this general comparison is found within LAM (Pradhan 2013; Pham et al. 2016; Steger et al. 2016; Youssef et al. 2016; Chen et al. 2017), and a brief discussion of our experience with these will follow hereupon. For particular details about these (popular) techniques in general, we suggest Witten et al. (2011), Mitchel (1997), and Hastie et al. (2009), wherein most of these are explained in detail.

The obtained values of the overlay measure (AUC) indicate a good performance for all four models with slight differences (Table 3 and Fig. 7). The values of TP (true positive) rate indicate that all four models have similar capability to correctly predict landslide instances. The RF-based model has the smallest distribution of misclassification of landslide (FPrate) and the largest distribution of correctly classified non-landslide instances (TNrate) compared to other models. Conversely, the EL-based model has the largest value of FPrate and the smallest value of TNrate compared to other models (Table 3). RF model gives the least overestimations of target landslide class, while other methods perform similarly in this respect (Fig. 7). However, RF does have significantly higher FNrate (Table 3), which indicate that it

Table 2 Pros and cons overview of some ML classifiers from our perspective

ML technique	Pros	Cons
DT[a]	Relatively fast (in both learning and classifying), deals well with nominal and numeric data, deals well with NoData, easily optimized, easy to interpret (especially when pruned)	Lower accuracy, sensitive to sampling strategy (tends to create too complex models with large training sets, and has low performance with small training sets), tends to overfit, performs poor with high-dimensional data
LR	Relatively fast (in both learning and classifying), deals well with high-dimensional data, no optimization required, outputs class probability for interpretation	Lower accuracy, sensitive to sampling strategy, deals badly with nominal data (requires indirect transformation to scaled/scored numeric), handles only linearly separable relation, tends to overfit and overestimate (high FP)
ANN[b]	Accurate, deals well with nominal (indirectly) and numeric data, suitable for high-dimensional data	Relatively slow, sensitive to sampling strategy (deals badly with redundant and noisy data in training), tends to overfit, requires extensive optimisation, not interpretable
EL	Accurate, fast, deals well with nominal (indirectly) and numeric data, suitable for high-dimensional data, no optimization required	Sensitive to sampling strategy, tends to overfit and overestimate, not interpretable
SVM[c]	Accurate, deals well with nominal (indirectly) and numeric data, deals well with noisy and redundant data, works well with high-dimensional data, works well (better) with small training sets	Relatively slow, deals badly with NoData, not easy to optimize (2–3 fitting parameters), tends to overfit, not interpretable
RF	Accurate, fast, deals well with nominal and numeric data, deals well with NoData, as well as with redundant and noisy data in training, easy to optimize, somewhat interpretable	Works badly with small training sets, works badly with high-dimensional data (due to random choice of attributes at the individual trees' node level)

[a]C4.5
[b]Perceptron feed-forward
[c]with radial-basis-function kernel

underestimates the landslide class. A tradeoff threshold between contingency table indices is needed to select the model of choice. As a rule of thumb, it is usually better to choose the model that has less FN, but other aspects of the model (e.g. spatial and geomorphic plausibility) also need consideration (Steger et al. 2016).

Some of the advantages of SVM compared to other ML techniques, which make this technique very popular are: the capability to provide a good generalization if the parameters of kernel function are appropriately optimized. It is also robust and provides unique classification rule when using the same training samples (unlike ANN, wherein randomization effectively influences weights at the node level and results in slightly different outcomes while using the same training sample). RF is

Table 3 Performance of various models implemented in Ljubovija Municipality case study

Model	TPrate	TNrate	FPrate	FNrate	AUC
ANN	0.83	0.54	0.46	0.17	0.77
SVM	0.86	0.60	0.40	0.14	0.73
RF	0.73	0.78	0.22	0.27	0.76
EL	0.89	0.39	0.61	0.12	0.76

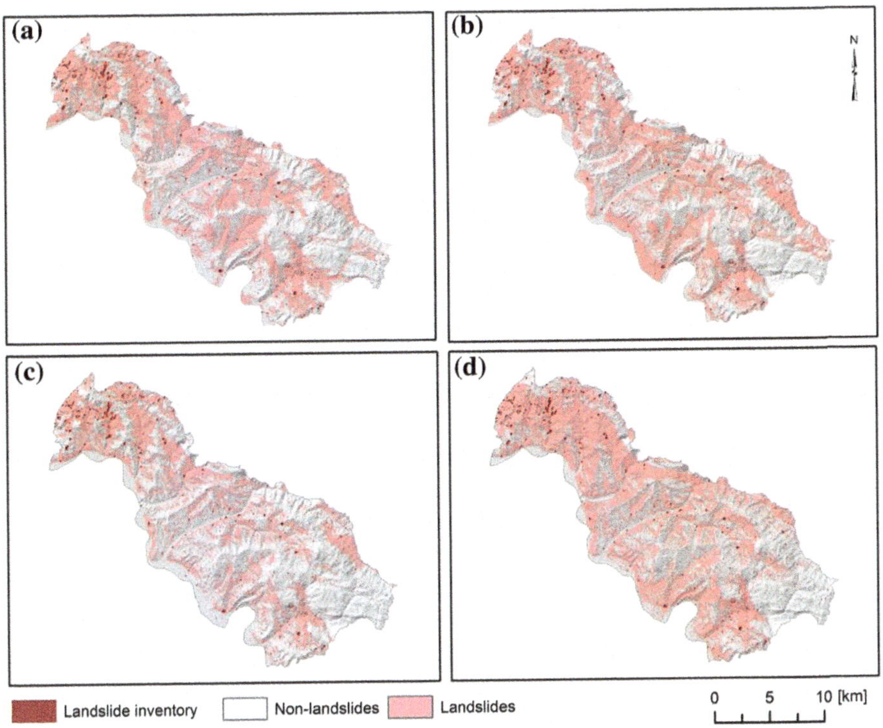

Fig. 7 Landslide (binary) predictions, for Ljubovija Municipality case study, derived by: **a** ANN, **b** SVM, **c** RF, and **d** EL

also a very popular technique in LAM because it avoids overfit, due to the *Law of Large Numbers*. Another advantage is RF's capability to handle less informative attributes. Recently introduced EL technique was also identified as one of the best performing and fastest models in terms of processing time (even faster than RF, and much faster than SVM and ANN).

6 Sampling Strategy

As indicated before, ML concept implies training and testing protocols. Training protocol is used for learning a classification or regression function based on the sample dataset, i.e. the training dataset. The two most commonly used strategies for sapling are random and uniform strategies. However, the nature of the landsliding process indicates the spatial variability and heterogeneity, demonstrating that landslides do not occur uniformly and usually cover only a small percentage of the whole study area. Therefore, using the training data set that contains randomly chosen samples or that contains uniformly distributed samples (e.g. 100 × 100 m grid) is not appropriate. Considering the relatively small number of landslide instances and the nature of landslides regarding their spatial distribution, the more suitable, two-step, sampling procedure is herein proposed. The first step includes the definition of all landslide areas as landslide polygons (if raster data are used), and all areas that are highly unlikely to host landslide (based on the expert knowledge) as non-landslide polygons. Second step randomly selects a half of all landslide instances as training landslide samples, ensuring that they are distributed across all landslide polygons from the inventory, considering the size of each landslide polygon, and selecting the same number of instances from non-landslide polygons in the same manner. By using this proposed balanced sampling strategy, both classes (landslide—1 and non-landslide—0) are equally represented, and thus the possibility that the ML model favors the majority classes (non-landslide—0) during the learning/training protocol is avoided. The rest of the instances that have not been sampled for training dataset are used for testing protocol as testing/validating dataset. These are subjective (expert-driven) criteria, which are dependent on the particular case study and experience of the practitioner. Beside expert knowledge, another way to sample non-landslide instances, recommended by Tsangaratos and Benardos (2014), is to use Mahalanobis distance metric (Mahalanobis 1936). This objective approach would involve the sampling that is based on the distance from the existing (recorded) landslides, wherein furthest areas are potentially interesting for sampling non-landslide instances (Kornejady et al. 2017). Namely, using Mahalanobis distance, which is a probabilistic distance (that considers the spatial variance, following a general geographic law that closer entities are similar), can result in successful grouping of non-landslide instances. Thereby, distant pixels are less likely to interfere and confuse the learning algorithm, and can be regarded as "safe" non-landslides. This procedure is considered superior to purely random sampling (Kornejady et al. 2017).

In order to determine how the proposed sampling strategy can contribute to the improvement of the model the following experiment was conducted in Ljubovija case study. Dataset (S), used to represent the entire Ljubovija territory, contains the total of 387,207 instances (30 × 30 m grid cells), out of which only 1475 (0.38%) instances represent landslides. Originally, the number of landslide instances was even smaller, but after introducing a 15 m tolerance buffer (see Section "Landslide inventory enhancements") their number has been increased. The entire dataset was sampled to determine the training split using all three sampling methods separately, random, uniform and customized, labeled S^r_{tr}, S^u_{tr} and S^c_{tr}, respectively. Remaining instances, i.e. instances that were not included in the training datasets, were used for testing/validating (separately for every sampling method), labeled S^r_{te}, S^u_{te} and S^c_{te}, respectively. Apart from enhancement of the inventory for increasing the number of landslide instances, the proposed strategy implies limiting the area for sampling non-landslide instances to areas that are arbitrarily chosen by the expert as least-likely landslide-prone areas (ridge zones, alluvial deposits, exclusion of the landslide upslope area, exclusion of the areas that are not suitable for landslide progression spatially or geologically, gentle slopes with less than 5°, etc.). It is also suggested to group non-landslide samples in clusters to achieve better performance (Conoscenti et al. 2016). Contents and spatial distribution of training and testing datasets are presented in Table 4 and Fig. 8.

Based on the training datasets, S^r_{tr} and S^u_{tr} contents, it is evident that the use of random and uniform strategy resulted in very small numbers of landslide instances (almost none), which was expected given that the probability of their selection during the sampling was small due to the very small percentage of landslide instances and their heterogeneous spatial distribution. On the other hand, the use of proposed strategy has contributed to balance dataset, with the original spatial distribution of landslide instances.

RF technique with default parameters was used to build models for this experiment. Therein, the LAM is considered as a classification task (landslide—1 and non-landslide—0). The measures of validation that were used include TPrate, FPrate, FNrate, TNrate and AUC (Bradley 1997). Results are shown in Table 5.

When the models were built with S^r_{tr} and S^u_{tr} datasets and then tested on corresponding S^r_{tr} and S^u_{tr} datasets, the results for the TNrate and FPrate were excellent (Table 5). Contrarily, the values of TPrate and FNrate were very bad. Further, the comprehensive measure (AUC) indicates that the same models would have been derived even if the random class scoring was performed. This result reflected the nature of both training datasets in which almost all of the instances are non-landslides. Thus, during the training protocol, RF was not fed by the data on

Table 4 Number of training instances in various sampling strategies (indexes next to sample lable S: $_{tr}$—training, $_{te}$—testing, ^{r}—random, ^{u}—uniform, ^{c}—custom)

Dataset	Sampling strategy	# of landslide instances	# of non-landslide instances	Total # of instances
S^r_{tr}	Randomly	5	1469	1474
S^r_{te}	$S - S^r_{tr}$	1470	384,263	385,733
S^u_{tr}	Uniformly	3	1471	1474
S^u_{te}	$S - S^u_{tr}$	1472	384,261	385,733
S^c_{tr}	Custom sampling	737	737	1474
S^c_{te}	$S - S^c_{tr}$	738	384,995	385,733

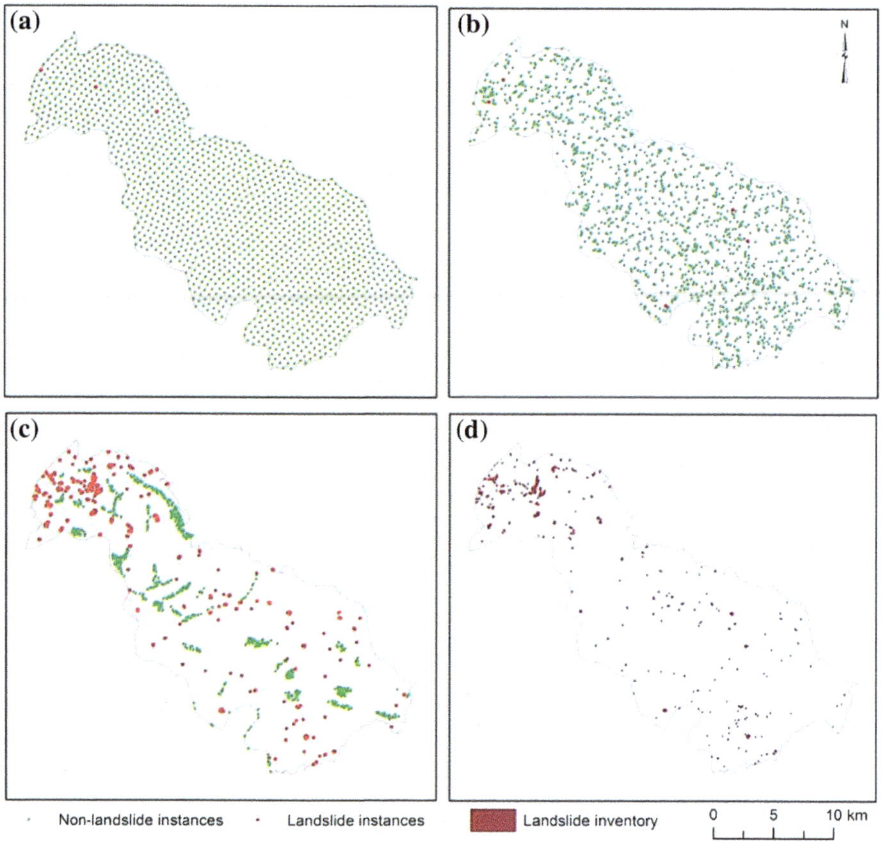

Fig. 8 Distribution of sampling instances by various strategies: **a** random, **b** uniform, **c** custom sampling strategy (expert-defined areas for sampling non-landslide instances), and **d** landslide inventory

Table 5 Performance of RF model using different sampling strategies for Ljubovija Municipality case study (indexes next to sample lable S: $_{tr}$—training, ^{r}—random, ^{u}—uniform, ^{c}—custom)

Training set	Testing set	TPrate	TNrate	FPrate	FNrate	AUC
S^r_{tr}	S^r_{te}	0.001	1.000	0.000	0.999	0.565
S^u_{tr}	S^u_{te}	0.001	1.000	0.000	0.999	0.566
S^c_{tr}	S^c_{te}	0.936	0.391	0.609	0.064	0.825

the landslides and consequently could not learn to classify instances of landslides. When more balanced training dataset, S^p_{tr}, was used for training protocol, the model learned to classify landslide instances to a much greater extent. In addition, the AUC measure value indicates that these are "good" models. The TNrate and FPrate values indicate that the model has misclassified a great number of instances as landslides. This problem can be solved to a certain degree, as explained in Section "Classification versus Regression", by taking more classes for landslide probability or by switching to a regression task.

Each study can be specific in some way. During the sampling procedure, researchers need to be led by the fact that the ML relay on the data. Furthermore, the applied sampling strategy needs to include significant information as much as possible required for learning.

7 Cross-Scaling Concept

Cross-scaling is a recently proposed concept (Marjanović 2014), that has not yet received enough attention. It relies on the fact that input data are introduced as rasters of different resolution and support, wherein the relationship between resolution to support is very delicate and often neglected. From ML point of view, an algorithm cannot learn well if the input resolution is too coarse, as it would learn from large-sized pixel (pixel $\gg$ support) that give average value of target attribute that is otherwise much more spatially variable within the large-sized pixel area. On the other hand, if the resolution is too fine (pixel $\ll$ support), the algorithm might learn redundant detailed information, that can distract it from learning a good general classification/regression rule. Both cases might lead to overfit, so it is usually proposed to optimize the operative resolution to best fit the original data support. However, cross-scaling concept goes a step further. It exploits the possibility to learn from data resolutions that are slightly coarser than optimal (pixel > support), because it can introduce subtle generalizations of attributes' values and help the ML algorithm avoid the overfit. It is understood that only upscaling is allowed, i.e. sampling should go as much as the highest data resolution allows (if the best resolution within the dataset is 25 m, only enlarging the pixel sampling grid is allowed, e.g. 50 or 100 m).

In the City of Belgrade case study three different spatial resolutions of input datasets were used: 25, 50 and 100 m. The highest resolution was 25 m and included morphological and geological conditioning factors, as well as all interpolated factors (Fig. 3), whereas environmental factors were only available at 100 m resolution (CORINE land cover). Landslide class label was divided into: 0—non-landslides, 1—dormant landslides, and 2—active landslides. Suggested cross-scaling concept was implemented as follows:

1. training on 25 m set and testing on 25 m set,
2. training on 50 m set and testing on 25 m set,
3. training on 100 m set and testing on 25 m set,
4. training on 50 m set and testing on 50 m set,
5. training on 100 m set and testing on 50 m set.

All these combinations were tested by two different ML techniques, SVM and RF, wherein prediction of target landslide label classes was placed as a typical classification task. Result support the cross-scaling hypothesis, as they showed that regardless of the ML technique (SVM or RF) performance is improving for most cross-scaled variants, but especially combination 5 (training on 100 m set and testing on 50 m set). What is additionally important, class 2 (landslides) witnesses significant AUC improvement, from about 0.5 to 0.7 in SVM variants, while in RF variants the improvement is less drastic (Table 6). Figure 9, visually illustrates these findings on a detail regarding the right Danube River bank near Grocka (Serbia). Reduction of false positives is apparent in combination 100–25 cross-scaled RF model, while true positive improvement is demonstrated in 100–25 cross-scaled SVM model. The prediction rate curves (for initial 50–50 and cross-scaled 50–100 variants) also indicate that predictive power is slightly increased, while success rate curves are clearly showing better fitting skill of the cross-scaled variant (Fig. 10).

Similarly, for Halenkovice area cross-scaling also gave improvements in the implementation of SVM algorithm for landslide prediction. Training over 10 m data and testing also over 10 m data initially gave poor performance (AUC = 0.57) with highly underestimated landslide class (Fig. 11a). Training over 30 m data and testing on 10 m gave significant improvements (AUC = 0.70), with slight over-estimation of landslide class (Fig. 11b). Even better results (AUC = 0.74) were achieved with the combination that trains on 20 m and tests on 10 m data (Fig. 11c), meaning that fine tuning is necessary to find the best resolution combination.

Table 6 AUC ROC model performance comparison of original and cross-scaled models in the validation area (only) for The City of Belgrade case study

Model, resolution combo	ROC area			
	Class 0	Class 1	Class 2	Average AUC
SVM, 25–25	0.52	0.49	0.55	0.52
SVM, 50–25	0.52	0.49	0.54	0.51
SVM, 100–25	0.61	0.51	0.67	0.60
SVM, 50–50	0.51	0.49	0.53	0.51
SVM, 100–50	**0.63**	**0.53**	**0.72**	**0.63**
RF, 25–25	0.64	0.50	0.67	0.60
RF, 50–25	0.37	0.46	0.46	0.43
RF, 100–25	0.69	0.56	0.72	0.66
RF, 50–50	0.65	0.52	0.64	0.60
RF, 100–50	**0.69**	**0.57**	**0.72**	**0.66**

Class 0 = non-landslide, class 1 = dormant landslide, class 2 = active landslide

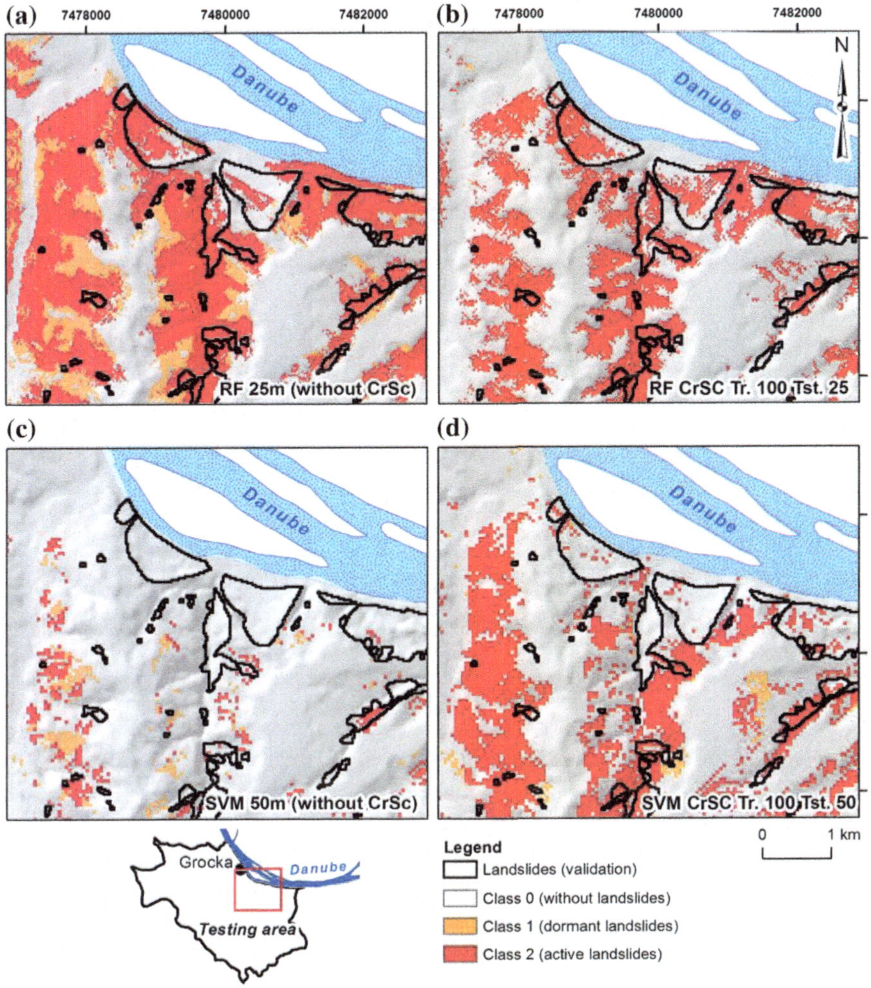

Fig. 9 Details on landslide prediction for the City of Belgrade case study: **a** RF model trained on 25 and tested on 25 m resolution dataset, **b** RF model trained on 100 and tested on 25 m resolution dataset, **c** SVM model trained on 25 and tested on 25 m resolution dataset, **d** SVM model trained on 100 and tested on 25 m resolution dataset

8 Quasi-hazard Concept

As mentioned before (see Section "Landslide inventory enhancements"), LAM is usually practiced as a landslide susceptibility, i.e. spatial probability of landslide occurrence. We are supporting the view that susceptibility should remain as static as possible, and include conditioning factors that do not change drastically over time, although there are opposite views which involve even rainfall data within the

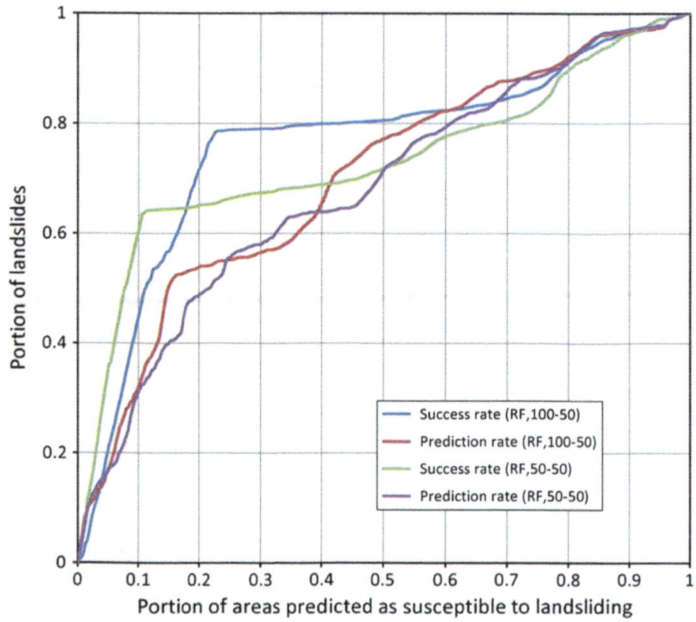

Fig. 10 Success and prediction rates curves for RF, 100–50 model

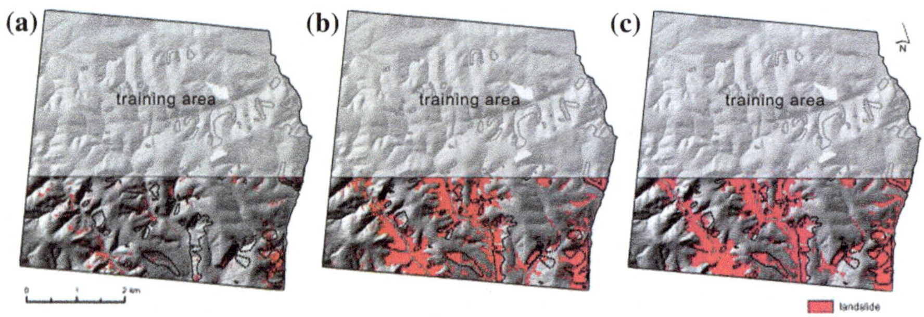

Fig. 11 Landslide prediction for Halenkovice case study: **a** SVM model trained on 10 and tested on 10 m resolution dataset, **b** SVM model trained on 30 and tested on 10 m resolution dataset, **c** SVM model trained on 20 and tested on 10 m resolution dataset (after Marjanović 2014)

susceptibility analysis (Marjanović et al. 2011). When attempting to extend the analysis temporally, practitioners come across incomplete historical data regarding the landslide frequency/magnitude for longer return periods (e.g. several decades). Herein, it is proposed to overcome this problem by introducing the temporal base of the trigger. This is justified only if the triggering process is unambiguous. For instance, when the field evidence supports that shallow landslides within a specified area are directly mobilized by rainfall that exceeds specified threshold (intensity/duration), it can be expected that the dynamics of the landslides will correspond to the dynamics of the rainfall, and that the rainfall is the likely trigger. In most of our case studies, the triggering mechanism was more complex, except for Tuzla and partly Ljubovija. Hereinafter, a quasi-hazard concept will be elaborated for the Tuzla case study.

Landslides in the Tuzla area are mainly shallow translational slides or debris flows, and it is justified to assume their direct link to landslide triggering factor. Some calculations based on 2010 rainfall event in Tuzla (Mumic et al. 2013), suggest that 72 h rainfall that exceeds the threshold of 100 mm triggers landslides. It is first necessary to perform a standard susceptibility assessment based on some modeling approach, preferably ML-based. Therein, standard geological, morphometric and environmental attributes were used as input conditioning factors and landslide susceptibility, with a relative scale 0–1, i.e. low to high landslide susceptibility was created (Fig. 12a). Secondly, the procedure required detailed studies and prediction of rainfall patterns for specific return periods. Historical rainfall from 1981 to 2010 period were analyzed on annual level, and given as relative rainfall intensity in 0–1 range, interpolated over the entire study area (Fig. 12b). Finally, multiplying susceptibility and rainfall intensity map in their relative scales gave the quasi-hazard map, which highlights the areas of high spatial probability of landslides and areas of high annual rainfall intensity that can trigger landslides (Fig. 12c). Overlapping of high susceptibility and high trigger intensity (calculated for a specific period of time) highlights high quasi-hazard, thereby answering: *where* (overlap of high susceptibility and high trigger intensity), *when* (within a return period specified by the trigger), and (partly) *what magnitude* of landslides can be expected (entirely relative 0–1 scale) within a given area. It is still far from the exact probabilities that can be only quantified when all essential information is supplied. The latter should include the true landslide frequency in specified return period (which would require process monitoring for longer time and over a larger area), estimations of volume of transported material, the velocity of the process (even more detailed monitoring), separately mapped source and accumulation zones, and so forth, which is all usually missing in regional scale landslide inventories.

9 Objective Model Evaluation

One of the most important and a mandatory step is to clarify the validity of the ML-based model results. Which of the measures or analyzing techniques should be used for selecting the best performing model is an issue for many scientists. In most of the landslide modeling studies the authors compare the model outcome map with the original landslide inventory map. Usually the first approach is to perform a visual comparison and examination of similarity between those two maps, which is then followed by application of some of the statistical measures. The simplest and a very popular way to quantify similarity between two raster maps is to do a cell by cell (pixel by pixel) spatial match to get the total number of matching cells or to get the

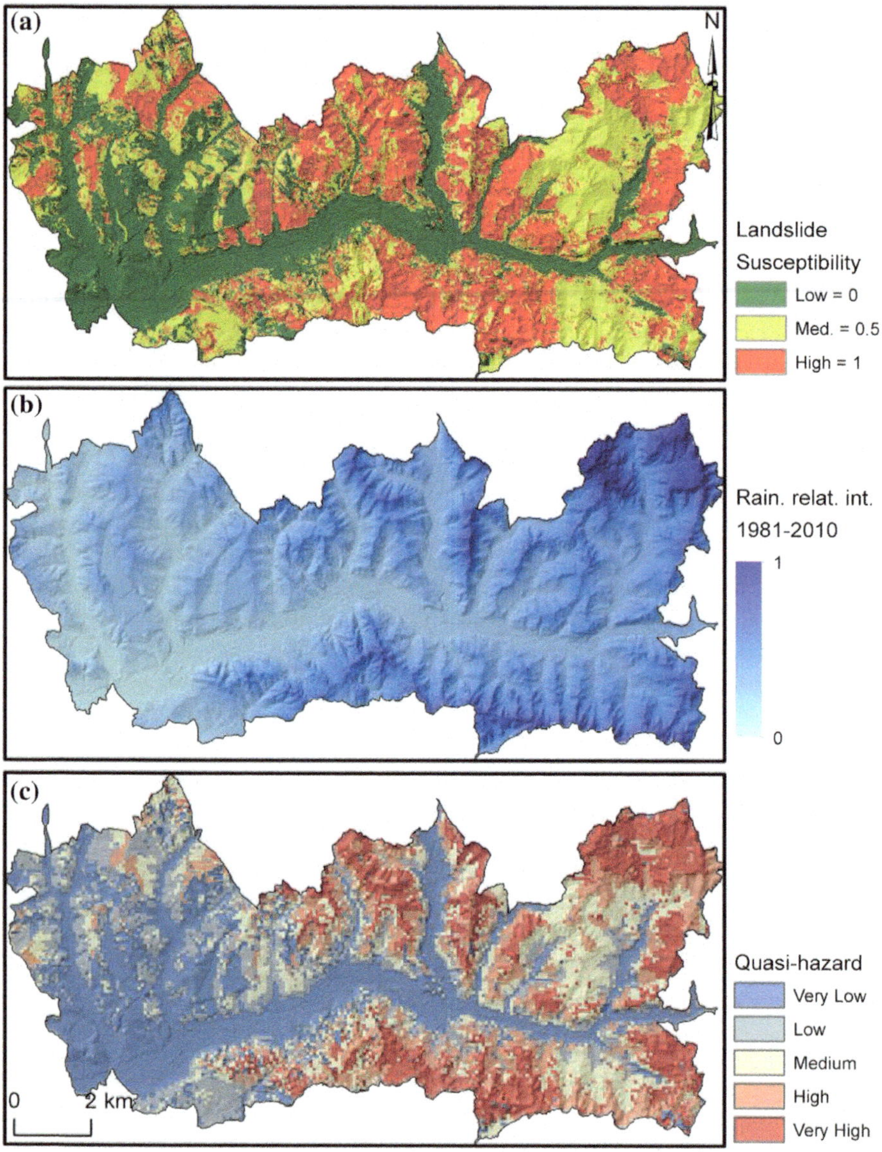

Fig. 12 Quasi-hazard procedure: **a** Landslide susceptibility model, **b** Landslide trigger model (annual rainfall intensity for 1981–2010), **c** Quasi-hazard model

proportion of observations classified correctly, i.e. accuracy. Considering the great importance of the model evaluation, many statistical measures are introduced and used for that purpose. Some of the most commonly used for landslide model evaluation are Relative Operating Characteristic (ROC) curve, sensitivity, specificity, area under the ROC curve (AUC) and kappa statistics and its variations. ROC curve is well-established measure used for validation of ML classification, which is derived (constructed) from the confusion matrix. The confusion matrix (contingency table) depicts how the distribution of classes in a map derived from ML-based model differs from the original map. In the two-class problem (positive—landslide and negative—non-landslide) it contains four values: True Positive (TP), False Negative (FN), False Positive (FP), and True Negative (TN). Those values are also known as hits, misses, false alarm references and correct rejections, respectively. In an *n*-class task ($n > 2$), ROC curve is constructed for each class using a class contingency table, one class versus all other classes. Generally, when model outcomes are discrete class labels (landslide—1 and non-landslide—0), validation produces only one point in the ROC space of a class defined with values of TPrate—sensitivity and FPrate—(1 —specificity) (Fig. 13a). In the case of probabilistic (regression) ML-based models, the number of points that define a ROC curve depends on the number of considering decision thresholds (Fig. 13b). The AUC is used as a single measure of model evaluation, approaching a value of 1 for good models, while a value of 0.5 denotes a random guess model (Bradley 1997).

The common and well-established raster map comparison measure is kappa statistic. The kappa statistic measures the differences between the observed agreement among two maps and the agreement that might be achieved solely by chance due to the alignment of those two maps (Aronoff 2005; Cohen 1960). The standard kappa index that was introduced by Cohen (1960) is, as previously mentioned measures, derived from the confusion matrix and it is calculated as:

$$Kappa = \frac{P(O) - P(E)}{1 - P(E)} \tag{1}$$

where $P(O)$ represents observed agreement and $P(E)$ is the proportion of agreement that may be expected to arise by chance.

Pontius and Millones (2011) were one of the many authors which criticized the kappa statistics (Brennan and Prediger 1981; Feinstein and Cicchetti 1990; Foody 2002; Spitznagel and Helzer 1985). They indicated that the main issues with the kappa and its variations is that they attempt to compare accuracy to a baseline of

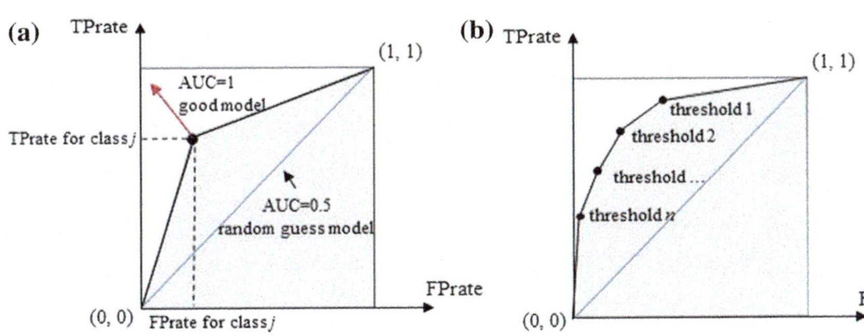

Fig. 13 ROC curve for **a** discrete two class models, and **b** regression models

randomness, which is not a reasonable in the case of map construction/classification. However, even though kappa statistic and its variations are criticized), they are still very popular as evaluation measures for landslide models (Baeza et al. 2016; Bui et al. 2016; Moosavi and Niazi 2016; Shirzadi et al. 2017).

Considering the importance of a choice of evaluation measures, the objective of the following experiment (examples) is to provide insight into the most common used measures in landslide modeling and to discuss on the factors that must be considered when performing model evaluation with respect to the map similarity assessment.

Some of the possible landslide ML-based model outcomes derived from common landslide testing datasets which represent the study area, where landslides cover only a small percentage (approximately 1%) over the whole study area, are:

- ML model significantly better classifies class non-landslide (0),
- ML model equally well classifies both classes (landslide-1 and non-landslide - 0), and
- ML model significantly better classifies class landslide (1).

Confusion matrices for all three examples are presented in (Table 7).

All three models are further evaluated using the previously described measures and the results are presented in Table 8.

Some of the advantages of the accuracy measure are that it is very easy to interpret and is simple to calculate, which are the main reasons why it is very popular. Nevertheless, this measure does not consider the distribution/proportion of each class in the dataset. This is obvious in example 1), where the values of accuracy indicated that the model derived almost perfect results, despite the fact that it is not capable to correctly classify landslides. Therefore, during the testing protocol, in most of the published landslide-related studies, this measure is usually used together with measures such as Kappa and AUC.

Table 7 Confusion matrices

(1) Example/model	Model outcomes class	1	0	
Original class	1	9900	100	10,000
	0	90	10	100
	$\sum$	9990	110	10,100
(2) Example/model	Model outcomes class	1	0	
Original class	1	9000	1000	10,000
	0	10	90	100
	$\sum$	9010	1090	10,100
(3) Example/model	Model outcomes class	1	0	
Original class	1	1000	9000	10,000
	0	1	99	100
	$\sum$	1001	9099	10,100

Table 8 Comparison of different evaluation measures

Example/Model	Accuracy	Kappa	TPrate	TNrate	FPrate	FNrate	AUC
1	0.98	0.06	0.10	0.99	0.01	0.900	0.54
2	0.90	0.14	0.90	0.90	0.10	0.100	0.90
3	0.11	0.00	0.99	0.10	0.9	0.01	0.54

Considering that Kappa has a range from −1 to +1 and AUC from 0 to +1, it can be concluded that the derived Kappa and AUC values for examples 1 and 3 indicate almost the same. The agreement of model outcome classification and the original map is almost equivalent to chance, i.e. random guess. Contrarily, in example 2, Kappa and AUC have a disagreement. According to Landis and Koch (1977), which proposed arbitrary ranges of kappa values to be used as "benchmarks" for the interpretation of Kappa values (for 0.8–1.0 almost perfect, 0.6–0.8 substantial, 0.4–0.6 moderate, 0.2–0.4 fair, 0–0.2 slight, and ≤ 0 poor), the results of example 2 indicated that the performance of the model can be evaluated as "slight". According to the common interpretation find in the literature, for AUC values higher then 0.90, the performance of model 2 can be evaluated as "excellent". Therefore, which measures should be used remains arbitrary. In order to answer this question, it is necessary to look at the confusion matrix or TPrate, TNrate, FPrate and FNrate. The example 2 is specific because it emphasizes the non-landslide class (almost 99%) in testing dataset. The ML model for this example equally well classifies both classes, 90% instances of both classes are classified correctly, which can be observed based on values of TPrate and TNrate. Since that AUC values present the overlay measure of the performance of the model, giving the equal importance to each class, the high values of TPrate and TNrate will produce the high value of AUC. The low values of kappa, derived for example 2, can be explained by one of the two kappa paradoxes defined by Feinstein and Cicchetti (1990). It states that if the expected agreement $P(E)$ is large, then the chance correction process can convert a relatively high value of observed agreement $P(O)$ into a relatively low value of kappa (Eq. 1). Therefore, the high obtained agreement value ($P(O) = 0.90$) and the high expected agreement value ($P(E) = 0.89$) for example 2 produced low value of kappa. This paradox is caused by high prevalence, i.e. by the large disagreement between the distributions of data across the classes.

To give a detailed insight into the behavior of the derived landslide models, many authors used two additional performance measures, success rate curve (SRC) and prediction rate curve (PRC) (Chung and Fabbri 2003; Jaafari et al. 2015; Kavzoglu et al. 2014; Wang et al. 2015). The success rate curve assesses how many landslides are correctly classified (detected) during the training protocol (process) of modeling and measures a goodness of fit assuming that the model is "correct". The prediction rate assesses how many landslides are correctly classified (detected) during the validation (test) process and provides the validation of the prediction regardless of the prediction model (Chung and Fabbri 2003). Same as ROC curve, both curves have a corresponding value of the area under the rate curve, the area under the success rate curve (AUSRC) and the area under the prediction rate curve

(AUPRC), which provide information on model's accuracy (performance) and prediction power (generalization), respectively (Kornejady et al. 2017).

10 Conclusions

To conclude this chapter, a short recapitulation of essential points arising from the addressed issues will be given hereinafter.

Landslide inventories can be enhanced by locally increasing resolution of polygon to raster transformation or introducing buffer zone around converging polygons on behalf of precision tolerance and geometry issues that can happen during conversion. This will provide more landslide instances to work with, either for training protocol, or for validation. Although it can create redundant instances (grid pixel < support) it is useful for learning, especially with algorithms that are not sensitive to redundant training instances (see Table 2). However, such intervention is not recommended for all data in general (for non-landslide instances, or for general data downscaling below the support limits of the input data).

Operative attributes are to be chosen wisely, wherein attribute extraction can be considered mostly if there are large numbers of attributes (e.g. >20, when synthetic attributes are introduced). Special caution is advised when using discrete attributes (such as lithology or land cover), and includes their disaggregation to binary sub-attributes.

Once the inputs are resolved (class label and attributes), one needs to decide whether to run classification or regression task, depending on what type of the analysis would be preferable, exclusive (e.g. predicting landslide propagation for planning purposes) or probabilistic (e.g. defining acceptable landslide risk in the management process). Practitioners should be encouraged to explore possibilities of ML techniques they are most comfortable with, but should remain tuned for important breakthroughs in ML community. Even when routinely applying their preferable techniques, they are advised to strategize with sampling landslide and non-landslide instances before the training procedure. Presented experiment indicates that two-stepped balanced sampling strategy (in which equal number of landslide and non-landslide instances is included to learn both classes equally well) provides better model outcomes.

Optimizing operative resolution, and trying to intermix them is further encouraged, because some better generalization capacity of cross-scaled ML algorithms was experimentally confirmed.

If the research objective requires not only spatial, but also the temporal distribution of landslide zones, it is possible to use good knowledge of the dynamics of the trigger to support a quasi-hazard outcomes. However, these should be only used as a preliminary quantitative assessment, with relative (0–1) scale of the landsliding magnitude.

Finally, in order to obtain more accurate interpretation of model evaluation, beside the accuracy, Kappa and AUC, which are largely used, it is necessary to observe visual interpretation and additional values such as values from the confusion matrix (TP, TN, FP and FN) or values of TPrate, TNrate, FPrate and FNrate, as well as success and prediction rates of their models Using those additional measures, practitioners can gain more insight into the model performance and its validation. These measures put together provide a convergence of evidence to improve reliability in the model evaluation.

Acknowledgements This work was supported by the Ministry of Education, Science and Technological Development of the Republic of Serbia, Project No TR 36009.

References

Aronoff S (2005) Remote sensing for GIS managers. ESRI Press, Readlands

Baeza C, Lantada N, Amorim S (2016) Statistical and spatial analysis of landslide susceptibility maps with different classification systems. Environ Earth Sci 75(19):1318

Bíl M, Müller I (2008) The origin of shallow landslides in Moravia (Czech Republic) in the spring of 2006. Geomorphology 99:246–253

Božović B, Lazić M, Sunarić D, Todorović B (1981) Prikaz stepena istraženosti i kritička analiza metodologije dosadašnjih istraživanja stabilnosti terena beogradskog područja, Simpozijum Istraživanje i sanacija klizišta, Bled, Slovenija, Knjiga 1. Zbornik radova, pp 107–118 (in Serbian)

Bradley AP (1997) The use of the area under the ROC curve in the evaluation of machine learning algorithms. Pattern Recogn 30(7):1145–1159

Brennan RL, Prediger DJ (1981) Coefficient kappa: some uses, misuses, and alternatives. Educ Psychol Measur 41(3):687–699

Bui DT, Tuan TA, Klempe H, Pradhan B, Revhaug I (2016) Spatial prediction models for shallow landslide hazards: a comparative assessment of the efficacy of support vector machines, artificial neural networks, kernel logistic regression, and logistic model tree. Landslides 13 (2):361–378

Cama M, Conoscenti C, Lombardo L, Rotigliano E (2016) Exploring relationships between grid cell size and accuracy for debris-flow susceptibility models: a test in the Giampilieri catchment (Sicily, Italy). Environ Earth Sci 75:238

Caruana R, Niculescu-Mizil A (2006) An empirical comparison of supervised learning algorithms. ICML '06 Proceedings of the 23rd international conference on machine learning, Pittsburgh, June 2006, pp 161–168

Chen W, Pourghasemi HR, Kornejady A, Zhang N (2017) Landslide spatial modeling: introducing new ensembles of ANN, MaxEnt, and SVM machine learning techniques. Geoderma 305:314–327

Choi J, Oh HJ, Lee HJ, Lee C, Lee S (2012) Combining landslide susceptibility maps obtained from frequency ratio, logistic regression, and artificial neural network models using ASTER images and GIS. Eng Geol 124:12–23

Chung CJF, Fabbri AG (2003) Validation of spatial prediction models for landslide hazard mapping. Nat Hazards 30(3):451–472

Ciurleo M, Cascini L, Calvello M (2017) A comparison of statistical and deterministic methods for shallow landslide susceptibility zoning in clayey soils. Eng Geol 223:71–81

Cohen J (1960) A coefficient of agreement for nominal scales. Educ Psychol Measur 20(1):37–46

Conoscenti C, Rotigliano E, Cama M, Caraballo-Arias NA, Lombardo L, Agnesi V (2016) Exploring the effect of absence selection on landslide susceptibility models: a case study in Sicily, Italy. Geomorphology 261:222–235

Dumlao AJ, Victor JA (2015) GIS-aided statistical landslide susceptibility modeling and mapping of Antipolo Rizal (Philippines). IOP Conf Ser: Earth Environ Sci 26:12031. https://doi.org/10.1088/1755-1315/26/1/012031

Đurić D, Mladenović A, Pešić-Georgiadis M, Marjanović M, Abolmasov B (2017) Using multiresolution and multitemporal satellite data for post-disaster landslide inventory in the Republic of Serbia. Landslides, https://doi.org/10.1007/s10346-017-0847-2

Erener A, Mutlub A, Düzgün HS (2016) A comparative study for landslide susceptibility mapping using GIS-based multi-criteria decision analysis (MCDA), logistic regression (LR) and association rule mining (ARM). Eng Geol 203:45–55

Feinstein AR, Cicchetti DV (1990) High agreement but low kappa: I. the problems of two paradoxes. J Clin Epidemiol 43(6):543–549

Foody GM (2002) Status of land cover classification accuracy assessment. Remote Sens Environ 80(1):185–201

Foumelis M, Lekkas E, Parcharidis I (2004) Landslide susceptibility mapping by GIS-based qualitative weighting procedure in Corinth Area, Bulletin of the Geological Society of Greece, vol XXXVI, 2004 Proceedings of the 10th International Congress, Thessaloniki, April 2004, pp 904–912

Gallus D, Abecker A (2008) Classification of landslide susceptibility in the development of early warning systems. 11th AGILE International Conference on Geographic Information Science, University of Girona, Spain, pp 1–17

Goetz JN, Brenning A, Petschko H, Leopold P (2015) Evaluating machine learning and statistical prediction techniques for landslide susceptibility modeling. Comput Geosci 81:1–11. https://doi.org/10.1016/j.cageo.2015.04.007

Gojgić D, Petrović N, Komad Z (1995) Katastar klizišta i nestabilnih padina u funkciji prostornog i urbanističkog planiranja, projektovanja i građenja, II Simpozijum Istraživanje i sanacija klizišta, D. Milanovac, Srbija, pp 103–111 (in Serbian)

Guzzetti F, Mondini AC, Cardinali M, Fiorucci F, Santangelo M, Chang KT (2012) Landslide inventory maps: new tools for an old problem. Earth-Sci Rev 112:42–66

Hall MA, Holmes G (2003) Benchmarking attribute selection techniques for discrete class data mining. IEEE Trans Knowl Data Eng 15(6):1437–1447

Hastie T, Tibshirani RI, Frieman J (2009) The elements of statistical learning: data mining, inference, and prediction. Springer, New York

Heymann Y, Steenmans C, Croissille G, Bossard M (1994) CORINE Land Cover. Technical Guide, Official Publications of the European Communities

Jaafari A, Najafi A, Rezaeian J, Sattarian A, Ghajar I (2015) Planning road networks in landslide-prone areas: a case study from the northern forests of Iran. Land Use Policy 47:198–208

Jones R (2002) Algorithms for using a DEM for mapping catchment areas of stream sediment samples. Comput Geosci 28:1051–1060. https://doi.org/10.1016/S0098-3004(02)00022-5

Kavzoglu T, Sahin EK, Colkesen I (2014) Landslide susceptibility mapping using GIS-based multi-criteria decision analysis, support vector machines, and logistic regression. Landslides 11 (3):425–439. https://doi.org/10.1007/s10346-013-0391-7

Kircher K, Krejčí O, Máčka Z, Bíl M (2000) Slope deformations in Eastern Moravia, Vsetín District (Outer Western Carpathians). Acta Universitas Carolinae 35:133–143

Kornejady A, Ownegh M, Bahremand A (2017) Landslide susceptibility assessment using maximum entropy model with two different data sampling methods. CATENA 152:144–162

Kukemilks K, Saks T (2013) Landslides and gully slope erosion on the banks of the Gauja River between the towns of Sigulda and Līgatne. Est J Earth Sci 62:231. https://doi.org/10.3176/earth.2013.17

Landis JR, Koch GG (1977) The measurement of observer agreement for categorical data. Biometrics 3:159–174

Lokin P, Pavlović R, Trivić B, Lazić M, Batalović K, Đurić U (2012) Belgrade landslide cadastre, XIV simpozijum iz Inženjerske geologije i Geotehnike, proceedings, Belgrade, Serbia, pp 389–403 (in Serbian)

Lombardo L, Cama M, Maerker M, Rotigliano E (2014) A test of transferability for landslides susceptibility models under extreme climatic events: application to the Messina 2009 disaster. Nat Hazards 74:1951–1989

Mahalanobis PC (1936) On the generalized distance in statistics. Proc Natl Inst Sci (Calcutta) 2:49–55

Malamud B, Turcotte DL, Guzzetti F, Reichenbach P (2004) Landslide inventories and their statistical properties. Earth Surf Proc Land 29:687–711

Marjanović M, Kovačević M, Bajat B, Voženílek V (2011) Landslide susceptibility assessment using SVM machine learning algorithm. Eng Geol 123(3):225–234

Marjanović M (2013) Comparing the performance of different landslide susceptibility models in ROC space. In: Margottini C et al (eds) Landslide science and practice, vol 1, Springer, Berlin. https://doi.org/10.1007/978-3-642-31325-7_76

Marjanović M (2014) Conventional and machine learning methods for landslide assessment in GIS. Palacky University, Olomouc, Czech Republic

Marjanović M, Đurić U, (2016) From landslide inventory to landslide risk assessment: methodology, current practice and challenges. III Congress of Geologists of the Republic of Macedonia, 30 Sept–2 Oct 2016, Struga Macedonia, pp 199–208

Mitchell TM (1997) Machine learning. McGraw Hill, New York

Moosavi V, Niazi Y (2016) Development of hybrid wavelet packet-statistical models (WP-SM) for landslide susceptibility mapping. Landslides 13(1):97–114

Mumic E, Glade T, Hasel S (2013) Analysis of landslides triggered in 2010 in Tuzla, Bosnia and Herzegowina. Geophysical Research Abstracts EGU General Assembly 2013, Abstract #13016

Oommen T, Baise LG, Vogel RM (2011) Sampling bias and class imbalance in maximum-likelihood logistic regression. Math Geosci 43:99–120

Park S, Choi C, Kim B, Kim J (2013) Landslide susceptibility mapping using frequency ratio, analytic hierarchy process, logistic regression, and artificial neural network methods at the Inje area, Korea. Environ Earth Sci 68:1443. https://doi.org/10.1007/s12665-012-1842-5

Pham BT, Pradhan B, Bui DT, Prakash I, Dholakia MB (2016) A comparative study of different machine learning methods for landslide susceptibility assessment: a case study of Uttarakhand area (India). Environ Model Softw 84:240–250

Pontius RG Jr, Millones M (2011) Death to kappa: birth of quantity disagreement and allocation disagreement for accuracy assessment. Int J Remote Sens 32(15):4407–4429

Pradhan B (2013) A comparative study on the predictive ability of the decision tree, support vector machine and neuro-fuzzy models in landslide susceptibility mapping using GIS. Comput Geosci 51:350–365. https://doi.org/10.1016/j.cageo.2012.08.023

Shahabi H, Khezri S, Ahmad BB, Hashim M (2014) Landslide susceptibility mapping at central Zab basin, Iran: a comparison between analytical hierarchy process, frequency ratio and logistic regression models. CATENA 115:55–70. https://doi.org/10.1016/j.catena.2013.11.014

Sharma LP, Patel N, Debnath P, Ghose MK (2012) Assessing landslide vulnerability from soil characteristics-a GIS-based analysis. Arab J Geosci 5:789–796. https://doi.org/10.1007/s12517-010-0272-5

Shirzadi A, Bui DT, Pham BT, Solaimani K, Chapi K, Kavian A, Shahabi H, Revhaug I (2017) Shallow landslide susceptibility assessment using a novel hybrid intelligence approach. Environ Earth Sci 76(2):60

Spitznagel EL, Helzer JE (1985) A proposed solution to the base rate problem in the kappa statistic. Arch Gen Psychiatry 42(7):725–728

Steger S, Brenning A, Bell R, Petschko H, Glade T (2016) Exploring discrepancies between quantitative validation results and the geomorphic plausibility of statistical landslide susceptibility maps. Geomorphology 262:8–23

Tsangaratos P, Ilia I (2017) Landslide assessments through soft computing techniques within a GIS-based framework. Am J Geogr Inf Syst 6(1A), https://doi.org/10.5923/s.ajgis.201701

Tsangaratos P, Benardos A (2014) Estimating landslide susceptibility through an artificial neural network classifier. Nat Hazards 74(3):1489–1516

Varnes DJ (1984) Landslide hazard zonation: a review of principles and practice. International Association for Engineering Geology, Paris, France, p 63

Wang Q, Li W, Chen W, Bai H (2015) GIS-based assessment of landslide susceptibility using certainty factor and index of entropy models for the Qianyang County of Baoji city, China. J Earth Syst Sci 124(7):1399–1415

Witten IH, Frank E, Hall MA (2011) Data mining practical machine learning tools and techniques. Elsevier, Burlington

Yalcin A, Reis S, Aydinoglu AC, Yomralioglu T (2011) A GIS-based comparative study of frequency ratio, analytical hierarchy process, bivariate statistics and logistics regression methods for landslide susceptibility mapping in Trabzon, NE Turkey. CATENA 85:274–287

Yao X, Tham LG, Dai FC (2008) Landslide susceptibility mapping based on support vector machine: a case study on natural slopes of Hong Kong, China. Geomorphology 101(4):572–582. https://doi.org/10.1016/j.geomorph.2008.02.011

Yilmaz I (2009) Comparison of landslide susceptibility mapping methodologies for Koyulhisar, Turkey: conditional probability, logistic regression, artificial neural networks, and support vector machine. Environ Earth Sci 61(4):821–836. https://doi.org/10.1007/s12665-009-0394-9

Youssef AM, Pourghasemi HR, Pourtaghi ZS, Al-Katheeri MM (2016) Landslide susceptibility mapping using random forest, boosted regression tree, classification and regression tree, and general linear models and comparison of their performance at Wadi Tayyah Basin, Asir Region, Saudi Arabia. Landslides 13:839–856

Assessment of the Contribution of Geo-environmental Factors to Flood Inundation in a Semi-arid Region of SW Iran: Comparison of Different Advanced Modeling Approaches

Davoud Davoudi Moghaddam, Hamid Reza Pourghasemi and Omid Rahmati

Abstract Floods are a hazard for artificial structures and humans. From natural hazard management point of view, present the new techniques to assess the flood susceptibility is considerably important. The aim of this research is on one hand to evaluate applicability of different machine learning and advanced techniques (MLTs) for flood susceptibility analysis and on the other hand to investigate of the contribution of geo-environmental factors to flood inundation in a semi-arid part of SW Iran. Here, we compare the performance of six modeling techniques namely random forest (RF), maximum entropy (ME), multivariate adaptive regression splines (MARS), general linear model (GLM), generalized additive model (GAM), and classification and regression tree (CART)for first time to spatial predict the flood prone-area at Tashan Watershed, southwestern Iran. In the first step of study, a flood inventory map with 169 flood events was constructed through field surveys. These flood locations were then spatially randomly split into train, and validation sets with two different proportions of ratio 70 and 30%. Ten flood conditioning factors such as landuse, lithology, drainage density, distance from roads, topographic wetness index (TWI), slope aspect, distance from rivers, slope angle, plan curvature and altitude were considered in the analysis. In addition, learning vector quantization (LVQ) was used as a new supervised neural network algorithm to analyse thevariable importance. The applied models were evaluated for performance appliyng the area under the receiver operating characteristic curve (AUC). The result demonstrated that CART had the AUC value of 93.96%. It was followed by ME (88.58%), RF (86.81%), GAM (81.35%), MARS (75.62%), and GLM (73.66%).

D. Davoudi Moghaddam · O. Rahmati
Department of Watershed Management Engineering, Faculty of Natural Resources Management and Agriculture, Lorestan University, Lorestan, Iran

H. R. Pourghasemi (✉)
Department of Natural Resources and Environmental Engineering, College of Agriculture, Shiraz University, Shiraz, Iran
e-mail: hr.pourghasemi@shirazu.ac.ir; hamidreza.pourghasemi@yahoo.com

H. R. Pourghasemi and M. Rossi (eds.), *Natural Hazards GIS-based Spatial Modeling Using Data Mining Techniques*, Advances in Natural and Technological Hazards Research 48, https://doi.org/10.1007/978-3-319-73383-8_3

Keywords Machine learning techniques · Flood prone areas
Learning vector quantization · Iran

1 Introduction

As a global phenomenon, flooding causes loss of human lives, socio-economic damages, and widespread devastation. The development of techniques can help managers and decision makers greatly in reducing damages through flood hazard management methods focusing on non-structural measures (Liu et al. 2015). This includes improved landuse planning, flood warning and insurance, the planning of a flood resilient environment, flood forecasting, and detecting the susceptible areas to flooding (Cherqui et al. 2015; Kazakis et al. 2015; Lumbroso et al. 2011). Regarding the last item, several index-based models such as weights-of-evidence (Rahmati et al. 2015a; Tehrany et al. 2014b), multi criteria evaluation (Rahmati et al. 2015b), frequency ratio (Lee et al. 2012; Rahmati et al. 2015a), and logistic regression (Pradhan 2009) have been used for flood susceptibility mapping. In these years, the subject of flood hazard modeling has grown rapidly, resulting in the application of some machine learning techniques (MLTs). For instance, Tehrany et al. (2013) applied decision tree algorithm in flood susceptibility analysis at Kelantan, Malaysia. Also, in Johor River Basin, Malaysia, Kia et al. (2012) investigated the performance of artificial neural network (ANN) to identify the flood susceptible zones. Lohani et al. (2012) used a neural network-based method with several geo-environmental factors to recognize the flood-prone sites in Bhakra region, India. Tehrany et al. (2015b) utilized different kernel functions of support vector machine (SVM) method to assess relations among flood occurrences and several predictor variables for producing a flood susceptibility zonation map. The mentioned studies demonstrated that flood modeling can be precisely carried out using MLTs in comparison with non-linear and simple multivariate and bivariate methods such as logistic regression and frequency ratio due the dynamic of ecosystems and non-linear and complex structures of watersheds. Although the precision of aforementioned techniques is still debated; however, more reliable, accurate, rapid and less expensive methods are needed to improve prediction accuracy of the flood susceptibility zonation map.

Base on the aforementioned literature, the performance of machine learning techniques is better than statistical multivariate and bivariate approaches in various researches (Witten et al. 2011). Thus, the principal target of this research is to use CART, GAM, GLM, MARS, ME and RF models for flood susceptibility mapping, and for this aim, Tashan Watershed in southwestern Iran was chosen. The considerable difference between current study and previous flood modeling studies is that six machine learning techniques were used and the modeling results are compared in the Tashan Watershed. The proposed models have not been applied before for preparing the flood susceptibility map (FSM). Therefore, application of the CART, GAM, GLM, MARS, ME and RF models in flood susceptibility mapping belongs novelty to the this research. Furthermore, because no such studies have been performed based on the machine learning and/or statistical models so far in the Tashan Watershed, therefore, this study is an essential work to sustainable development planning. The particular aims of current research are to (1) investigate of the contribution of geo-environmental factors to flood inundation in a semi-arid part of SW Iran, (2) explore and evaluate the prediction capability of CART, GAM, GLM, MARS, ME, and RF models for generating FSM, (3) compare the accuracy of mentioned techniques to find the best of them that is more accurate to recognize flood-prone areas.

2 Study Area

The authors selected the Tashan Watershed in SW Iran as case study for preparing FSMs. This region is located in the Khuzestan Province, between 30° 37′ 53″ N and 30° 56′ 11″ N latitude and 50° 03′ 37″ E and 50° 15′ 56″ E longitude (Fig. 1). The area of this watershed is about 369 km^2. The altitude of the Tashan Watershed changes between 238 and 1968 m.a.s.l. 317.3 mm is recorded as the mean annual point precipitation in the weather station (IMO 2014). Regarding to the geological survey of Iran (GSI 1997), about 57% of the lithology of Tashan Watershed falls within the low level pediment fan and other Quaternary deposits (Q_{t2}). More than half of the area (about 56%) is covered by rangelands areas. Because of physiographic conditions, the Tashan Watershed is always exposed to flood risk. For example, Fig. 2 displays the severity of the flash flood that occurred in December 2013.

3 Methods

The adapted methodology of this research is indicated in Fig. 3 as a flowchart.

3.1 Flood Inventory Map

In an area, the future flooding can be estimated using and analyzing the records of past flood inundation events (Pradhan 2009). Therefore, a flood inventory map is necessary to study the dependence among the conditioning factors and the flooding. The preparation of an accurate flood-inundation inventory database is required (Jebur et al. 2013). In this study, from the field surveys and available information (e.g. documentary sources of Iranian Water Resources Department), a flood inundation database/inventory containing 169 inundation occurrences was prepared for the study area. We randomly partition the database into a calibration phase (i.e. 118 (70%) of the inundation occurrences), and a validation phase (i.e. 51 (30%) of the inundation occurrences) (Oh and Pradhan 2011; Ohlmacher and Davis 2003).

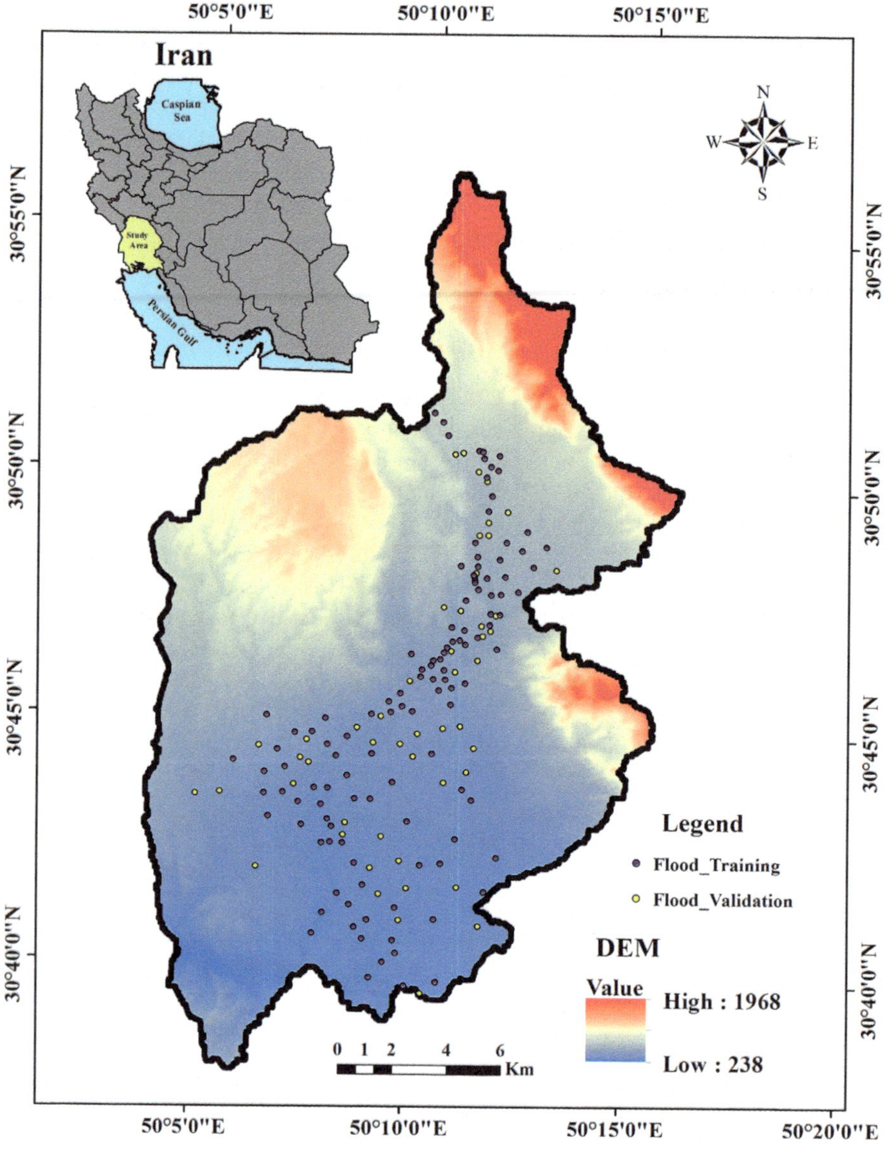

Fig. 1 Flood locations map with the hill-shaded map of Tashan Watershed, SW Iran

4 Flood Conditioning Factors

For modelling the flood susceptibility, the role of factors on the flood inundation events must be specified (Kia et al. 2012). Thus, using the literature review ten conditioning factors was selected. These thematic data layers are landuse, lithology,

Fig. 2 Photographs showing the severity of the flood that occurred in study area

drainage density, distance from roads, topographic wetness index (TWI), slope aspect, distance from rivers, slope angle, plan curvature and altitude.

By applying topographic maps (1:50,000-scale), the digital elevation model (DEM) was produced in resolution of 30 m. Some of these factors such as altitude, aspect, and slope angle maps were constructed on the basis of DEM in ArcGIS 10.3 software and illustrated in Figs. 4a–c.

For characterizing the curvature of watersheds and studying the convergence and divergence of surface flow, the plan curvature map can be applied (Fig. 4d).

Using the prepared database, the distance from river and road maps were constructed. The buffers of road and river were generated in 1500 and 500 m intervals, respectively, as shown in Fig. 4e–f.

The drainage density map was created using topographic maps and was classified applying quantile classification scheme into four classes (Fig. 4g).

Ln ($\alpha/\tan\beta$) is defined as the TWI (Fig. 4h), where β and α are the slope angle at the point andcumulative upslope area draining through a point, respectively (Beven and Kirkby 1979).

The 1:100,000-scale geology map of Khuzestan Province was utilized for preparing the lithology map of study area (Fig. 4i). Different types of lithological formations cover in the Tashan Watershed which classified into five classes (Table 1).

Regarding the maximum likelihood algorithm and supervised classification technique and applying Landsat 7/ETM + images, the landuse map was created.

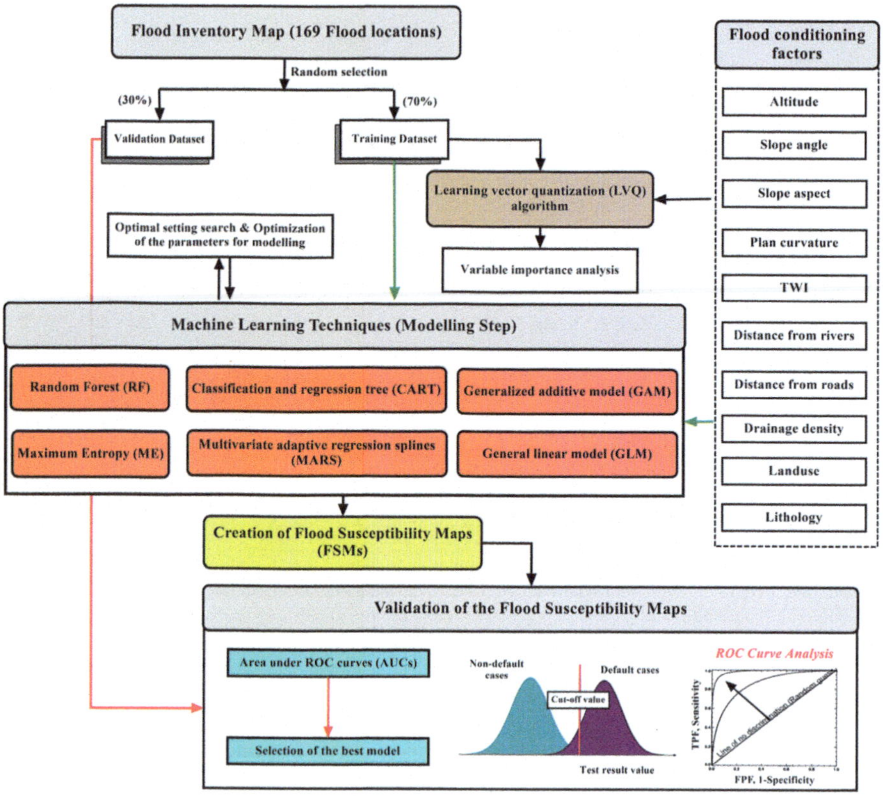

Fig. 3 Methodological flow chart employed in this study

Agriculture, rangeland, residential area and forest are the land-use types of Tashan Watershed (Fig. 4j).

5 Variable Importance Analysis

Variable importance analysis is aimed at finding the contributions by the inputs conditioning factors to the accuracy in a model output. To quantitatively assesse the relative contribution, learning vector quantization (LVQ) algorithm was used. LVQ is a supervised learning technique and was developed by Kohonen (1995) which allows analyzing the variable importance. In this study R package *lvq* is used to find most important independent variables.

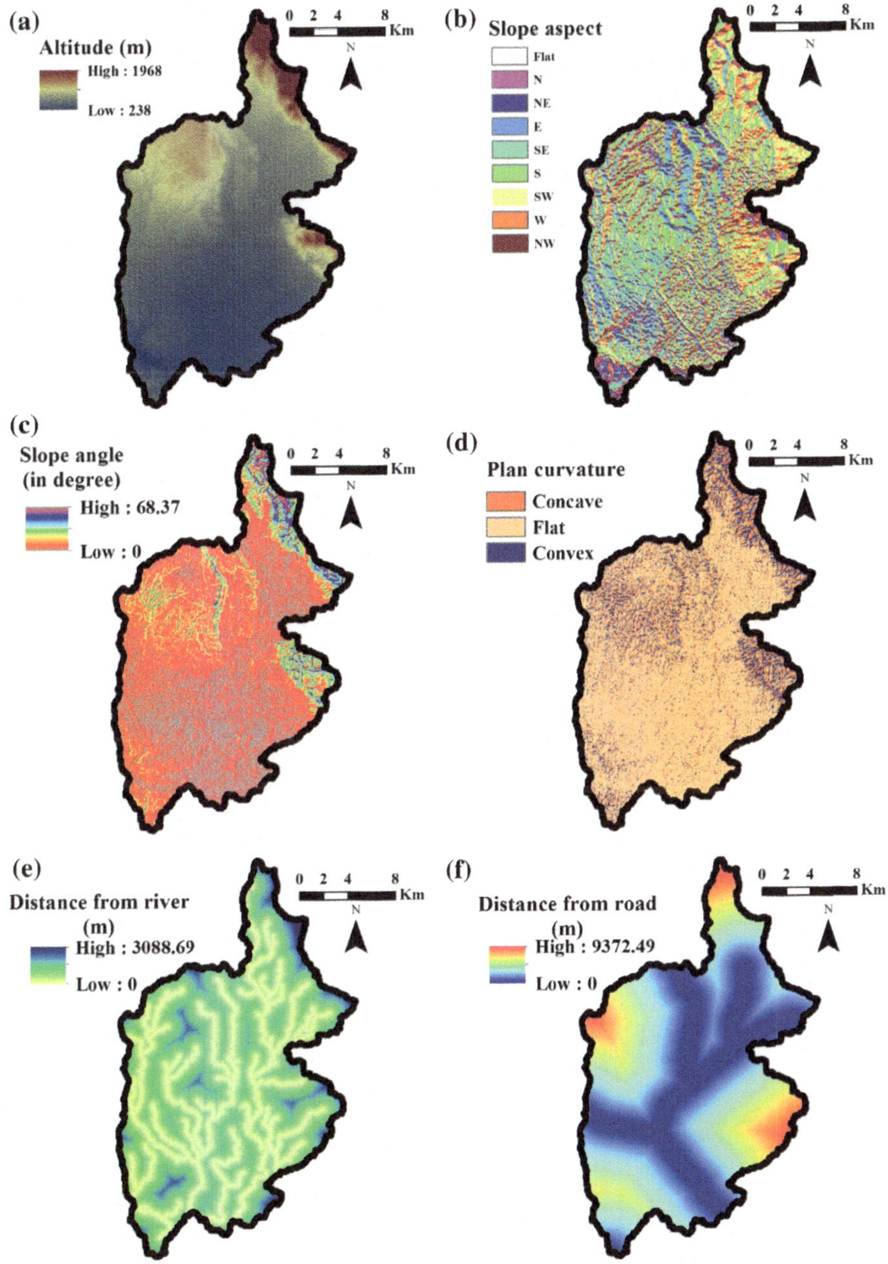

Fig. 4 Input predictor variables: **a** altitude, **b** slope aspect, **c** slope angle, **d** plan curvature, **e** distance from river, **f** distance from road, **g** drainage density, **h** TWI, **i** lithology, and **j** landuse

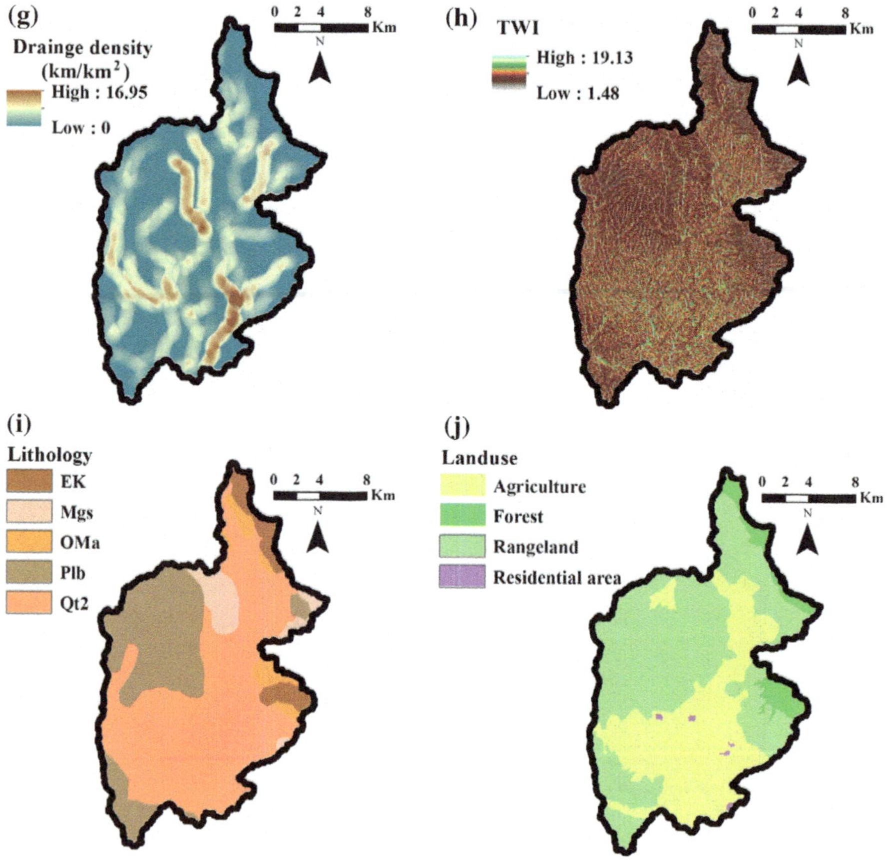

Fig. 4 (continued)

Table 1 Lithology of the study area

Code	Formation	Lithology	Geological age
Ek	Bangestan Group	Limestone and shale	Cretaceous
Mgs	Gachsaran	Anhydrite, salt, grey and red marl	Miocene
OMa	Asmari	Jointed limestone with intercalations of shale	Miocene
Plb	Bakhtyari	Conglomerate and sandstone	Pliocene
Qt2	–	Pediment fan and valley terrace deposits	Quaternary

6 Application of Models

Classification and Regression Tree (CART)

CART is a 'decision tree' algorithm to be applied both for classification and regression. It creates a tree-like structure using all inundation conditioning factors to make two child nodes repeatedly. Appling a difference of diversity or impurity measures, the best predictor is chosen. With respect to the given factor, the aim is to create groups of the data which are as homogeneous as possible (Kurt et al. 2008). A full detail of the CART model is provided in Breiman et al. (1984).

Generalized Additive Model (GAM)

One of the most important extensions of the GLMs is GAM that makes it easy to check nonlinear relations among explanatory and response variables so long as being less lead to overfitting in natural hazard analysis than the other models (Brenning 2009; Goetz et al. 2011; Hastie and Tibshirani 1990). Unlike most machine learning techniques, the model fit of the GAM can be understandable and explainable so easily (Brenning 2008; Goetz et al. 2011). The fundamental concept of a GAM model is applying the linear function of any co-variate as applied in a general linear model with an empirically fitted smooth function to reveal a general trend and to choose the proper functional form (Hastie and Tibshirani 1990). Hence a GAM utilizes a combination of nonlinear and linear reconstruction in an additive manner.

General linear model (GLM)

The regression methods in compare with logistic, linear, and log-linear regression models have been usually applied to study of flood hazard analysis. The main aim of the logistic regression is to determine the best method to show the relation among multiple independent variables and a dependent variable (i.e. response variable) (Ozdemir and Altural 2013). The simplest form of LR model can be demonstrated as:

$$L = \frac{1}{1 + e^2} \tag{1}$$

where, L is the estimated possibility of a flood happening. Because R can vary from $-\infty$ to $+\infty$, the flood possibility changes between 0 and 1 as a sigmoid curve. Parameter R is described as:

$$R = B_0 + B_1X_1 + B_2X_2 + \cdots + B_nX_n \tag{2}$$

where, B_0 and n are the intercept and the number of flood conditioning factors, respectively. Values of B_i (i = 0, 1, 2, …, n) denote the slope coefficients and X_i are the flood conditioning factors. Considering Eqs. (1) and (2), the logistic regression can be shown in the following form:

$$Logit(L) = \frac{1}{1 + e^{-B_0 + B_1 X_1 + B_2 X_2 + \cdots + B_n X_n}} \quad (3)$$

Multivariate adaptive regression splines (MARS)

The MARS is relatively an efficient machine learning model that combines mathematical structure of splines, classical linear regression, and binary recursive partitioning to construct a local analysis/method in which the relations among the factors/predictors (i.e. independent factors) and response (i.e. dependent factor) are either non-linear or linear (Felicísimo et al. 2013). The general expression of MARS can be described as a sum of basic functions (Eq. 4):

$$Y = \beta_0 + \sum_{m=1}^{M} \beta_m H_m(x) \quad (4)$$

where y, n, $H_m(x)$, and β_0 is the predicted value by the MARS model, the number of basic functions in the model, constant basis function, and initial constant of equation, respectively. The basic functions and complete algorithm of MARS are presented by Friedman (1991) and Hastie et al. (2001).

Maximum Entropy (ME)

Maximum entropy (ME) model illustrates the distribution of floods as a probability distribution (PD) that assigns a non-negative value to each pixel of the study area (Phillips et al. 2006). Graham et al. (2008) indicated that this method is quite robust to spatial errors in occurrence data and utilizes presence only (i.e. flood locations) datasets to predict the flood susceptibility. ME model applies a Gibbs probability distribution for spatial prediction of the phenomena, which completely discussed in Phillips et al. (2006). In current study, maximum entropy modeling was conducted applying of free "MaxEnt" software (version 3.3.3) (Phillips et al. 2004).

Random Forest (RF)

Random forest (RF) is flexible ensemble classifiers based upon decision trees algorithm which fits a big number of regression trees to each of subsamples, the first developed by Breiman (2001). RF considers the predictions of every single regression tree model applying a rule based method.

In flood susceptibility modeling based on RF technique, the samples (i.e. flood locations) which are not applied for the calibration of the k-th tree in the bagging process are distinguished as a separate subset called out-of-bag (OOB). The OOB indicators can be utilized by the k-th-tree to assess model accuracy (Peters et al. 2007). Therefore, this model can prepare an unbiased estimation of the prediction error without performing an external validation (Breiman 2001). Additionally, RF presents other important advantages which make it interesting for its application in flood hazard modeling:

- It is relatively robust to outliers, noise and spurious data.
- It can be run efficiently on large databases with different types of data.
- It determines most vital variables in the prediction.
- It is computationally easier than other common machine learning models such as ANNs.
- It can surmount the limitation of black-box models such as SVM and ANNs.

RF needs two parameters to be determined for modeling: (1) the number of factors/variables (i.e. m_{try}), (2) the number of trees (i.e. n_{tree}), to be chosen from the accessible set of features stochastically. The average of the outcomes of all the trees is considered as a last result of RF model (Breiman 2001; Cutler et al. 2007). For this research, the "randomForest" package of R software (R Development Core Team 2015) was applied for performing RF model, and then the last constructed map was transported into GIS environment to prepare the FSM.

Validation and Comparison of the FSMs

The flood susceptibility maps (FSMs) must be validated and it is an essential step in modeling process. The capability of the machine learning methods was evaluated applying a non-dependent threshold approach: the receiver operating characteristic (ROC) curve. The area under ROC curve (AUC) has been generally used in various studies to appraise the accuracy of FSMs (Rahmati et al. 2015a; Tehrany et al. 2014a, b, 2015a). The curve of prediction-rate can be used to the validation and illustrates how well the model predicts the flood occurrences (Lee and Pradhan 2007; Naghibi and Pourghasemi 2015; Tien Bui et al. 2012).

7 Results

Assessing the contribution of predictor variables in identifying flood-prone areas

Result from LVQ method is shown in Fig. 5. This indicated that landuse (VI = 77.1%), distance from road (VI = 72.9%), distance from river (VI = 70.7%), lithology (VI = 69.4%), and drainage density (VI = 66.3%) are the most significant factors, followed by slope angle (VI = 65.4%), altitude (VI = 65.3%), TWI (VI = 64.3%), slope aspect (VI = 48.8%), and plan curvature (VI = 46.7%). Thus, these factors were selected as input variables to modeling and produce the FSMs, because they have significant contribution on flooding in the study area.

Comparison of flood susceptibility maps

In this research, six flood susceptibility maps from CART, GAM, GLM, MARS, ME and RF machine learning models were prepared in ArcGIS 10.3 software (Figs. 6a–f). There are many schemes for classifying susceptibility levels (Ayalew et al. 2004; Suzen and Doyuran 2004). The natural break technique can determine break points by picking the category limits which maximize the differences between categories and minimize the differences within category. Consequently, the flood susceptibility maps were divided to four categories and according to natural break

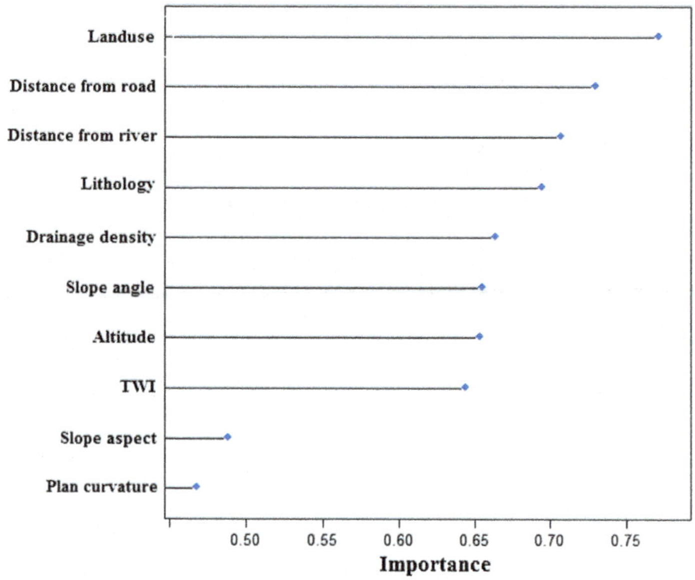

Fig. 5 Variables importance analysis using LVQ algorithm

Table 2 The distribution of the flood susceptibility classes and areas with respect to the flood occurrence

Susceptibility class	Area (%)					
	CART	GAM	GLM	MARS	ME	RF
Low	62.70	51.64	52.56	52.56	58.22	46.82
Moderate	9.40	14.61	15.18	15.17	17.18	18.22
High	4.01	14.13	13.03	13.04	14.22	19.08
Very high	23.90	19.61	19.23	19.23	10.38	15.88

classification scheme into low, moderate, high, and very high susceptible groups (Table 2) (Pourghasemi et al. 2012). Regarding to the FSMs of CART, GAM, GLM, MARS, ME and RF, low class of FSMs covered 62.70,51.64, 52.56, 52.56, 58.22, and 46.82% of the Tashan Watershed, respectively, while the total of very high and high categories for CART, GAM, GLM, MARS, ME and RF are 27.91, 33.74, 32.26, 32.27, 24.6 and 34.96%, respectively. Therefore, it is clear that ME considered the lowermost value of area corresponding to high and very high susceptibility, while RF and GAM have highest value for these two classes.

Performance and validation of flood susceptibility techniques

The outcomes of the six models of flood susceptibility were validated applying the validation dataset of the flood inventory with the application of the AUC value (Devkota et al. 2013). It can be observe in Fig. (7a–f), the AUC values for the

CART, GAM, GLM, MARS, ME and RF models are 93.96, 81.35, 73.66, 75.62, 88.58, and 86.81%, respectively. The logical relations among AUC value and model precision can be described into the following groups: 90–100% (excellent); 80–90% (very good); 70–80% (good); 60–70% (average); and 50–60% (poor) (Yesilnacar 2005). Thus, it can be concluded that CART technique has the highest prediction capability (AUC = 93.96%) and excellent accuracy in predicting the flood-prone areas in the Tashan Watershed, while the GLM technique has the lowest prediction capability (AUC = 73.66%). Furthermore, the results indicate that the GAM (AUC = 81.35%), ME (AUC = 88.58%) and RF (AUC = 86.81%) techniques have very good accuracy and MARS technique (AUC = 75.62%) has reasonably good accuracy for identifying flood-prone areas.

8 Discussions

The maps of flood susceptibility zone are considered as a final export that would be beneficial for future planning of urban development projects and sustainable land use management and also these exports are the first significant step in flood risk and hazard evaluations (Pradhan and Youssef 2011; Tehrany et al. 2014a, b). So it is needed to accurately recognize flood prone areas with high proficiency based on several techniques and past flood events. Because the precision of different proposed approaches for analyzing flood susceptibility is debated as yet (Rahmati et al. 2015a), investigations of new MLTs for the evaluation of floods are extremely essential. These researches will assist to provide a sufficient scientific background and then to gain some useful outcomes (Tien Bui et al. 2015). Progress in the interdisciplinary field of the GIS and machine learning has presented very new MLTs that have been distinguished as having superior overall efficiency (Witten et al. 2011). Some advanced MLTs such as the RF, GLM, MARS, and CART have been applied in other subjects with great precision (Liu et al. 2013; Naghibi and Pourghasemi 2015; Trigila et al. 2015; Vorpahl et al. 2012; Youssef et al. 2015), howsoever, the assessment of these techniques for flood susceptibility mapping has still not been carried out. In the present study, we addressed for first time this issue with the investigation and comparison of the six MLTs (i.e. CART, GAM, GLM, MARS, ME and RF) for flood susceptibility analysis.

The comparison of several MLTs has allowed us to better assess limits and strengths of each technique and the statistical reliability of the flood prone areas. According to validation results, all FSMs are considered to have acceptable and representable appearance (AUC > 70%). CART technique exposed the totally foremost cross-validated performance, followed by ME technique. Conversely, both visual assessment and quantitative validation—through ROC curve—agreed on GLM technique to be the minimum performing model approach. This may be due to the fact that GLM model has a linear predictor system which consequently leads to lower capability in comparison with other MLTs (Agresti 1996; Crawley

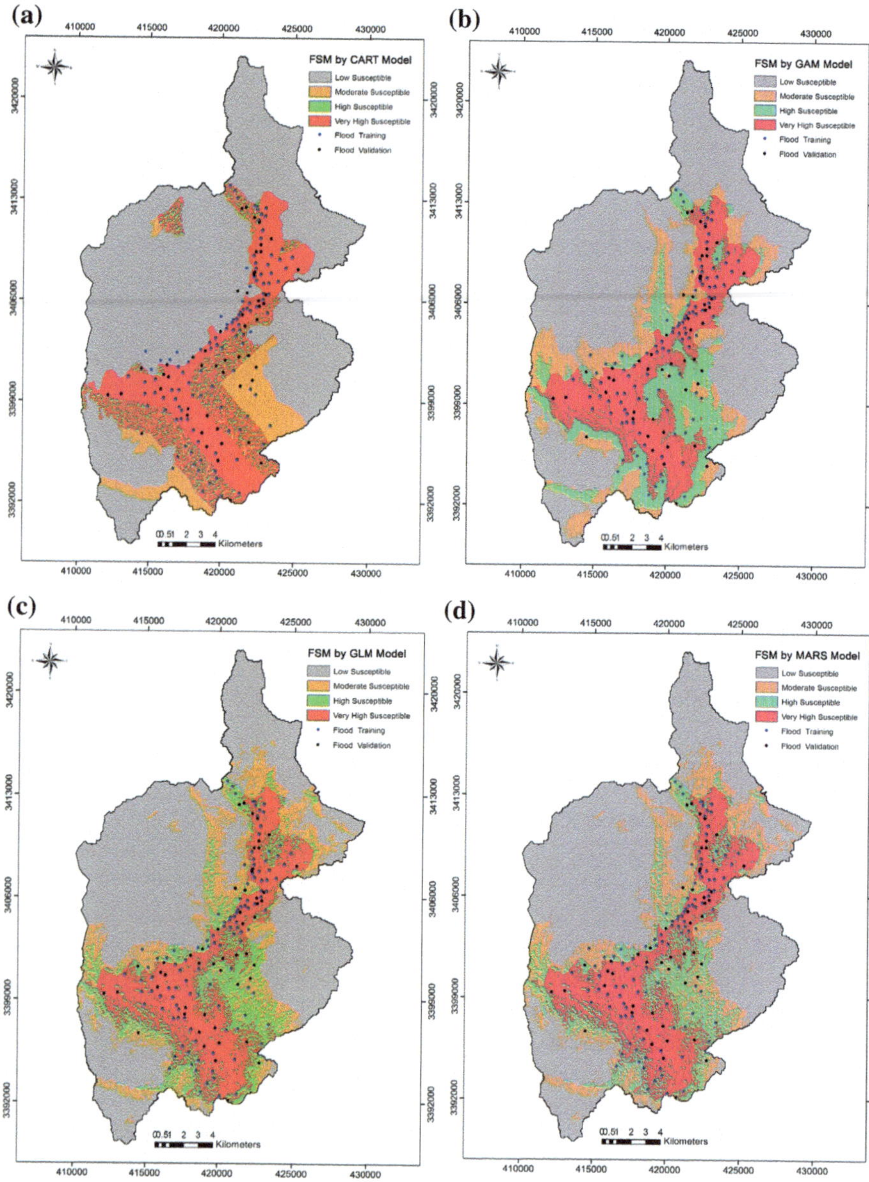

Fig. 6 Flood susceptibility map produced from: **a** CART, **b** GAM, **c** GLM, **d** MARS, **e** ME, and **f** RF

1993). Nevertheless, the GLM technique presents a remarkable simplicity of application and also has acceptable results for flood susceptibility mapping.

It is a obvious reality that the accuracy of the resulting MLTs is impressed by the conditioning factors which are applied to generate the FSMs, hence the analysis of contribution of predictor variables (i.e. importance of each conditioning factors) is

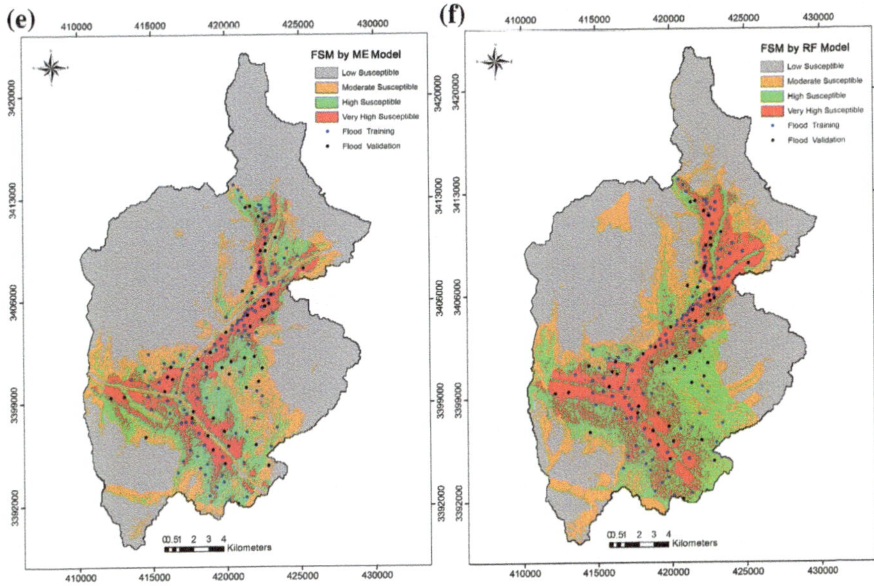

Fig. 6 (continued)

considered a key point (Tehrany et al. 2015a). Overall, our findings showed that the most influencing variables/factors on flooding were land use, distance from road, distance from river, lithology, and drainage density. This agrees with Tehrany et al. (2013) and Tehrany et al. (2015a) in that landuse is most important predictor in flood susceptibility mapping.

In flood hazard analysis, accuracy and time are two key parameters of modeling that are needed for flood control/mitigation measures and flood warning programs (Mustafa et al. 2015; Tehrany et al. 2015a). From a computational time of MLTs viewpoint, Tehrany et al (2015b) stated that the SVM model (with various kernel functions) requires considerable time for the data analysis, which can be considered as one of the disadvantages of SVM (Pourghasemi et al. 2013). In this study, similar to SVM, a disadvantage of RF technique is its long run time. This result also is in very concurrence with study of Rahmati et al. (2016).

Lohani et al. (2012) demonstrated that ANFIS model needed a large amount of parameters to delineate the flood prone areas. Hence, application of ANFIS model in flood susceptibility analysis needs a large amount of parameters, which is very inconvenient to use, especially in data-scarce regions. In contrast, based on our results, an advantage of the ME, CART techniques is that they don't require a large amount of parameters for learning.

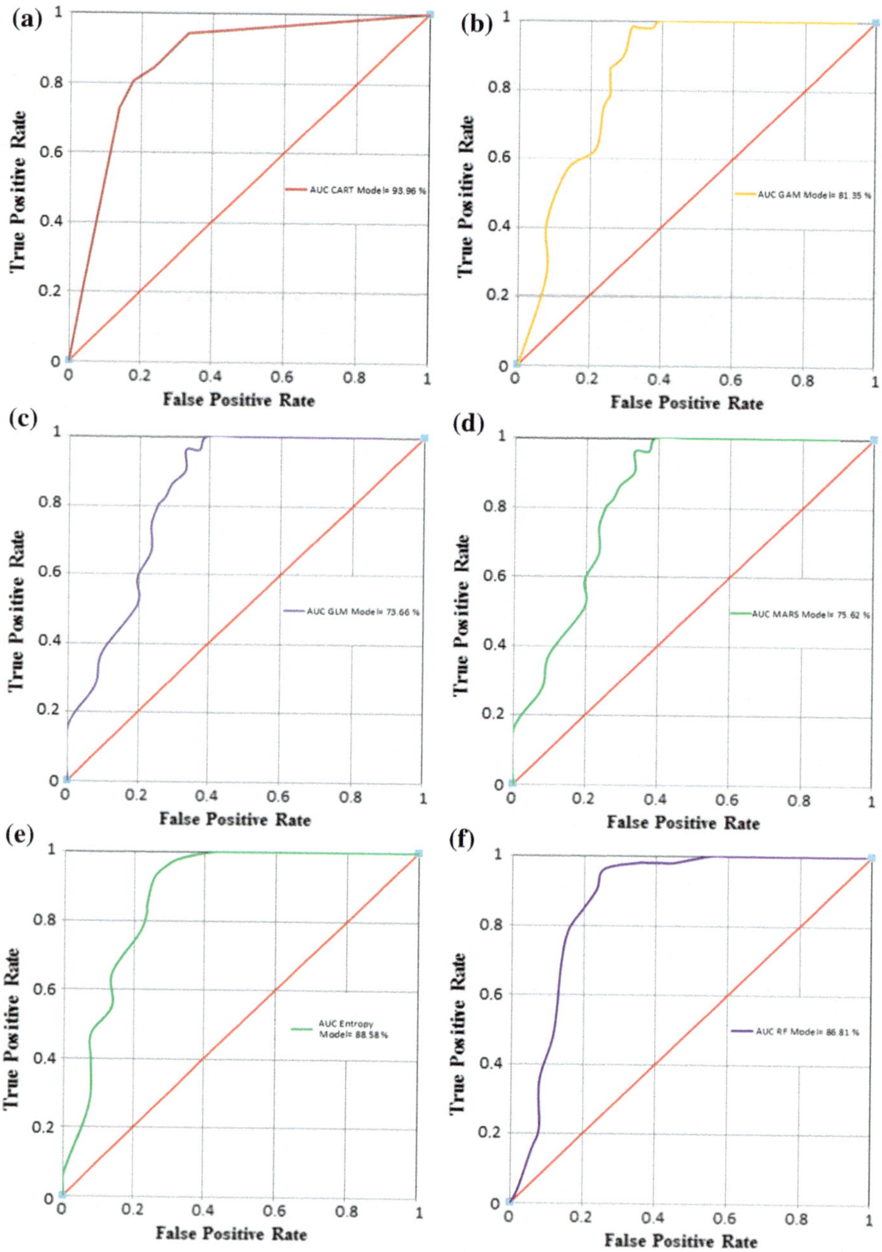

Fig. 7 ROC curve: **a** CART, **b** GAM, **c** GLM, **d** MARS, **e** ME, and **f** RF

9 Conclusion

Delineation of the flood susceptible areas, using advanced machine learning models is one of the main demands of research in natural hazard management. Our aim was to assess the applicability of six machine learning models namely, CART, GAM, GLM, MARS, ME, and RF, which have been widely applied for geo-hazards and environmental modeling, but which have not been wholly explored for flood susceptibility modeling. In first, a flood inventory dataset was prepared using the 169 flood locations (obtainable from extensive field surveys and Iranian Water Resources Department) that occured in the study area (Tashan Watershed in SW Iran).

Flooding is managed mostly by several geomorphological and geo-environmental factors. The relations among flood occurrence and these factors/predictors (altitude, slope angle, slope aspect, topographic wetness index (TWI), plan curvature, distance from roads, distance from rivers, drainage density, land use and lithology) have been assessed using the six advanced machine learning techniques in a flood susceptibility map. Validation results showed that the CART model displayed significantly better predictive efficiency than other applied machine learning techniques.

Since there is no guideline with respect to most influencing conditioning factors on flooding, our research prepares a quantitative assessment of the effect of the factor contribution on model performance using learning vector quantization (LVQ) algorithm to evaluate the variable importance analysis. Based on this analysis, landuse, distance from road, distance from river, lithology, and drainage density were recognized as the vital factors affecting the accuracy of flood susceptibility models.

One of the benefits of MLTs is that there is no classification needed for conditioning factors, whilst for common bivariate statistical techniques (e.g. FR, WOE), reclassification before flood modeling is needed. Furthermore, unlike the other machine learning models such as artificial neural networks (ANNs), the CART and ME techniques can prepare good information (such as factor importance analysis, response curves) for understanding of the flood occurrence process. These results can produce real interpretations. In addition, CART and ME techniques could handle huge geospatial data and perform minimum time which is an effective parameter of early warning system in flood hazard studies. Hence, as a final conclusion of our work, we could say that the applied MLTs (particularly CART and ME techniques) provided cost effective and accurate results.

References

Agresti A (1996) An introduction to categorical data analysis. Wiley, New York

Ayalew L, Yamagishi H, Ugawa N (2004) Landslide susceptibility mapping using GIS-based weighted linear combination, the case in Tsugawa area of Agano River, Niigata Prefecture, Japan. Landslides 1(1):73–81

Beven KJ, Kirkby MJ (1979) A physically based, variable contributing area model of basin hydrology/Un modèle à base physique de zone d'appel variable de l'hydrologie du bassin versant. Hydrolog Sci Bull 24(1):43–69

Breiman L (2001) Random forests. Mach Learn 45:5–32

Breiman L, Friedman JH, Olshen RA, Stone CJ (1984) Classification and regression trees. Wadsworth, Belmont, CA

Brenning A (2008) Statistical geocomputing combining R and SAGA: the example of landslide susceptibility analysis with generalized additive models. In: Böhner J, Blaschke T, Montanarella L (eds) SAGA—seconds out (=Hamburger Beiträge zur Physischen Geographie und Landschaftsökologie, 19), pp 23–32

Brenning A (2009) Benchmarking classifiers to optimally integrate terrain analysis and multispectral remote sensing in automatic rock glacier detection. Remote Sens Environ 113 (1):239–247

Cherqui F, Belmeziti A, Granger D, Sourdril A, Gauffre PL (2015) Assessing urban potential flooding risk and identifying effective risk-reduction measures. Sci Total Environ 514:418–425

Crawley MJ (1993) GLIM for ecologists. Blackwell Scientific Publications, Oxford

Cutler DR, Edwards TC, Beard KH, Cutler A, Hess KT, Gibson J, Lawler JJ (2007) Random forests for classification in ecology. Ecology 88(11):2783–2792

Devkota KC, Regmi AD, Pourghasemi HR, Yoshida K, Pradhan B, Ryu IC, Dhital MR, Althuwaynee OF (2013) Landslide susceptibility mapping using certainty factor, index of entropy and logistic regression models in GIS and their comparison at Mugling-Narayanghat road section in Nepal Himalaya. Nat Hazards 65:135–165

Felicísimo Á, Cuartero A, Remondo J, Quirós E (2013) Mapping landslide susceptibility with logistic regression, multiple adaptive regression splines, classification and regression trees, and maximum entropy methods: a comparative study. Landslides 10:175–189

Friedman JH (1991) Multivariate adaptive regression splines. Ann Stat 19:1–141

Geology Survey of Iran (GSI) (1997) http://www.gsi.ir/Main/Lang_en/index.html

Goetz JN, Guthrie RH, Brenning A (2011) Integrating physical and empirical landslide susceptibility models using generalized additive models. Geomorphology 129:376–386

Graham CH, Elith J, Hijmans RJ, Guisan A, Peterson AT, Loiselle BA The NCEAS predicting Species Distributions Working Group (2008) The influence of spatial errors in species occurrence data used in distribution models. J Appl Ecol 45:239–247

Hastie TJ, Tibshirani RJ (1990) Generalized additive models, 2nd edn. Chapman and Hall, London

Hastie T, Tibshirani R, Friedman J (2001) The elements of statistical learning: data mining, inference, and prediction. Springer, New York

Iranian meteorological organization (IMO) (2014) http://www.irimo.ir/eng/index.php

Jebur MN, Pradhan B, Tehrany MS (2013) Using ALOS PALSAR derived high- resolution DInSAR to detect slow-moving landslides in tropical forest: Cameron Highlands, Malaysia. Geomatics Nat Hazards Risk 6(8):1–19

Kazakis N, Kougias I, Patsialis T (2015) Assessment of flood hazard areas at a regional scale using an index-based approach and analytical hierarchy process: application in Rhodope-Evros region, Greece. Sci Total Environ 538:555–563

Kia MB, Pirasteh S, Pradhan B, Mahmud AR, Sulaiman WNA, Moradi A (2012) An artificial neural network model for flood simulation using GIS: Johor River Basin, Malaysia. Environ Earth Sci 67:251–264

Kohonen T (1995) Learning vector quantization; self-organizing maps. Springer, Berlin, pp 175–189

Kurt I, Ture M, Kurum AT (2008) Comparing performances of logistic regression, classification and regression tree, and neural networks for predicting coronary artery disease. Expert Syst Appl 34(1):366–374

Lee S, Pradhan B (2007) Landslide hazard mapping at Selangor, Malaysia using frequency ratio and logistic regression models. Landslides 4:33–41

Lee MJ, Kang JE, Jeon S (2012) Application of frequency ratio model and validation for predictive flooded area susceptibility mapping using GIS. In: 2012 IEEE international geoscience and remote sensing symposium, pp 895–898

Liu C, White M, Newell G, Griffioen P (2013) Species distribution modelling for conservation planning in Victoria, Australia. Ecol Model 249:68–74
Liu X, Li N, Yuan S, Xu N, Shi W, Chen W (2015) The joint return period analysis of natural disasters based on monitoring and statistical modeling of multidimensional hazard factors. Sci Total Environ 538:724–732
Lohani A, Kumar R, Singh R (2012) Hydrological time series modeling: a comparison between adaptive neuro-fuzzy, neural network and autoregressive techniques. J Hydrol 442:23–35
Lumbroso D, Stone K, Vinet F (2011) An assessment of flood emergency plans in England and Wales, France, and the Netherlands. Nat Hazards 58:341–363
Mustafa D, Gioli G, Qazi S, Waraich R, Rehman A, Zahoor R (2015) Gendering flood early warning systems: the case of Pakistan. Environ Hazards 14(4):312–328
Naghibi A, Pourghasemi HR (2015) A comparative assessment between three machine learning models and their performance comparison by bivariate and multivariate statistical methods for groundwater potential mapping in Iran. Water Resour Manag 29:5217–5236
Oh HJ, Pradhan B (2011) Application of a neuro-fuzzy model to landslide-susceptibility mapping for shallow landslides in a tropical hilly area. Comput Geosci 37:1264–1276
Ohlmacher GC, Davis JC (2003) Using multiple logistic regression and GIS technology to predict landslide hazard in northeast Kansas, USA. Eng Geol 69:331–343
Ozdemir A, Altural T (2013) A comparative study of frequency ratio, weights of evidence and logistic regression methods for landslide susceptibility mapping: Sultan Mountains, SW Turkey. J Asian Earth Sci 64:180–197
Peters J, Baets BD, Verhoest NEC, Samson R, Degroeve S, Becker PD, Huybrechts WH (2007) Random forests as a tool for ecohydrological distribution modelling. Ecol Model 207:304–318
Phillips S, Dudík M, Schapire R (2004) A maximum entropy approach to species distribution modeling. In: Proceedings of the 21th International conference on machine learning. Association for Computing Machinery (ACM), Banff, Canada
Phillips S, Anderson R, Schapire R (2006) Maximum entropy modelling of species geographic distributions. Ecol Model 190:231–259
Pourghasemi HR, Pradhan B, Gokceoglu C (2012) Application of fuzzy logic and analytical hierarchy process (AHP) to landslide susceptibility mapping at Haraz watershed, Iran. Nat Hazards 63:965–996
Pourghasemi HR, Goli Jirandeh A, Pradhan B, Xu C, Gokceoglu C (2013) Landslide susceptibility mapping using support vector machine and GIS at the Golestan Province, Iran. J Earth Sys Sci 122(2):349–369
Pradhan B (2009) Flood susceptible mapping and risk area delineation using logistic regression, GIS and remote sensing. J Spatial Hydrol 9(2):1–18
Pradhan B, Youssef AM (2011) A100-year maximum floood susceptibility mapping using integrated hydrological and hydrodynamic models: Kelantan River Corridor, Malaysia. J Flood Risk Manage 4(3):189–202
R Development Core Team (2015) R: a language and environment for statistical computing. R Foundation for Statistical Computing, Vienna
Rahmati O, Pourghasemi HR, Zeinivand H (2015a) Flood susceptibility mapping using frequency ratio and weights-of-evidence models in the Golastan Province, Iran. Geocarto Int. https://doi.org/10.1080/10106049.2015.1041559
Rahmati O, Zeinivand H, Besharat M (2015b) Flood hazard zoning in Yasooj region, Iran, using GIS and multi-criteria decision analysis, Geomatics. Nat Hazards & Risk. https://doi.org/10.1080/19475705.2015.1045043
Rahmati O, Pourghasemi HR, Melesse AM (2016) Application of GIS-based data driven random forest and maximum entropy models for groundwater potential mapping: a case study at Mehran Region, Iran. Catena 137:360–372
Suzen ML, Doyuran V (2004) A comparison of the GIS based landslide susceptibility assessment methods: multivariate versus bivariate. Environ Geol 45:665–679

Tehrany M, Pradhan B, Jebur MN (2013) Spatial prediction of flood susceptible areas using rule based decision tree (DT) and a novel ensemble bivariate and multivariate statistical models in GIS. J Hydrol 504:69–79
Tehrany M, Lee MJ, Pradhan B, Jebur MN, Lee S (2014a) Flood susceptibility mapping using integrated bivariate and multivariate statistical models. Environ Earth Sci 72:4001–4015
Tehrany M, Pradhan B, Jebur MN (2014b) Flood susceptibility mapping using a novel ensemble weights-of-evidence and support vector machine models in GIS. J Hydrol 512:332–343
Tehrany MS, Pradhan B, Jebur MN (2015a) Flood susceptibility analysis and its verification using a novel ensemble support vector machine and frequency ratio method. Stoch Environ Res Risk Assess. https://doi.org/10.1007/s00477-015-1021-9
Tehrany MS, Pradhan B, Mansor S, Ahmad N (2015b) Flood susceptibility assessment using GIS-based support vector machine model with different kernel types. CATENA 125:91–101
Tien Bui D, Pradhan B, Lofman O, Revhaug I, Dick OB (2012) Spatial prediction of landslide hazards in Hoa Binh province (Vietnam): a comparative assessment of the efficacy of evidential belief functions and fuzzy logic models. CATENA 96:28–40
Tien Bui D, Tuan TA, Klempe H, Pradhan B, Revhaug I (2015) Spatial prediction models for shallow landslide hazards: a comparative assessment of the efficacy of support vector machines, artificial neural networks, kernel logistic regression, and logistic model tree. Landslides. https://doi.org/10.1007/s10346-015-0557-6
Trigila A, Iadanza C, Esposito C, Scarascia-Mugnozza G (2015) Comparison of logistic regression and random forests techniques for shallow landslide susceptibility assessment in Giampilieri (NE Sicily, Italy). Geomorphology 249:119–136
Vorpahl P, Elsenbeer H, Marker M, Schroder B (2012) How can statistical models help to determine driving factors of landslides? Ecol Model 239:27–39
Witten IH, Frank E, Mark AH (2011) Data mining: practical machine learning tools and techniques, 3rd edn. Morgan Kaufmann, Burlington, USA
Yesilnacar EK (2005) The application of computational intelligence to landslide susceptibility mapping in Turkey [PhD thesis]. Department of Geomatics the University of Melbourne, Melbourne, p 423
Youssef AM, Pourghasemi HR, Pourtaghi ZS, Al-Katheeri MM (2015) Landslide susceptibility mapping using random forest, boosted regression tree, classification and regression tree, and general linear models and comparison of their performance at Wadi Tayyah Basin, Landslides, Asir Region, Saudi Arabia. Springer, Berlin. https://doi.org/10.1007/s10346-015-0614-1

Land Subsidence Modelling Using Data Mining Techniques. The Case Study of Western Thessaly, Greece

Paraskevas Tsangaratos, Ioanna Ilia and Constantinos Loupasakis

Abstract The main objective of the present study was to investigate land subsidence phenomena and prepare a land subsidence map using spatio-temporal analysis of groundwater resources, remote sensing techniques and data mining methods. The methodology was implemented at the wider plain area extending northwest of Farsala town, Thessaly, Greece, covering an area of approximately 145 Km^2. In order to estimate the spatio-temporal trend concerning groundwater level the non-parametric Mann–Kendall test and Sen's Slope estimator were applied, whereas a set of Synthetic Aperture Radar images, processed with the Persistent Scatterer Interferometry technique, were evaluated in order to estimate the spatial and temporal patterns of ground deformation. In a test site where ground deformation rate values derived by the analysis of SAR images, Support Vector Machines was utilized to predict the subsidence deformation rate based on three variables, namely: thickness of loose deposits, the Sen's Slope value of groundwater trend and the Compression Index of the formation covering the area of research. Based on the Support Vector Machine model, a land subsidence map was then produced for the entire research area. The outcomes of the study indicated a strong relation between the thickness of the loose deposits and the deformation subsidence rate and a clear trend between the subsidence deformation rate and the groundwater fluctuation. The *r square* value for the validation dataset within the test site was estimated to be 0.75. The land subsidence map produced by the Support Vector Machine model was validated by field surveys and measurements and showed good predictive performance. In conclusion, the subsidence model proposed in this study allows the accurate identification of surface deformations and can be helpful for the local

P. Tsangaratos (✉) · I. Ilia · C. Loupasakis
Department of Geological Studies, School of Mining and Metallurgical Engineering, National Technical University of Athens, Zografou Campus: Heroon Polytechniou 9, 15780 Zografou, Greece
e-mail: ptsag@metal.ntua.gr

I. Ilia
e-mail: gilia@metal.ntua.gr

C. Loupasakis
e-mail: cloupasakis@metal.ntua.gr

H. R. Pourghasemi and M. Rossi (eds.), *Natural Hazards GIS-based Spatial Modeling Using Data Mining Techniques*, Advances in Natural and Technological Hazards Research 48, https://doi.org/10.1007/978-3-319-73383-8_4

authorities and government agencies to take measures before the evolution of severe subsidence phenomena and therefore for timely protection of the affected areas.

Keywords Land subsidence · Remote sensing techniques · Water table fluctuation Support vector machine

1 Introduction

Land subsidence is considered among the most frequent geological hazards worldwide that usually occurs as a consequence of a number of physical and human induced phenomena, namely: natural compaction of unconsolidated fine-grained deposits, groundwater over-exploitation, collapse of natural or manmade cavities, oxidation of peat-rich materials and tectonic activity (Galloway and Burbey 2011). In general, land subsidence involves a gradual settling or a sudden sinking of discrete segments of ground surfaces causing extensive deformations over large areas. Land subsidence may lead to damages to private and public buildings, bridges, roads, railroads, storm drains, sewer and canals systems, the failure of well casings and changes in the morphology of streams, canals and drains. Land subsidence incidents due to aquifers over–exploitation have been reported worldwide with immense socio-economic impacts, mainly in coastal areas and urbanized deltas, with notable examples located in Mexico, China, Thailand, Italy, Spain, Japan and the USA (Galloway et al. 1998; Tomás et al. 2005; Stramondo et al. 2007; Raspini et al. 2013). Several cases showing areas affected by land subsidence due to aquifers over–exploitation have also been reported in Greece. Kalochori and Sindos villages at the west Thessaloniki plain (Stiros 2001; Psimoulis et al. 2007; Raucoules et al. 2008; Loupasakis and Rozos 2009; Raspini et al. 2014; Svigkas et al. 2016), Anthemountas basin at the east of Thessaloniki (Koumantakis et al. 2008; Raspini et al. 2013), Thessaly plain in central Greece (Soulios 1997; Kaplanidis and Fountoulis 1997; Marinos et al. 1997; Salvi et al. 2004; Ganas et al. 2006; Apostolidis and Georgiou 2007; Kontogianni et al. 2007; Rozos et al. 2010; Vassilopoulou et al. 2013; Apostolidis and Koukis 2013; Ilia et al. 2016), the area extending at the west-northwest of the Amyntaio opencast coal mine at Florina Prefecture, Northern Greece (Tzampoglou and Loupasakis 2016; Loupasakis et al. 2014; Soulios et al. 2011), Messara valley in Crete (Mertikas and Papadaki 2009) and Thriasio plain at the west of Athens (Kaitantzian et al. 2014) are some of the well-known cases.

Despite the fact that there is a well established theory that describes land subsidence phenomena, modelling of the phenomenon is often a challenging task. The exact mechanism responsible for land subsidence is defined in each case by complex interrelations between geological, hydro-geological, morphological and tectonic settings and also human activities; therefore the full understanding of these phenomena requires the intervention of multiple scientific specialties. The most common approach for simulating land subsidence involves the use of deterministic methods that apply either the conventional consolidation theory or more complicated soil deformation constitutive laws (Brinkgreve et al. 2006). Such methods require highly accurate geotechnical and hydrological data, and produce 2-D or 3-D simulation providing quantitative estimations of the expected deformation rates (Gambolati et al. 2005; Loupasakis and Rozos 2009; Raspini et al. 2014). However, in most cases such data are not available and numerous assumptions have to be adopted.

Another approach that quite recently has been used for modelling land subsidence is based on the evaluation of the relation between the distribution of past ground deformations and land subsidence related variables which, as a result, provides a series of susceptible, hazardous and risk maps. Key aspect in this approach is the assumption that future land subsidence is likely to occur in situations that have lead to land subsidence in the past and present. This approach has been applied in medium scale studies enabling knowledge driven or data driven methods to assess data providing both qualitative and quantitative results (Teartisup and Kerdsueb 2013).

Data driven methods and specifically data mining methods (e.g. artificial neural network and decision trees), have been widely used to assess the land subsidence occurring as a consequence of underground mining (Kim et al. 2009; Choi et al. 2011; Oh et al. 2011; Lee et al. 2012; Malinowska 2014). Kim et al. (2009) evaluated ground subsidence hazard using artificial neural network within a GIS environment, whereas Galve et al. (2009) performed land subsidence susceptibility mapping in Ebro Valley (NE Spain) using nearest neighbour distance (NND) and probabilistic analysis. Choi et al. (2011) produced land subsidence susceptibility maps based on fuzzy relations in Taebaek, Korea, while similar maps were constructed by Oh et al. (2011) in Samcheok, Korea, using the frequency ratio, weight of evidence, logistic regression, and artificial neural network methods. Few cases have also been reported concerning data mining methods and land subsidence due to groundwater withdraw (Modoni et al. 2013; Zhu et al. 2013b, 2015).

In contrast to classical statistical approaches, data mining methods do not rely on the nature of data and on assumptions that data are drawn from a given probability distribution (Fayyad et al. 1996). Data mining methods involve processes that extract patterns from data sets which are then used to gain insight into relational aspects of the phenomena being studied and to predict outcomes to aid decision making (Flentje et al. 2007). As reported by many researchers the advantage of data mining methods for the analysis of land subsidence is their ability to handle non-linear problems and their robustness regarding noisy or incomplete data (Kim et al. 2009; Malinowska 2014).

The main objective of the present study was (i) to examine land subsidence phenomena due to groundwater withdraw, (ii) to define critical aspects for understanding their underlying mechanism and (iii) to provide a land subsidence susceptibility map using spatio-temporal analysis of groundwater resources, remote sensing techniques and data mining methods.

Specifically, the analysis presented in this study considers lithological and geotechnical data, water table measurements, land use information, distribution of human activities and spatio-temporal measurements of past displacement derived from remote sensing techniques. The analyses of the above data allowed predicting the deformation rate by implementing a Support Vector Machine (SVM). SVM was selected as a promising alternative since the presence of a regularization parameter, which controls the trade-off between training error and validation error, provides the tool to avoid over-fitting during the training phase. Additional, SVM is tolerant to "noisy" data and presents robustness towards small datasets. In our case, the remote sensing data measuring the evidence of past land subsidence were considered as dependent variable, whereas a number of land subsidence related variables were considered as the independent variables that influence and explain the evolution of the phenomena. In order to analyze the spatial and temporal trend of the groundwater resources and to implement the SVM model, the "e1071" package (Meyer et al. 2017) was used in R Studio (ver.0.99.489) (RStudio Team 2015), whereas ArcGIS 10.1 (ESRI 2013) was used for compiling the data and producing the land subsidence map.

Our research was focused in Thessaly plain, Central Greece, where land subsidence phenomena related to reservoir compaction have been observed since the early 90's (Apostolidis and Georgiou 2007; Kontogianni et al. 2007; Rozos et al. 2010; Vassilopoulou et al. 2013; Apostolidis and Koukis 2013; Modis and Sideri 2015; Ilia et al. 2016). Based on piezometric level measurements conducted between 1980 and 2005, the aquifer systems of the study area were subject to excessive over-exploitation (Apostolidis 2014). In particular, at Farsala—Stavros plain area, the consistent over-exploitation during the last three decades led to the complete draining of the overlaying shallow unconfined aquifer and the progressive drawdown of the successive confined—artesian aquifers. According to Rozos et al. (2010), this phenomenon resulted to the compaction of the compressible intercalated clayey horizons and the manifestation of intensive land subsidence since 2002.

A brief description of land subsidence mechanism and the detection and monitoring techniques is provided in the following paragraphs while a more detail analysis of the case of Thessaly plain will be presented.

2 Land Subsidence Modelling Due to Over-Exploitation of Aquifers

The mechanism behind land subsidence triggered by over-exploitation of aquifers and groundwater withdrawal is based on the principle of effective stress, a principle proposed by Karl Terzaghi in 1925 (Terzaghi 1925). Terzaghi's principle states that when a saturated soil is subjected to a total stress σ, this stress can be expressed by:

$$\sigma = \sigma' + u$$

where: u, the pressure acting on water and on granular structure and σ', the effective stress supported by granular structure only.

In general, excessive groundwater withdrawal from aquifer systems decreases the pore water pressure and increases the normal effective stress, which results in the compaction of the hydrostratigraphic units and eventually leads to land subsidence (Galloway and Burbey 2011).

In many aquifers system, the most prolific layers, in terms of water storage, represented by sand and gravel horizons, are intercalated by fine grain layers. These layers, when subjected to increasing geostatic loads tend to consolidate leading to the reduction of the overall thickness of the aquifer layers. The lowering of the water pressure in the sand and gravel causes slow release of water from clay and silt confining units, lens and interbeds. As these fine-grained deposits are particularly susceptible to consolidation, the total volume of the normally consolidated fine-grained deposits and weakly cemented sediments of the exploited aquifer is reduced. The ultimate effect is the non-reversible lowering of the land surface. So, land subsidence phenomena accompanying the groundwater over-exploitation can be attributed to deformation of porous matrix of aquitards (Fig. 1).

In a 1D model the groundwater level drawdown in aquifers and the soil deformation could be simulated by the consolidation of aquitards in vertical direction. Thus, the amount of land subsidence could be estimated based on the level drop of groundwater and the thickness and compressibility of the soil layers.

Concerning the detection and monitoring of land subsidence phenomena several methods have been used that could be separated into ground-based and remotely sensed geodetic surveys and techniques. Both techniques are based on the accurate measuring of the vertical and horizontal displacement of the land surface. During the last two decades, Earth Observation (EO) techniques, especially Global Positioning System (GPS) and Differential Interferometric Synthetic Aperture Radar (DInSAR) technologies have been widely applied in land subsidence (Galloway et al. 1998; Dixon et al. 2006; Herrera et al. 2009; Hu et al. 2009; Galloway and Burbey 2011; Osmanoglu et al. 2011; Chaussard et al. 2013; Raspini et al. 2013, 2014; Zhu et al. 2013a, 2015; Svigkas et al. 2016).

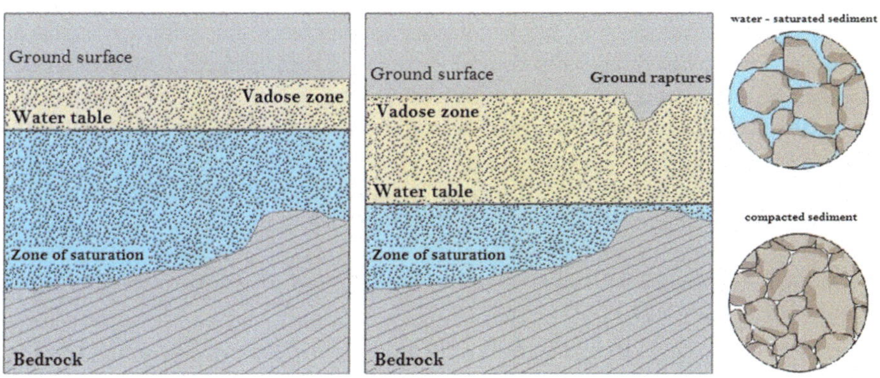

Fig. 1 Land subsidence mechanism due to the overexploitation of the aquifers

DInSAR technique analyzes phase variations or interference between two different radar images gathered over the same area at different times with the same acquisition mode and properties (Gabriel et al. 1989; Massonnet and Rabaute 1993; Massonnet and Feigl 1998; Rosen et al. 2000). The main objective of DInSAR techniques is to retrieve measurements of the surface displacement that occurred between the two different acquisitions. Within the DInSAR techniques, Permanent Scatterers Interferometry (PSI) technique was the first technique specifically implemented for the processing of multi-temporal radar imagery (Ferretti et al. 2001). The algorithm focuses on ground resolution elements containing a single dominant scatterer having stable radiometric characteristics, namely Permanent Scatterers (PS) points. These points, strongly correlated in time, are the points on the surface at which velocities along with the Line of Ssight (LOS) deformation rate are going to be estimated. Most of the times, PSs can be manmade constructions (rooftops, roads etc.), natural formations (protruding rocks etc.) or even more custom made reflectors. Areas with frequent surface changes (urban areas, cultivated areas) or areas with no PSs (lakes, forests etc.) are areas with low coherence. This is due to the fact that there is a different or low amount of micro-wave backscattering to the satellite at each acquisition. For those areas there is no information regarding the deformation rates. Over urban areas, where many PS can be identified, LOS (Line of Sight) deformation rate can be estimated with accuracy theoretically better than 0.1 mm/yr (Colesanti et al. 2003). In the present study PSI technique was utilized to detect and monitor past LOS deformation rates that would serve as evidence of past land subsidence, while land subsidence was estimated based on the level drop of groundwater, the thickness and compressibility of the formations within the research area.

3 The Study Area

The study area is located in Thessaly basin, central Greece, at the wider plain area extending northwest of Farsala town, covering an area of approximately 145 Km^2 (Fig. 2). The morphology of the wider area appears very flat, with low landscape variation. Two major rivers Enipeas and Farsaliotis cross the area. The area has been undergoing intensive cultivation, mainly cotton, corn, sugar beet, tomato and cereals crops which consume large volumes of irrigation water (Dimopoulos et al. 2003).

Based on the Köppen climate classification system (Aguado and Burt 2012), the climate of the wider area of research is characterized as Mediterranean type (Csa) having hot dry summer and a mild winter. The rainy season is from October to May accounting to almost 90% of the total amount of annual rainfall which approximately reaches 31.7–87.7 mm/month. December appears to be the rainiest month (87.7 mm) followed by November (86.2 mm), while the driest month appears to be August (10.4 mm) followed by July (14.1 mm). The annual average mean temperature is 15.13 °C with the highest and lowest average temperature being 20.94 and 9.39 °C respectively. The climate data were obtained from the

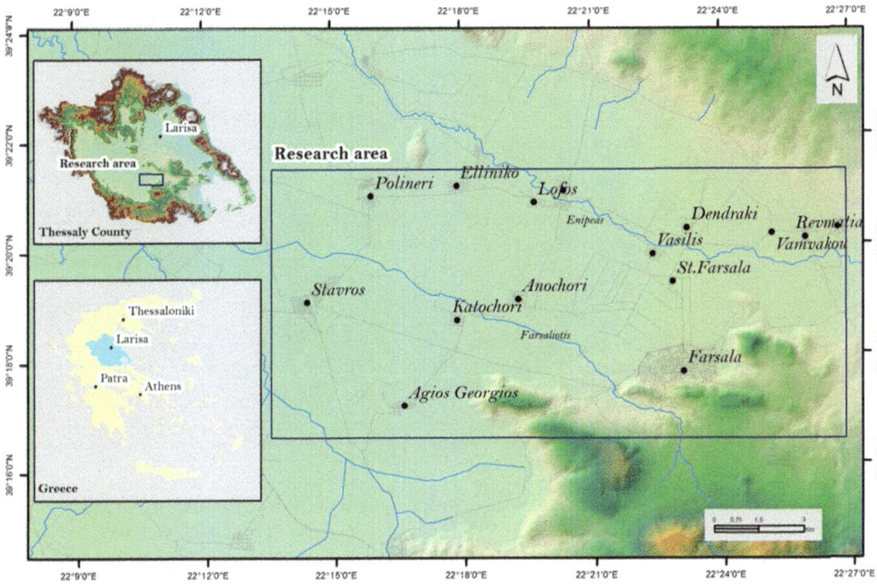

Fig. 2 The study area

University of East Anglia Climate Research Unit (CRU) and referred to a period over 107 years between 1901 and 2008 (Jones and Harris 2008).

The wider area of west Thessaly plain belongs mainly to the Pindos geotectonic zone and is characterized by a variety of geological formations. Mesozoic Alpine formations belonging to the Pelagonian, Sub-Pelagonian and Pindos geotectonic zone, occupy most of the area while post-Alpine deposits cover the lowlands of the basin. The Mesozoic Alpine formations that constitute the bedrock of the Quaternary deposits consist of Schist—chert formation (sh), Ophiolites (o) and Limestones (Le) while the post-Alpine deposits include Neogene (Ne), Pleistocene and Holocene deposits (Mariolakos et al. 2001; Rozos and Tzitziras 2002; Apostolidis 2014).

As presented in Fig. 3, the coarser deposits consisting of sands and gravels (sd–gr, gr–sd) occupy the riverbeds, while the rest of the plain is covered by the finer clayey silts and silty clays (cl–sl, with ranging percentage of intercalated sands and gravels. Normal faults in E–W direction have mainly structured the basin (Bornovas et al. 1969; Katsikatsos et al. 1983).

Considering the hydro-geological setting, highly productive aquifers are developed in the Quaternary deposits which constitute of Pleistocene sand and gravel horizons and brown and grey clayey silt to silty clay intercalations. These alternations of permeable coarse-grained deposits (aquifers) with impermeable to low permeability strata (aquitards) create shallow unconfined aquifers and a number of successive semi-confined to confined aquifers, sometimes artesian (Paleologos and Mertikas 2013). A great number of wells exploit the unconfined aquifers for

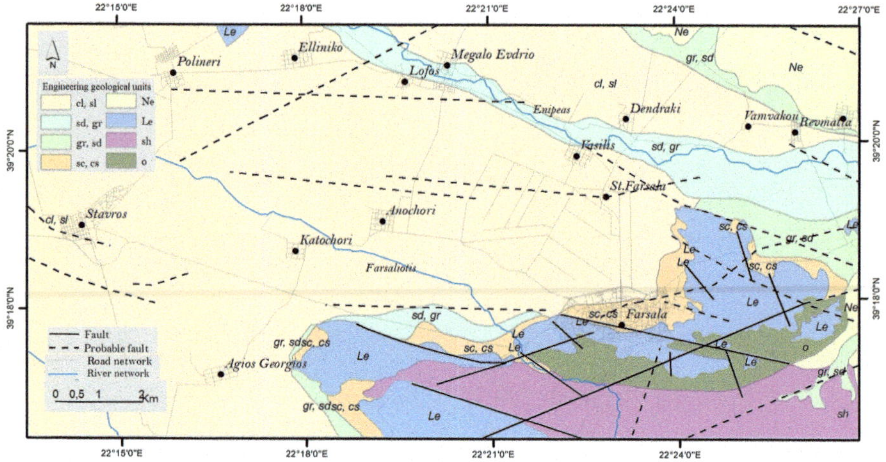

Fig. 3 Engineering geological settings of the wider study area. Modified after Apostolidis (2014)

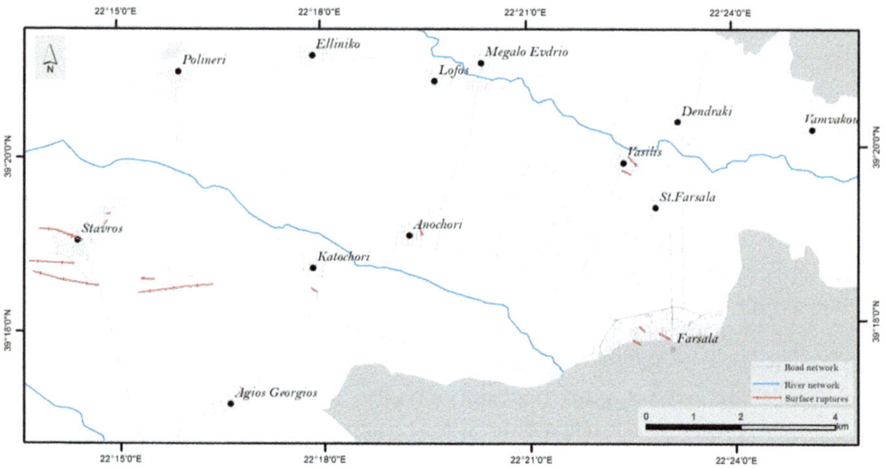

Fig. 4 Surface ruptures

irrigation purposes due to the intense agricultural activity in the area. The discharge rates of the wells range between 50 and 100 m^3/h, with groundwater flowing from eastwards to westwards of the basin (Kallergis 1971, 1973; Mariolakos et al. 2001). The recharge of these systems are mainly from the infiltration of the surface water but also through the lateral infiltration of the karstic aquifers developed in the carbonate formations of the Narthaki Mountain (Rozos et al. 2010).

Concerning the reported damages due to land subsidence, the majority of them affected roads and private buildings in the town of Farsala and the villages of Agios Georgios, Stavros and Anochori. Figure 4 presents the spatial distribution of the surface raptures while Fig. 5 illustrates some characteristic photos.

Fig. 5 Damages in Farsala, Agios Georgios, Stavros, Anochori

4 Data and Methods

The applied methodology could be distinguished into three phases: (a) the first phase involved defining the geological, hydro-geological and tectonic settings of the study area, the estimation of the physical and geo–technical properties of the geological formations and also the analysis of the spatial and temporal trend of groundwater level, (b) the second phase involved the analysis of the PSI data, while the final phase (c) involved a normalization process and the construction of the land subsidence map, which was conducted by implementing the SVM method. Figure 6 illustrates the flowchart of the applied methodology, whereas details of each phase are described in the following paragraphs.

During the first phase, the geological, hydro-geological and tectonic settings of the study area, the physical and geo–technical properties of the geological

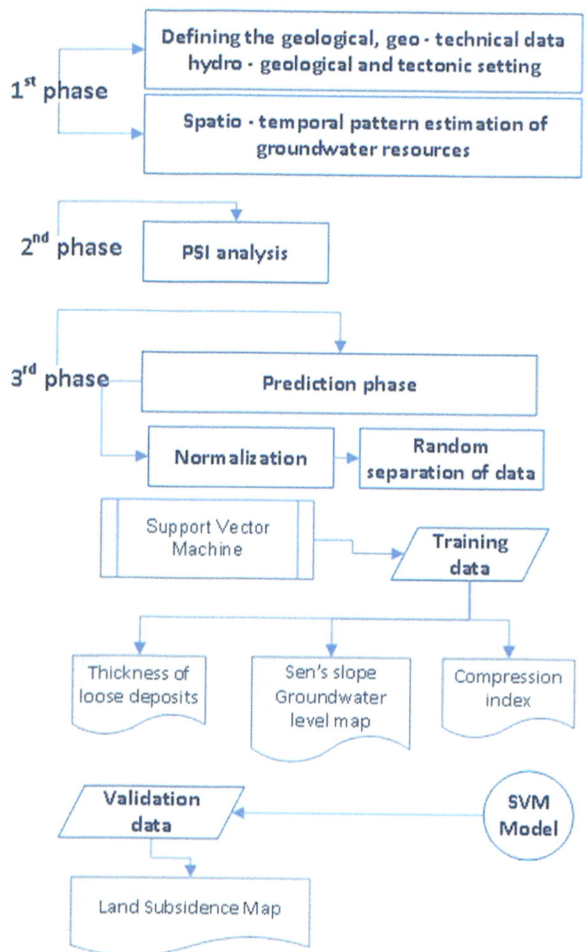

Fig. 6 Flowchart of the followed methodology

formations and also the spatial and temporal trend of groundwater level were investigated. In many studies the thickness of the susceptible deposits has been directly linked with the amount of land subsidence (Xu et al. 2008; Zhu et al. 2013a, 2015). Also the Compression Index, which is a measure of the volume decreases due to increase in load, influences significantly the extent of surface deformability. The two variables were estimated by investigating previous studies and numerous borehole data (Fig. 7) (SOGREAH 1974; Apostolidis and Koukis 2013; Apostolidis 2014).

During this phase, the data series of Sen Slope's value, a metric able to provide the fluctuation per unit time and estimate the magnitude of the detected trend in the groundwater level was derived (in our case Q m/year) (Sen 1968; Hirsch et al. 1982). The yearly groundwater table fluctuation was estimated by processing nine groundwater monitoring wells, with data obtained from the department of Hydrology of the Thessaly Prefecture referring to the time period from 1980 to 2005 and for the low–level season (September). Only the yearly records of September were analyzed, since in most of the groundwater monitoring wells it was the month with the lowest groundwater table level. The trend of this month concerning the groundwater table could serve as good indicator of the overall pressure the aquifer is subject to. Before the application of the Sen's Slope estimator the common and modified Mann–Kendall method (Mann 1945; Kendall and Stuart 1967; Hamed and Rao 1998) was applied in order to identify trends in the groundwater level time—series data.

For the spatialization of the land subsidence related variables (thickness of loose deposits, Compression Index, and trend of groundwater level) Kriging interpolation technique was utilized (David 1977). Specifically, ordinary and universal kriging were used while experimental semivariograms were fitted with various theoretical models like spherical, exponential and Gaussian in order to select the most appropriate (Kumar and Remadevi 2006; Gundogdu and Guney 2007). The least

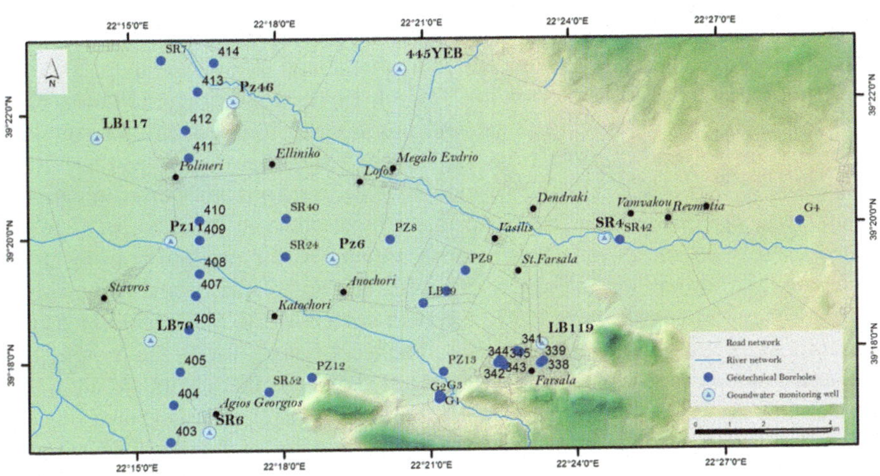

Fig. 7 Geotechnical boreholes and groundwater monitoring wells

root mean square error (RMSE) value, for every semivariogram model was estimated and the one with the lowest value was selected as more accurate (Johnson et al. 2001).

During the second phase, the analysis of the available PSI data, the necessary training and validation datasets were obtained. According to Raspini et al. (2013) SAR interferometry and in particular PSI techniques are considered as valuable tools for the early stage detection of the vertical deformations caused by the overexploitation of the aquifers. PSI data were derived from a descending data set provided by the German Space Agency (DLR) acquired in 1995–2003 by the European Space Agency (ESA) satellites ERS1 and ERS2. This set of data were processed within the framework of the Terrafirma project, that was supported by the Global Monitoring for Environment and Security (GMES) Service element Program, promoted and financed by the European Space Agency (ESA) (Adam et al. 2011). The negative displacement rate values indicate a movement away from the sensor (down lift), while the positive values represent a movement towards the sensor (up lift).

The next phase involved as a first step, the normalization of all variables in order to receive equal attention during the training process. The normalized values ranged between 0.1 and 0.9, using the Max-Min normalization procedure as follows:

$$vnew = \frac{v - Min(v)}{Max(v) - Min(v)} \times (u - l) + l$$

where $vnew$ is the normalized data matrix, v is the original data matrix, and u and l are the upper and lower normalization boundaries.

The next step involved the separation of the entire database randomly into a training dataset, for training the model and a validation dataset for evaluating the predictive power of the developed model. The training dataset included 70% of the total data, while the remaining 30% was included into the validation dataset.

As mentioned earlier, SVM was chosen among data mining techniques in order to model land subsidence phenomena and construct a land subsidence map. SVM is a non-parametric kernel-based technique (Vapnik 1998; Moguerza and Munoz 2006). It is really efficient in solving linear and non-linear classification and regression problems in a sophisticated robust manner (Cherkassky and Mulier 2007). According to Yao et al. (2008) SVM modelling differs from other discriminant type solutions since it utilizes an optimum linear hyperplane in order to separate data patterns and uses kernel functions in order to convert the original non-linear data patterns into a format that is linearly separable in a high-dimensional feature space.

In SVM regression (Vapnik 1995), the main objective is to find a function $f(x)$ that has at most ε deviation from the actually obtained targets y_i for all the training data, and at the same time is as “flat” as possible. This means that errors are allowed as long as they are less than ε. The ε—*epsilon* intensive loss function ensures existence of global minimum and at the same time optimization of reliable generalization bound.

In the case of linear analogous, the linear function $f(x)$ is given by the following equation:

$$f(x) = \langle w, x \rangle + b$$

where, the vector x presents the independent variables and vector w represents the vector of weights which has to be estimated by the model, and b is the "bias" term.

"Flatness" in the case of linear function $f(x)$ means that w has to be small. This can be achieved by minimizing the norm $\|w\|^2 = \langle w|w \rangle$ and the problem could be solved by handling it as a convex optimization problem:

minimize $\frac{1}{2}\|w\|^2$

$$subject \begin{cases} y_i - \langle w, x_i \rangle - b \leq \varepsilon \\ \langle w, x_i \rangle + b - y_i \leq \varepsilon \end{cases}$$

The above assumes that function f exists and approximates all pairs (x_i, y_i) with ε precision. In order to allow for some errors, one could introduce slack variables ξ_i, ξ_i^* to cope with otherwise infeasible constraints of the optimization problem (Smola and Schölkopf 2004). The convex optimization problem formulates into (Vapnik 1995):

minimize $\frac{1}{2}\|w\|^2 + C\sum_{i=1}^{l}(\xi_i, + \xi_i^*)$

$$\text{subject} \begin{cases} y_i - \langle w, x_i \rangle - b \leq \varepsilon + \xi_i \\ \langle w, x_i \rangle + b - y_i \leq \varepsilon + \xi_i^* \\ \xi_i, \xi_i^* \geq 0 \end{cases}$$

The constant $C > 0$ determines the trade-off between the "flatness" of f and the amount up to which deviations larger than ε are tolerated. This corresponds to dealing with a so called ε-insensitive loss function $|\xi|_\varepsilon$ described by

$$|\xi|_\varepsilon = \begin{cases} 0 & if\,|\xi| \leq \varepsilon \\ |\xi| - \varepsilon & otherwise \end{cases}$$

Extension of linear classification formulation to non-linear SVM can be achieved using the Kernel trick (Schölkopf and Smola 2002). Although many kernel functions have been proposed, the most commonly used are linear, polynomial, radial basis function and sigmoid.

In the present study the radial basis function (RBF) Gaussian kernel was implemented, which is influenced by the kernel width (γ) and the regularization (C) parameters (Tien Bui et al. 2016). The RBF Gaussian kernel is a defined by the following equation:

$$K(x_i, x_j) = \exp\left(-\gamma\|x_i - x_j\|^2\right)$$

Parameter C determines the tradeoff between the model complexity ("flatness") and the degree to which deviations larger than ε are tolerated in optimization formulation for example, if C is too large (infinity), then the objective is to minimize the empirical risk only, without regarding model complexity part in the optimization formulation.

Parameter ε controls the width of the ε-insensitive zone, used to fit the training data. The value of ε can affect the number of support vectors used to construct the regression function. The bigger ε, the fewer support vectors are selected. On the other hand, bigger ε-values result in more 'flat' estimates. Hence, both C and ε-values affect model complexity, however in a different manner.

In the present study, a grid search algorithm along with cross validation to find the optimum value for C, γ and ε were used. The leave-one-out cross-validation was employed to determine the optimal parameters and the set of values with the best leave-one-out cross-validation performance was selected for further analysis.

The next phase was to apply the SVM model that has been trained within a test area, to the entire research area and produce the land subsidence map, with the optimal parameters found in the previous step. Finally, the third phase ends with the validation of the predictive performance of the model. This was achieved by measuring two statistical metrics, the root mean squared error (*RMSE*) and the r square (R^2) (Willmott et al. 1985). *RMSE* is a quadratic scoring rule that measures the average magnitude of error, which is estimated by the differences between prediction and actual observations, whereas *r square*, provides a measure of how well observed outcomes are replicated by the model, based on the proportion of total variation of outcomes explained by the model.

5 Results

Based on the analysis of the geotechnical borehole data that provided information about the depth of bedrock formations that cover the area, a map which illustrates the spatial distribution of the thickness of the loose deposits was produced by applying ordinary Kriging method (Penížek and Borůvkav 2006) with the Gaussian model resulted in the minimum RMSE error and so considered the best-fit model. The maximum thickness was estimated to be 270 m and the minimum 30 m. Figure 8 shows the normalized values. Higher values were estimated in the area close to Stavros and Polineri villages, while lower values were estimated close to the town of Farsala.

From the same geotechnical database and the analysis of over 60 oedometer tests (Fig. 7), the Compression Index of the Quaternary deposits was estimated (Apostolidis 2014). The value of Compression Index was equal to the average value (when more than one oedometer test have been conducted in the same borehole) of the dominant sediment layer in that location. The red brown to black brown clays and the silty clays-clayey silts present higher values ranging between 0.060 and 0.835, while the alternating loose sandy clay or silty sand horizons present lower values ranging between 0.040 and 0.450. Figure 9 illustrates the normalized values of the Compression Index that has been produced after the implementation of an ordinary Kriging method, fitted to a Gaussian function that produced the lowest

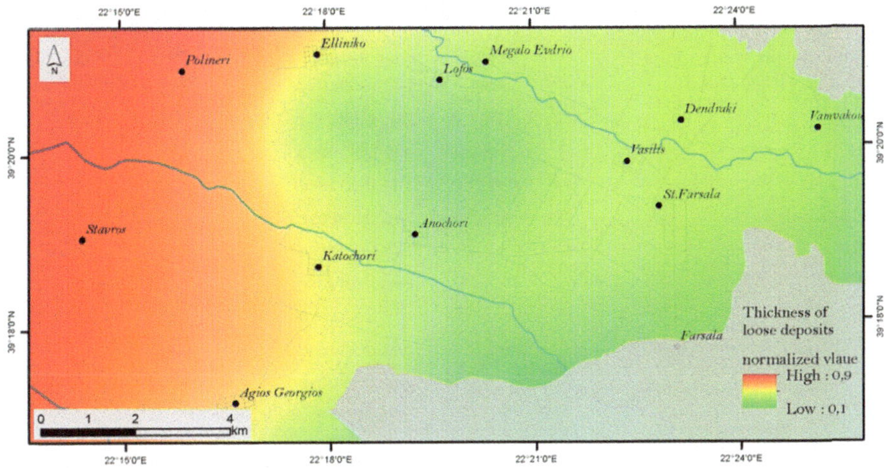

Fig. 8 Spatial distribution of the thickness of loose deposits

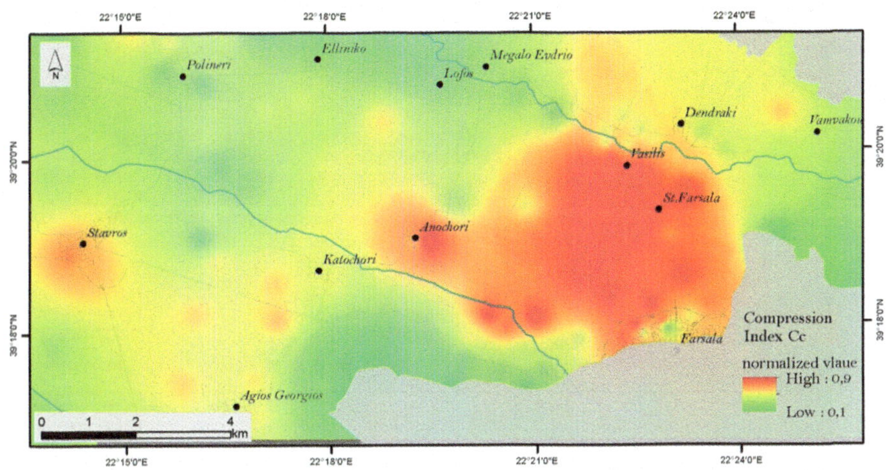

Fig. 9 Spatial distribution of the compression index (Cc)

RMSE. Higher values are located between the town of Farsala and the village of Vasilis while lower values are estimated at the northwest of the area.

Concerning the fluctuation per unit time and the magnitude of the trend in the groundwater level the analysis showed significant downward trend (Table 1).

The 445YEB monitoring well showed the highest values, −1.699 m/year, followed by SR4 (−1.504 m/year) and PZ46 (−1.400 m/year). The lowest values with statistical significant downward trend were recorder in LB117 (−0.595 m/year) and in LB119 (−1.080 m/year). The spatial distribution of the normalized Sen's Slope values is illustrated in Fig. 10.

Table 1 Results of trend analysis

Groundwater wells monitoring network (1980–2005)	Mann–Kendal trend low—level season		Sen's slope estimator low —level season
	Test Z	Significance	Q m/year
LB70	−5.14	sig.decrease	−1.196
PZ11	−2.11	sig.decrease	−1.225
SR4	−5.73	sig.decrease	−1.504
SR6	−4.45	sig.decrease	−1.146
LB117	−3.20	sig.decrease	−0.595
LB119	−5.67	sig.decrease	−1.080
445YEB	−2.24	sig.decrease	−1.699
PZ6	−5.95	sig.decrease	−1.310
PZ46	−2.42	sig.decrease	−1.400

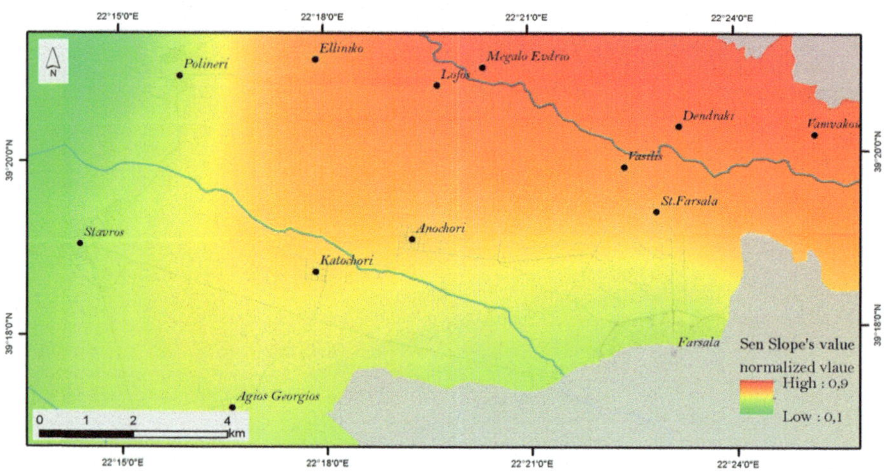

Fig. 10 Spatial distribution of Sen Slope's value

Based on the performed PSI analysis, 5848 PSs were identified in the data set acquired at the descending orbit of ERS1/2 from 19/6/1995 to 18/10/2003. Based on the mean deformation rate values and a threshold of ±1.50 mm/year, the 94.19% of the total PSs were stable. About 5.06% of the PSs show a downwards rate greater than −1.50 mm/year, indicating subsiding movement. Figure 11 presents the spatial distribution of the deformation rates at the test area. It is clear that the subsidence phenomena affect the town of Farsala as well as the plain area extending to the north. The area north of Vasilis village presented the highest recorded deformation rate reaching up to values of −20.339 mm/year, whereas the majority of the PSs had deformation rates ranging between −8.00 and −12.00 mm/year.

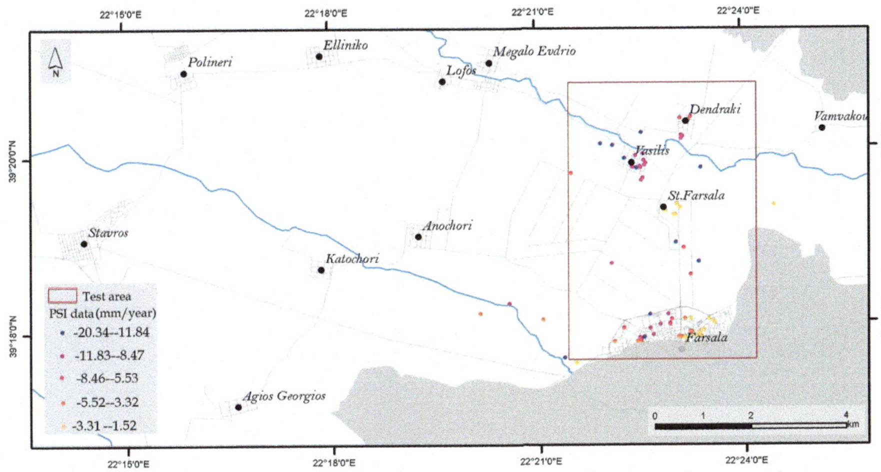

Fig. 11 PSI deformation rate measurements

Passing to the predictive phase, a number of 119 PSs that presented subsiding movements greater than −1.50 mm/year were selected within the test area and separated into a training and validation dataset. PSs located at areas covered by bedrock formations were excluded from the analysis. Following the developed methodology, for each PS point of the training and validation dataset, the values of thickness of the loose deposits, Sen's Slope value and Compression Index were obtained by applying the tool Extract Multi Values to Points, a tool found within the Extraction toolset, component of the Spatial Analyst toolbox (ESRI 2013). The next step was the implementation of the SVM method in order to predict deformation rate based on the three variables. During the implementation and after performing a tuning process, the optimal values C, ε and γ were estimated to be 1.55, 0.14 and 0.46, respectively.

A multi-linear regression analysis (MLR) (Montgomery et al. 2012) was also conducted in order to provide a base regression model and compare it with the one produced by the SVM model. Table 2 illustrates the results from the implementation of SVM and MLR, in which the SVM model outperforms the MLR model. The R^2 value in the training dataset was estimated for the SVM model to be 0.81 and the *RMSE* 1.93, while the R^2 for the same dataset and the MLR model was estimated to be 0.42 and the *RMSE* 3.16. The same pattern of accuracy was identified in the validation dataset. Specifically, the R^2 value was estimated for the SVM model to be 0.75 and the *RMSE* 2.43, while the R^2 for the MLR model was estimated to be 0.20 and the *RMSE* 2.79.

The final step was to produce the land subsidence map, based on the SVM model for the entire research area. Figure 12 illustrates the spatial distribution of the land subsidence values, with higher values estimated close to Vasilis village, Farsala station, the villages of Stavros and Anochori and the town of Farsala.

Table 2 Results of the validation analysis

Model	Training dataset		Validation dataset	
	RMSE	R^2	RMSE	R^2
Multi-linear regression	3.16	0.42	2.79	0.20
Support vector machine	1.93	0.81	2.43	0.75

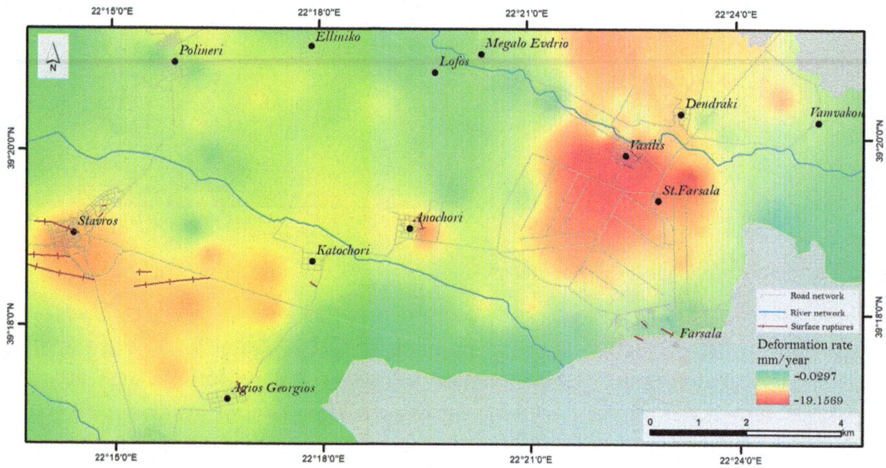

Fig. 12 Land subsidence map

6 Discussion

According to the US National Research Council (1991), the detailed mapping, characterization and the simulation of subsidence is the most essential phase that has to precede the design and implementation phase of mitigation methods. The most common practice concerning the mitigation measures in areas of land subsidence problems involves mainly the regulation of groundwater pumping systems, the design of alternative water supply and also the construction of maps that provide, in most cases, the spatial distribution of land subsidence (Raspini et al. 2013).

In this context, the motivation of the present study was to produce a land subsidence map in order to early detect surface deformations and serve as a helpful tool to local authorities. The results of the study indicated areas that although presented minor reported damages, exhibit high probability of land subsidence. Based on the fact that the present management of the groundwater resources will continue in the future, severe subsidence phenomena may appear in Vasilis, Dendraki and Anochori villages. The same phenomena will probably continue to be present in the village Stavros and would probably expand to the east. According to the SVM model, the town of Farsala seems to be less influenced by land subsidence; however it should be noticed that the model does not consider any other variables that could influence the phenomena, such as the building density, the infrastructure distribution and human interventions, which would certainly alter the observed deformation rates. The above mentioned areas are covered by formations with high Compression Index values and are also characterized by successive thick layers of loose deposits.

Based on the analysis of the groundwater level data, it appears that a significant drawdown takes place within the research area. Even though a natural recharge of aquifers takes place every year which is mainly provided by rainfall, infiltration of irrigation waters, leakage from the bed of the two major rivers and subsurface flow from the mountain ranges bounding the plain, a constant drawdown overall tendency occurs. According to Manakos (2010), the over-exploitation of the water resources in the wider Thessaly plain has resulted to the systematic groundwater level dropdown and also to the degradation of the quality of the water in the majority of the aquifer systems.

The high accuracy (*r square* value 0.75) achieved by the SVM model is an indication of the sufficiency and applicability of the conceptual model that is based on the thickness of loose deposits, the Sen's slope value and the Compression Index. These three variables could be considered as the independent variables that influence the evolution of the land subsidence and better describe the mechanism of the phenomena in the study area. Furthermore, the low accuracy of the MLR model (*r square* value 0.20) is an indication of the non-linear and complex nature of land subsidence phenomena that can be better modelled by more advanced methods such as SVM. The SVM model appears to be ideal for such complex problems since it performs better in regression problems with small number of training data and avoids over-fitting.

The outcomes of the study are also in agreement with previous studies that promote remote sensing data and EO techniques as valuable tools concerning the verification and validation of land subsidence and as a cost-efficient method for the management of land subsidence related hazards (Raspini et al. 2014).

Overall, the outcomes of the present study are in agreement with the theory concerning the mechanism of land subsidence evolution, suggesting that an excessive lowering of the groundwater level leads to the radical change of the geostatic loads triggering or accelerating the consolidation of compressible ground layers.

7 Conclusion

The present study provides a methodological approach for the investigation of land subsidence phenomena by utilizing spatio-temporal analysis of groundwater resources, remote sensing techniques and data mining methods, implemented at the wider Farsala plain located in western Thessaly, Greece.

The SVM method was utilized to predict the land subsidence deformation rate based on three related variables, namely: thickness of loose deposits, the Sen'Slope value of groundwater trend and the Compression Index of the formation covering the area of interest.

The high accuracy achieved by the SVM model (*r square* value 0.75) was an indication of the efficiency and applicability of the conceptual model that was based on those three variables to describe the mechanism of land subsidence in the study area.

The conducted analysis detected areas that exhibit deformation which, however, have no records of damages. It is most certain that the continuing over-exploitation of the water resources will trigger further land subsidence phenomena and expand the affected areas. This early detection of surface deformations allows taking mitigation measures before severe land subsidence phenomena occur and therefore allows for timely protection of the affected areas.

Acknowledgements The Terrafirma Extension project has funded the SAR imagery processing as well as the geological interpretation presented in this paper. The project is one of the many services supported by the Global Monitoring for Environment and Security (GMES) Service Element Program, promoted and financed by ESA. The project is aimed at providing civil protection agencies, local authorities and disaster management organisms with support in the process of risk assessment and mitigation by using the Persistent Scatterer Interferometry. The authors gratefully acknowledge the German Space Agency (DLR) for having processed the SAR data.

References

Adam N, Rodriguez Gonzalez F, Parizzi A, Liebhart W (2011) Wide area persistent scatterer Interferometry. In: Proceedings of IGARSS, Vancouver, Canada

Aguado E, Burt J (2012) Understanding weather and climate, 6th edn. Prentice Hall, Upper Saddle River, New Jersey, p 576

Apostolidis E (2014) Engineering-geological conditions in the western Thessaly basin: geomechanical characteristics of the quaternary deposits: analysis using geographic information systems. Ph.D Thesis, University of Patras, pp. 1199 (in Greek)

Apostolidis E, Georgiou H (2007) Engineering Geological of the surface ground ruptures in Thessalia basin sites. Recording and, documentation. Institute of Geology and Mineral Exploration (IGME), unpublished report (in Greek)

Apostolidis E, Koukis G (2013) Engineering-geological conditions of the formations in the Western Thessaly basin, Greece. Cent Eur J Geosci 5(3):407–422

Bornovas I, Filippakis N, Bizon JJG (1969) Geological map of Greece 1:50.000, Farsala Sheet. Publication IGME Athens

Brinkgreve RBJ, Broere W, Waterman D (2006) Plaxis, fine element code for soil and rock analyses, 2D—version 8. A.A. Balkema, Rotterdam Brookfield

Chaussard E, Amelung F, Abidin H, Hong SH (2013) Sinking cities in Indonesia: ALOS PALSAR detects rapid subsidence due to groundwater and gas extraction. Remote Sens Environ 128:150–161

Cherkassky V, Mulier F (2007) Learning from data: concepts, theory and methods. Wiley Inc., Hoboken, New Jersey

Choi JK, Won JS, Lee S (2011) Integration of a subsidence model and SAR interferometry for a coal mine subsidence hazard map in Taebaek, Korea. Int J Remote Sens 32(23):8161–8181

Colesanti C, Ferretti A, Prati C, Rocca F (2003) Monitoring landslides and tectonic motion with the permanent scatterers technique. Eng Geol 68:3–14

David M (1977) Geostatistical ore reserve estimation. Elsevier, Amsterdam

Dimopoulos M, Chalkiadaki M, Dassenakis M, Scoullos M (2003) Quality of groundwater in western Thessaly. The problem of nitrate pollution. Global Nest the Int J 5(3):185–191

Dixon TH, Amelung F, Ferretti A, Novali F, Rocca F, Dokka R, Sella G, Kim SW, Wdowinski S, Whitman D (2006) Subsidence and flooding in New Orleans—a subsidence map of the city offers insight into the failure of the levees during Hurricane Katrina. Nature 441:587–588

ESRI (2013) ArcGIS desktop: release 10.1. Environmental Systems Research Institute, Redlands, CA

Fayyad UM, Piatetsky-Shapiro G, Smyth P (1996) From data mining to knowledge discovery: an overview. In: Fayyad UM, Piatetsky-Shapiro G, Smyth P, Uthurasamy R (eds) Advances in knowledge discovery and data mining, MIT Press/AAAI Press

Ferretti A, Prati C, Rocca F (2001) Permanent scatterers in SAR interferometry. IEEE Geosci Remote S 39:8–20

Flentje P, Stirling D, Chowdhury R (2007) Landslide susceptibility and hazard derived from a landslide inventory using data mining—an Australian case study. In: Proceedings of the first North American Landslide Conference, Vail, Colorado, June 2007

Gabriel AK, Goldstein RM, Zebker HA (1989) Mapping small elevation changes over large areas: differential radar interferometry. J Geophys Res 94:9183–9191

Galloway DL, Burbey TJ (2011) Review: regional land subsidence accompanying groundwater extraction. Hydrogeol J 19:1459–1486

Galloway DL, Hudnut KW, Ingebritsen SE, Phillips SP, Peltzer G, Rogez F, Rosen PA (1998) Detection of aquifer system compaction and land subsidence using interferometric synthetic aperture radar, Antelope Valley, Mojave Desert, California. Water Resour Res. 34:2573–2585

Galve J, Gutiérrez F, Lucha P, Guerrero J, Bonachea J, Remondo J, Cendrero A (2009) Probabilistic sinkhole modelling for hazard assessment. Earth Surf Proc Land 34:437–452

Gambolati G, Teatini P, Ferronato M (2005) Anthropogenic land subsidence. In: Anderson MG (ed) Encyclopedia of hydrological sciences. Wiley, Wiley Online Library, Hoboken, p 17

Ganas A, Salvi S, Atzori S, Tolomei C (2006) Ground deformation in thessaly, central greece, retrieved from differential interferometric analysis of ERS-SAR data. In: 11th international symposium on natural and human induced hazards and 2nd workshop on earthquake prediction, Patras, Greece, vol 41, pp 22–25

Gundogdu KS, Guney I (2007) Spatial analyses of groundwater levels using universal kriging. J Earth Syst Sci 116(1):49–55

Hamed KH, Rao R (1998) A modified Mann-Kendall trend test for autocorrelated data. J Hydrol 204(1–4):182–196

Herrera G, Fernandez JA, Tomas R, Cooksley G, Mulas J (2009) Advanced interpretation of subsidence in Murcia (SE Spain) using A-DInSAR data modelling and validation. Nat Hazards Earth Syst Sci 9:647–661

Hirsch RM, Slack JR, Smith RA (1982) Techniques of trend analysis for monthly water quality data. Water Resour Res 18(1):107–121

Hu J, Shi B, Inyang HI, Chen J, Sui Z (2009) Patterns of subsidence in the lower Yangtze Delta of China: the case of the Suzhou-Wuxi-Changzhou Region. Environ Monit Assess 153:61–72

Ilia I, Loupasakis C, Tsangaratos P (2016) Assesing ground subsidence phenomena with persistent scatterer interferometry data in Western Thessaly, Greece. In: 14th international congress of the geological society of Greece, At Thessaloniki, Greece, vol volume L

Johnson K, Ver Hoef JM, Krivoruchko K, Lucas N (2001) Using ArcGIS geostatistical analyst. GIS by ESRI, Redlands, USA

Jones PD, Harris I (2008) Climatic Research Unit (CRU) time-series datasets of variations in climate with variations in other phenomena. University of East Anglia Climatic Research Unit, NCAS British Atmospheric Data Centre

Kaitantzian A, Loupasakis C, Rozos D (2014) Assessment of geo-hazards triggered by both natural events and human activities in rapidly urbanized areas. In: Lollino G et al (eds.) Proceedings of 12th international IAEG congress IAEG2014—engineering geology for society and territory, vol 5, pp 675–679

Kallergis G (1971) Ground subsidences during the drawdown of artesian aquifers due to their limited elasticity (in Greek). Technika Chronika 599–602

Kallergis G (1973) Hydrogeological study in sub-basin of Kalampaka (Western Thessaly). Institute of Geology and Mineral Exploration (IGME), unpublished report (in Greek), vol. XIV, No 1, Athens

Kaplanidis A, Fountoulis D (1997) Subsidence phenomena and ground fissures in Larissa, Karla basin, Greece: their results in urban and rural environment. In: Proceedings of the international symposium “engineering geology and environment”, vol 1, pp 729–735

Katsikatsos G, Mylonakis E, Triantaphyllis E, Papadeas G, Psonis K, Tsaila-Monopoli S, Skourtsi-Koroneou V (1983) Geological Map of Greece 1:50.000, Velestino Sheet. Publication IGME, Athens

Kendall MA, Stuart A (1967) The advanced theory of statistics, 2nd edn. Charles Griffin, Londres

Kim KD, Lee S, Oh HJ (2009) Prediction of ground subsidence in Samcheok City, Korea using artificial neural networks and GIS. Environ Geol 58:61–70. https://doi.org/10.1007/s00254-008-1492-9

Kontogianni V, Pytharouli S, Stiros S (2007) Ground subsidence, quaternary faults and vulnerability of utilities and transportation networks in Thessaly, Greece. Environ Geol 52:1085–1095

Koumantakis I, Rozos D, Markantonis K (2008) Ground subsidence in Thermaikos municipality of Thessaloniki County, Greece. In: International conference Gro-Pro—ground water protection—plans and implementation in a North European Perspective, vol 1, Korsør Denmark, 15–17 Sept 2008, pp 177–184

Kumar V, Remadevi (2006) Kriging of groundwater levels—a case study. J Spat Hydrol 6(1):81–94

Lee S, Park I, Choi JK (2012) Spatial prediction of ground subsidence susceptibility using an artificial neural network. J Environ Manage 49:347–358

Loupasakis C, Rozos D (2009) Land subsidence induced by water pumping in Kalochori Village (North Greece)—simulation of the phenomenon by means of the finite element method. Q J Eng GeolHydrogeol 42(3):369–382

Loupasakis C, Agelitsa V, Rozos D, Spanou N (2014) Mining geohazards—land subsidence caused by the dewatering of opencast coal mines: the case study of the Amyntaio coal mine, Florina, Greece. Nat Hazards 70:675–691. https://doi.org/10.1007/s11069-013-0837-1

Malinowska A (2014) Classification and regression tree theory application for assessment of building damage caused by surface deformation. Nat Hazards 73:317–334. https://doi.org/10.1007/s11069-014-1070-2

Manakos A (2010) Hydrogeological survey. Water district basin (08). In: 3rd community support framework, IGME, pp 547

Mann HB (1945) Non-parametric tests against trend. Econometrica 13(3):245–259

Marinos P, Perleros V, Kavvadas M (1997) Alluvial and karst aquifers in Thessaly plain. New data for their regime of overexploitation. In: Proceedings of the 2nd Hydrogeological Congress, Patras, Greece, pp 243–258 (in Greek)

Mariolakos H, Lekkas S, Papadopoulos T, Alexopoulos A, Spyridonos E, Mandekas I, Andreadakis E (2001) Underground tectonic structure in Farsala plain (Thessaly) as a determinative factor of the formation of the hydrogeological conditions of the area. In: Proceedings of the 9th congress of Greek Geological Society (in Greek)

Massonnet D, Feigl KL (1998) Radar interferometry and its application to changes in the earth's surface. Rev Geophys 36(4):441–500

Massonnet D, Rabaute T (1993) Radar interferometry: limits and potential. IEEE Trans Geosci Rem Sens 31:455–464

Mertikas SP, Papadaki ES (2009) Radar Interferometry for Monitoring Land Subsidence due to overpumping Ground Water in Crete, Greece, In: Proceedings of the Fringe Workshop, Frascati, Italy, 30 Nov–4 Dec 2009, p 4

Meyer D, Dimitriadou E, Hornik K, Weingessel A, Leisch F (2017) Package "e1071", pp. 62. Date/publication 02 Feb 2017

Modis K, Sideri D (2015) Spatiotemporal estimation of land subsidence and ground water level decline in West Thessaly basin, Greece. Nat Hazards 76:939–954

Modoni G, Darini G, Spacagna RL, Saroli M, Russo G, Croce P (2013) Spatial analysis of land subsidence induced by groundwater withdrawal. Eng Geol 167:59–71

Moguerza J, Munoz A (2006) Support vector machines with applications. Statist Sci 21(3): 322–336

Montgomery DC, Peck EA, Vining GG (2012) Introduction to linear regression analysis, 5th edn. Wiley, Hoboken, NJ, USA, p 672

Oh HJ, Ahn SC, Choi JK, Lee S (2011) Sensitivity analysis for the GIS-based mapping of the ground subsidence hazard near abandoned underground coal mines. Environ Earth Sci 64:347–358

Osmanoglu B, Dixon TH, Wdowinski S, Cabral-Cano E, Jiang Y (2011) Mexico City subsidence observed with persistent scatterer InSAR. J Appl Earth Obs Geoinf 13(1):1–12

Paleologos EK, Mertikas SP (2013) Evidence and implications of extensive groundwater overdraft-induced land subsidence in Greece. Eur Water 43:3–11

Penížek V, Borůvkav L (2006) Soil depth prediction supported by primary terrain attributes: a comparison of methods. Plant Soil Environ 52(9):424–430

Psimoulis P, Ghilardi M, Fouache E, Stiros S (2007) Subsidence and evolution of the Thessaloniki plain, Greece, based on historical levelling and GPS data. Eng Geol 90:55–70

Raspini F, Loupasakis C, Rozos D, Moretti S (2013) Advanced interpretation of land subsidence by validating multi-interferometric SAR data: the case study of Anthemountas basin (Northern Greece). Nat Hazards Earth Syst Sci 13:2425–2440

Raspini F, Loupasakis C, Rozos D, Adam N, Moretti S (2014) Ground subsidence phenomena in the Delta municipality region (Northern Greece): geotechnical modeling and validation with persistent scatterer interferometry. Int J Appl Earth Obs Geoinf 28:78–89

Raucoules D, Parcharidis I, Feurer D, Novali F, Ferretti A, Carnec C, Lagios E, Sakkas V, Le Mouelic S, Cooksley G, Hosford S (2008) Ground deformation detection of the greater area of Thessaloniki (Northern Greece) using radar interferometry techniques. Nat Hazards Earth Syst Sci 8:779–788. https://doi.org/10.5194/nhess-8-779-2008

Rosen PA, Hensley S, Joughin IR, Li FK, Madsen SN, Rodriguez E, Goldstein RM (2000) Synthetic aperture radar interferometry. IEEE Geosci Remote S 88:333–382. https://doi.org/10.1109/5.838084

Rozos D, Tzitziras A (2002) Report of the Engineering geological examination of ground water in Farsala area. Institute of Geology and Mineral Exploration (IGME), unpublished report (in Greek)

Rozos D, Sideri D, Loupasakis C, Apostolidis E (2010) Land subsidence due to excessive ground water withdrawal. A case study from Stavros—Farsala Site, West Thessaly Greece. In: Proceedings of the 12th international congress, Patras, Greece

RStudio Team (2015) RStudio: Integrated Development for R. RStudio Inc, Boston, MA URL

Salvi S, Ganas A, Stramondo S, Atzori S, Tolomei C, Pepe A, Manzo M, Casu F, Berardino P, Lanari R (2004) Monitoring long-term ground deformation by SAR interferometry: examples from the Abruzzi, Central Italy, and Thessaly, Greece. In: 5th international symposium on Eastern Mediterranean Geology, Thessaloniki, Greece, 14–20 Apr 2004, pp 1–4

Schölkopf B, Smola AJ (2002) Learning with kernels. MIT Press, Cambridge

Sen PK (1968) Estimates of the regression coefficient based on Kendall's tau. J Am Stat Assoc 63 (1968):1379–1389

Smola A, Schölkopf B (2004) A tutorial on support vector regression. Stat Comput 14(3):199–222

SOGREAH (1974) Project for the development of groundwater of Thessaly plain. Ministry of Agriculture, R 11971, Ministry of Agriculture, Athens

Soulios G (1997) Subsidence de terrains alluviaux dans le sud-est de la plaine de Thessalie, Grèce. In: Proceedings international symposium on engineering geology and the environment. Balkema, Rotterdam

Soulios G, Tsapanos T, Voudouris K, Kaklis T, Mattas C, Sotiriadis M (2011) Ruptures on surface and buildings due to land subsidence in Anargyri village (Florina Prefecture, Macedonia). In: Lambrakis N et al (eds) Proceedings of the advances in the research of aquatic environment, vol 2. Springer, Berlin, pp 505–512. https://doi.org/10.1007/978-3-642-24076-8

Stiros SC (2001) Subsidence of the Thessaloniki (northern Greece) coastal plain, 1960–1999. Eng Geol 61(4):243–256

Stramondo S, Saroli M, Tolomei C, Moro M, Doumaz F, Pesci A, Loddo F, Baldi P, Boschi E (2007) Surface movements in Bologna (Po Plain-Italy) detected by multitemporal DInSAR. Remote Sens Environ 110:304–316

Svigkas N, Papoutsis I, Loupasakis C, Tsangaratos P, An Kiratzi, Kontoes Ch (2016) Land subsidence rebound detected via multi-temporal InSAR and ground truth data in Kalochori and Sindos regions, Northern Greece. Eng Geol 209:175–186

Teartisup P, Kerdsueb P (2013) Land subsidence prediction in central plain of Thailand. Int J Environ Sci Dev 4(1):59–61

Terzaghi K (1925) Settlement and consolidation of clay. McGraw-Hill, New York, pp 874–878

Tien Bui D, Tuan TA, Klempe H, Pradhan B, Revhaug I (2016) Spatial prediction models for shallow landslide hazards: a comparative assessment of the efficacy of support vector machines, artificial neural networks, kernel logistic regression and logistic model tree. Landslides 13(2):361–378

Tomás R, Márquez Y, Lopez-Sanchez JM, Delgado J, Blanco P, Mallorqui JJ, Martinez M, Herrera G, Mulas J (2005) Mapping ground subsidence induced by aquifer overexploitation using advanced differential SAR Interferometry: Vega Media of the Segura River (SE Spain) case study. Remote Sens Environ 98(2–3):269–283

Tzampoglou P, Loupasakis C (2016) New data regarding the ground water level changes at the Amyntaio basin—Florina Prefecture, Greece. Proceedings of the 14th international congress of the geological society of Greece, Bulletin of the geological society of Greece, vol XLVIII

US National Research Council (1991) Mitigating losses from land subsidence in the United States. National Academy Press, Washington, DC

Vapnik V (1995) The nature of statistical learning theory. Springer, New York

Vapnik V (1998) Statistical learning theory. Wiley, New York

Vassilopoulou S, Sakkas V, Wegmuller U, Capes R (2013) Long term and seasonal ground deformation monitoring of Larissa plain (central Greece) by Persistent Scattering Inferferometry. Cent Eur J Geosci 5:61–76

Willmott CJ, Ackleson SG, Davis RE, Feddema JJ, Klink KM, Legates DR, O'Donnell J, Rowe CM (1985) Statistics for the evaluation of model performance. J Geophys Res 90 (C5):8995–9005

Xu YS, Shen SL, Cai ZY (2008) The state of land subsidence and prediction approaches due to groundwater withdrawal in China. Nat Hazards 45(123–135):2008

Yao X, Tham LG, Dai FC (2008) Landslide susceptibility mapping based on support vector machines: a case study on natural slopes of Hong Kong, China. Geomorphology 101:572–582

Zhu L, Chen Y, Gong HL, Liu C, Wang R (2013a) Spatial risk assessment on land subsidence in Beijing, China. In: 20th International congress on modelling and simulation, Adelaide, Australia, 1–6 Dec 2013

Zhu L, Gong H, Xiaojuan L, Yongyong L, Xiaosi S, Gaoxuan G (2013b) Comprehensive analysis and artificial intelligent simulation of land Subsidence of Beijing. China. Chin Geogra Sci 23 (2):237–248

Zhu L, Gong HL, Li XJ, Wang R, Chen BB, Dai ZX, Teatini P (2015) Land subsidence due to groundwater withdrawal in the northern Beijing plain, China. Eng Geol 193:243–255

Application of Fuzzy Analytical Network Process Model for Analyzing the Gully Erosion Susceptibility

Bahram Choubin, Omid Rahmati, Naser Tahmasebipour, Bakhtiar Feizizadeh and Hamid Reza Pourghasemi

Abstract Soil erosion is one of the most important processes in land degradation especially in semi-arid areas such as Iran. Awareness from susceptible areas to erosion is essential for decreasing the damages and restoration of the eroded areas and achieving the sustainable development goals. Thus, the main purposes of this study are prioritizing the effective variables in engender and extend of gully erosion and predicting the gully erosion susceptibility map in the Kashkan-Poldokhtar Basin, Iran. In order to achieve this purpose, the fuzzy analytical network process (Fuzzy ANP) was applied by means of considering the interrelationship network within the effective criteria on the gully erosion. The assessing step were conducted by the fuzzy approach in associate with the expert's opinions for determining the susceptible areas to gully erosion. Eventually, gully erosion susceptibility map was produced based on Fuzzy ANP weights and GIS aggregation functions. Results were validated by applying the known gullies collected in field surveys by GPS. The ROC curve was applied to investigate the susceptibility model's performance. Results of the Fuzzy-ANP was revealed that drainage density, soil texture, and lithology are most important factors for gully erosion. In addition, results delivered the accuracy of 90.4% for the study area which is very acceptable. This research

B. Choubin
Department of Watershed Management Engineering, Sari University of Agricultural Sciences and Natural Resurgences, Sari, Iran

O. Rahmati (✉)
Young Researchers and Elites Club, Khorramabad Branch, Islamic Azad University, Khorramabad, Iran
e-mail: Orahmati68@gmail.com

N. Tahmasebipour
Department of Watershed Management Engineering, Faculty of Agriculture, Lorestan University, Khorramabad, Iran

B. Feizizadeh
Department of Remote Sensing and GIS, University of Tabriz, Tabriz 51368, Iran

H. R. Pourghasemi
Department of Natural Resources and Environmental Engineering, Shiraz University, Shiraz, Iran

H. R. Pourghasemi and M. Rossi (eds.), *Natural Hazards GIS-based Spatial Modeling Using Data Mining Techniques*, Advances in Natural and Technological Hazards Research 48, https://doi.org/10.1007/978-3-319-73383-8_5

highlights that Fuzzy ANP as an efficient approach for producing the susceptibility map of gully erosion, especially in an environment with incomplete datasets.

Keywords Gully erosion · Susceptibility · Fuzzy ANP · GIS · Iran

1 Introduction

Soil is the most important resources of any country in the world, which disregarding it is caused that the achieving to the sustainable development goals would be impossible. Nowadays, soil erosion and soil degradation are serious challenges that persistently have an increasing trend. Soil erosion induced by water is one of the most important processes in land degradation, especially, in semi-arid environments (Kheir et al. 2007; Zucca et al. 2006; McCloskey et al. 2016; Choubin et al. 2018). In this context, gully erosion is the most effective occurrences in soil and land degradation among the various water erosion forms (Poesen et al. 2003; Fox et al. 2016).

Gully erosion known as natural hazard which is caused many damages, including; destruction of agricultural lands, contamination of the water quality (Chen et al. 2016; Mukai 2017), economic and environmental problems and destruction of the ecosystems both in aquatic and terrestrial habitats (Ibáñez et al. 2016; Zakerinejad and Maerker 2015). Therefore, predicting and determining susceptible areas to gully erosion is necessary for controlling gully and conservating soil and water resources (Conoscenti et al. 2013, 2014).

Previous studies have used different methods for representing the prone areas to gully erosion, for instance, logistic regression (Chaplot et al. 2005; Conoscenti et al. 2014), frequency ratio (Rahmati et al. 2016), weights-of-evidence (Dube et al. 2014), multivariate adaptive regression splines (Gómez-Gutiérrez et al. 2015), stochastic gradient treeboost (Angileri et al. 2016), conditional probability (Conoscenti et al. 2013; Rahmati et al. 2017), analytical hierarchy process (Svoray et al. 2012), and classification and regression trees (Geissen et al. 2007).

Since many drivers are effective on the gully erosion, including; rainfall, landuse, lithology, topography, and soil (Nazari Samani et al. 2010; Cui et al. 2012; McCloskey et al. 2016), multi-criteria decision making (MCDM) methods can be helped to ascertain of priority factors.

Analytic Network Process (ANP) is one the most effective GIS based MCDM technique, which is introduced (Saaty 1996) to overcome the shortcomings was associated with Analytic Hierarchy Process (AHP). Therefore, the substantial different between AHP and ANP is related to their structure (hierarchical in versus network, respectively), which ANP is proposed for decision making with dependencies and feedbacks (Saaty 2003). Problems in nature are not merely hierarchical and there are relationships between the components. Hence in contrast to the other MCDM methods such as AHP, the ANP method considers the internal and external relationships between criteria.

Within this article, we aim to represent a new technique based on the both expert's opinion and fuzzy approach for prioritizing effective variables in engender of gully erosion. Technically speaking, fuzzy decision making methods can be employed to find out solutions for complex and ambiguous problems such as gully erosion (Uygun and Dede 2016). In the present study, the integrated Fuzzy Analytical Network Process (Fuzzy-ANP) is proposed as a novel method for assessing the most effective parameters on gully erosion. In the first step, criteria and dimensions which are cause the gully erosion are discerned through both literature reviews and experts opinions. Then, interrelationship network obtained through the group ideas and Fuzzy ANP approach was implemented to determine the weights of variables associated with the network. based on this statement, the main objectives of this research are; (i) prioritizing the effective variables in engender and extend of gully erosion by considering the interrelationship network among the criteria, (ii) predicting the spatial distribution of gullies by fuzzy ANP method, and (iii) assessing the fuzzy approach associated with the expert's opinions in determining the prone areas to gully erosion.

2 Materials and Methods

2.1 *Study Area*

The study area is the Kashkan-Poldokhtar basin with the area of 245 km^2 which is located in Iran between 33° 2′ and 33° 13′N latitude, and 47° 23′ and 47° 37′E longitude. The elevation ranges from 461 to 2191 m (Fig. 1). Average precipitation in the study area is 385 mm/year. The temperature varies from 25 to 48 °C for summer, and from −5 to 11 °C for winter. Altitude changes in the study area causes changing in vegetation type. Shrublands and chestnut forests are mostly at altitudes above 1000 m, whereas grassland is prominence in the altitudes below 1000 m. Deforestation, overgrazing, anthropogenic infrastructures such as buildings, roads and other human activates are the most important reasons of land degradation in the study area. Figure 2 indicates examples of influences of these activates on land degradation in the study area.

2.2 *Methodology*

2.2.1 Data Collection

Data for this study divided in two sections. The first section was represented how field surveys to the construction of the gully erosion inventory map, and the second section aims to describe of the effective factors on gully erosion accordingly.

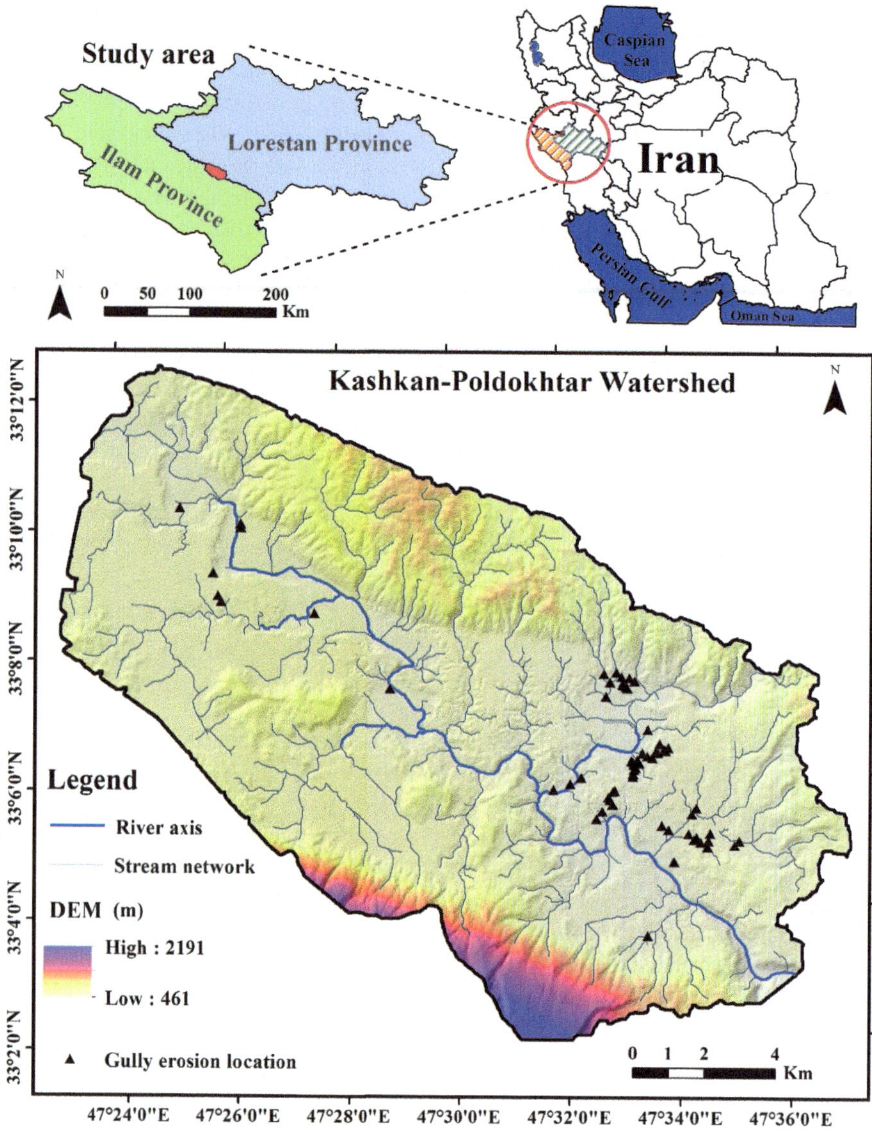

Fig. 1 Location of the study area and occured gullies

Gully Erosion Inventory Map

The preparation of the gully erosion inventory map is a fundamental step in geomorphological analyses and stochastic modelling. In our previous study (Rahmati et al. 2017), a verified gully inventory map was produced using multiple field surveys, Total Station (TS) survey, and the Geographic Object-Based Image

Fig. 2 Some gullies in the study area

Analysis (GEOBIA) method. According to results of monitoring reports of the watershed management department of Lorestan province, the gullies monitored almost occurred in the 2010–2016. The gullies which is exist in the study area are mostly formed by concentrated runoff (i.e. shear stress process). Locations of the recorded gullies were represented in Fig. 1. This gully erosion inventory was used for validation of gully erosion susceptibility model.

Effective Drivers on the Gully Erosion

To predicting the gully erosion map determining the effective drivers in the creation of the gully is important (De Vente et al. 2009; Rahmati et al. 2016). In the present study, using the literature review and data availability, effective factors were selected (Zakerinejad and Maerker 2015; Mukai 2017). The selected factors are; soil texture, drainage density, distance to streams, landuse, distance to road, lithology, steepness, slope aspect, plan curvature, altitude, and topographic wetness index (TWI) (Fig. 3). Details of the obtaining these factors are as follow:

Soil texture is one of the most important factors in the occurrence of gully erosion that was reported in many studies (e.g., Poesen et al. 2003; Dube et al. 2014; Deng et al. 2015). In order to obtain the soil texture map, we collected soil samples during field surveys. Totally, 75 samples (with weight about 0.5 kg in each location) were collected. The location of the each sample recorded by GPS. Soil samples in the laboratory were analyzed and soil textures were determined through hydrometer method. Soil texture types are including; sandy loam, silty loam, sand, sandy clay loam, silty clay, loamy sand, and salty sand. Eventually, the soil texture map was constructed based on the Zhao et al. (2009) procedure (Fig. 3a).

Drainage density and distance to streams (Fig. 3b,c) are also rest of factors which are effected the gully erosion (Dube et al. 2014). Line density and Euclidean distance tools in GIS were used to construct these layers.

According to the literature review (e.g., Serpa et al. 2015), the landuse types is one of the important factors in gully erosion occurrence, especially agricultural area. The ETM^{+} [1] satellite image was used to landuse mapping (in May 27th, 2014). To conduct the maximum likelihood supervised classification a total of 180 samples considered through field survey. Agriculture, fragmented forest, residential, and rangeland are classified landuses which are shown in Fig. 3d. Another factor is the distance to road, which is the cause of gully erosion because of the collecting and concentrating upstream runoff in the distinct locations (near bridges) to transferring it to the downstream. Distance to road prepared in GIS environment through Euclidean distance tool (Fig. 3e). Lithology is also one of the most important factors in investigating the natural hazard occurrences (Pourghasemi and Kerle 2016). Lithology classes in view of susceptibility to erosion was obtained from a geological map in the scale of the 1:100,000 based on Nekhay et al. (2009) (see Table 1 and Fig. 3f). Topographic wetness index also has been applied to quantify topographic control on hydrological processes for identifying hydrological flow paths and saturation zones in modeling (Moore et al. 1991). This index is calculated by two components of the upstream contributing area (A_s) and the slope gradient (β) as Eq. (1):

$$\mathrm{TWI} = \ln(\mathrm{A_s}/\tan\beta) \quad (1)$$

[1] Landsat Enhanced Thematic Mapper plus.

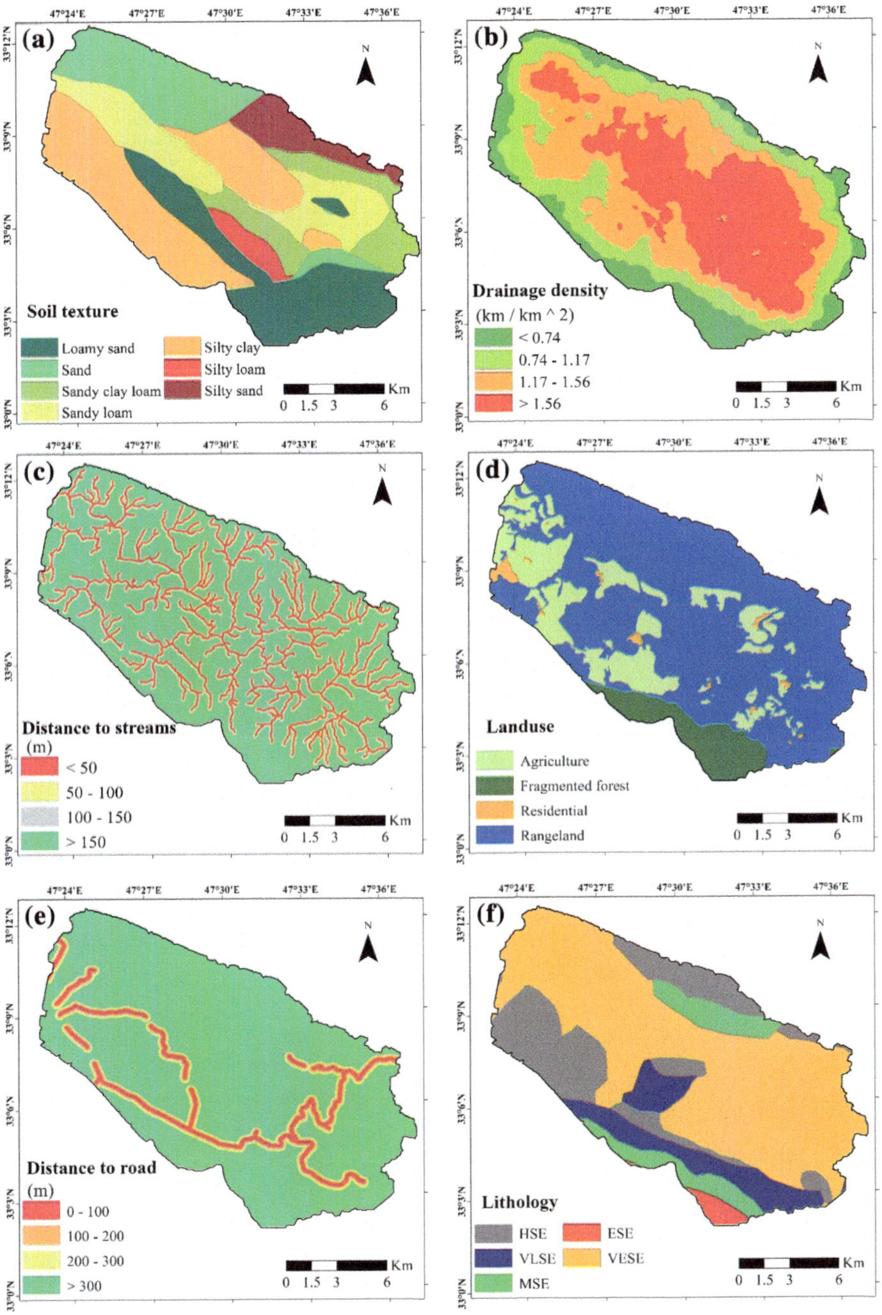

Fig. 3 Effective factors on the gully erosion: **a** soil texture, **b** drainage density, **c** distance to streams, **d** landuse, **e** distance to road, **f** lithology, **g** TWI, **h** altitude, **i** steepness, **j** slope aspect, and **k** plan curvature

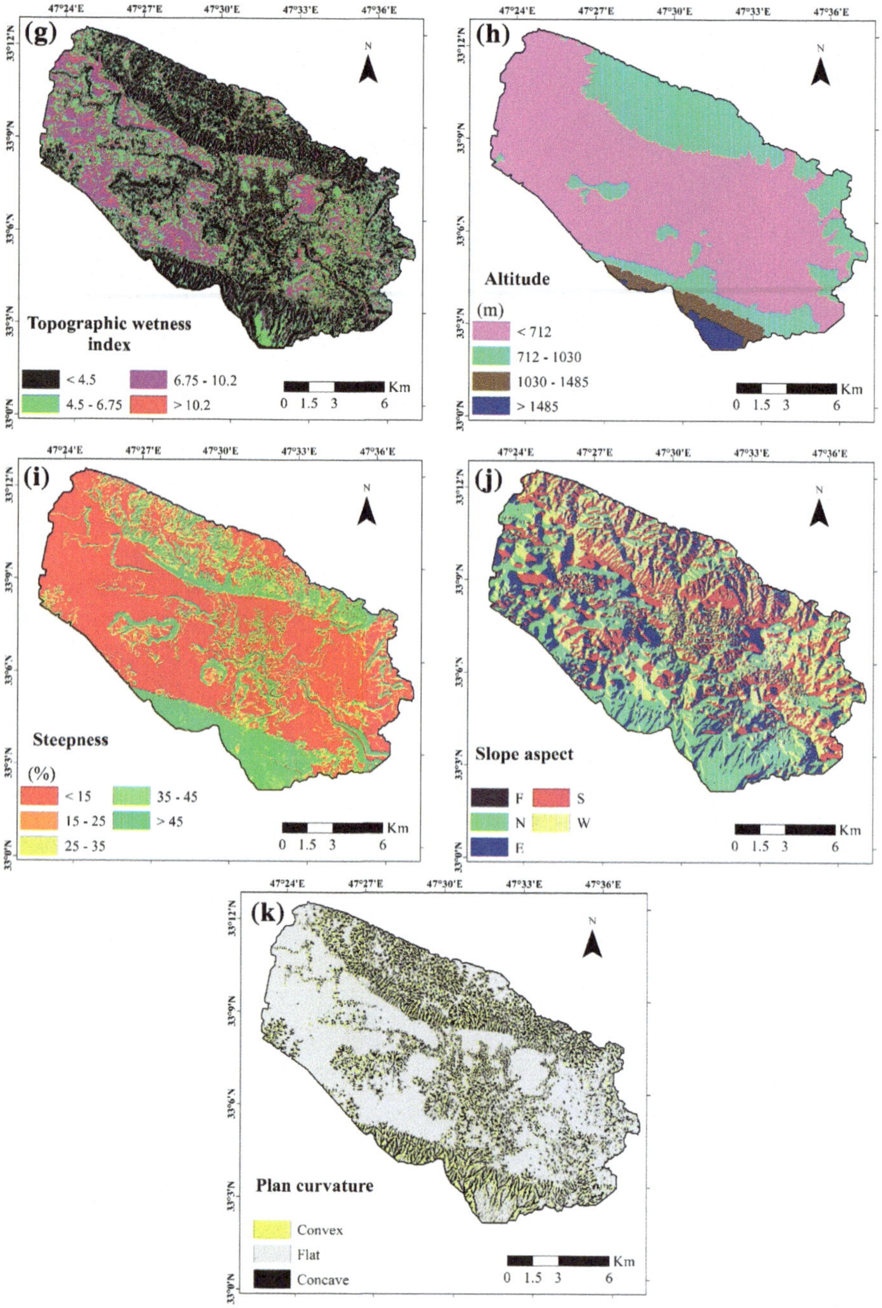

Fig. 3 (continued)

Table 1 Lithology classes in terms of susceptibility to erosion (Nekhay et al. 2009)

Classes	Geological materials
Hardly susceptible to erosion (HSE)	Igneous rock (volcanic, basic, acidic)
Very little susceptible to erosion (VLSE)	Well cemented rock: dolomite, limestone Compact siliceous rocks: phyllite, schist, quartzy sandstone, quartzite and similar
Moderately susceptible to erosion (MSE)	Rocks that are not as well consolidated: conglomerate, limestone, calcareous, gypsum, sandstone with little quartz, sand, and marl
Easy susceptible to erosion (ESE)	Soft formations: marl, gypsum, argillite, pelites, homogeneous clay, shale
Very easy susceptible to erosion (VESE)	Quaternary sediments: low level piedmont fan and valley terrace deposits, sand and clay

TWI map is shown in Fig. 3g.

Other considered indices that were applied to extract gully erosion susceptibility map are including topographic factors (i.e. steepness, slope aspect, altitude, and plan curvature). A 1/25,000 topographic map was applied to produce the Digital Elevation Model (DEM) with 10 m spatial resolution. The topographic factors were prepared in GIS environment using the DEM layer. Literature reviews are demonstrated the influences of the topographic attributes on the gully erosion initiation (Zakerinejad and Maerker 2015). Topographic factors are presented in Fig. 2h–k for altitude, steepness, slope aspect, and plan curvature, respectively.

2.2.2 Fuzzy Analytical Network Process (Fuzzy ANP)

Description

Analytic Network Process (ANP) is introduced by Saaty (1996) to overcome the on limitations of AHP method, which no considers mutual independence between criteria. Summarily, stages of ANP are in bellow (Saaty 1996):

1. Constructing f the network structure from a problem, and estimation of the relative weights of the criteria based on the pairwise comparison.
2. Creating the initial supermatrix based on the weights acquired from the previous step (including; weights of the clusters and nodes with consideration of inter-relationships among criteria).
3. Createing the weighted supermatrix by multiplying the initial supermatrix by cluster weights.
4. Finally, calculating the limit supermatrix by multiplying the weighted supermatrix, which obtained from step 3, n times by itself.

More details from stages of ANP represented in Saaty (1996).

Preferences scale for the pairwise comparison in ANP is alike the AHP, but subjective preferences have great influence on results. It is recognized that the linguistic assessment of human judgments and preferences are always subjective,

Table 2 Linguistic scale for relative importance in Fuzzy ANP (Uygun and Dede 2016)

Linguistic variables	Fuzzy number	Triangular fuzzy number	Triangular fuzzy reciprocal number
Equally important (EI)	~1	(1, 1, 1)	(1, 1, 1)
Weekly important (WI)	~3	(1, 3, 5)	(1/5, 1/3, 1)
Strongly important (SI)	~5	(3, 5, 7)	(1/7, 1/5, 1/3)
Very important (VI)	~7	(5, 7, 9)	(1/9, 1/7, 1/5)
Absolutely important (AI)	~9	(7, 9, 9)	(1/9, 1/9, 1/7)

vague, and uncertain. It is difficult and non-reasonable to provide numerical and exact values in pairwise comparison judgments. It feels which there is more confident to accept interval judgments than fixed value (Gholipour et al. 2014). So, conventional ANP seems to be insufficient to catch decision maker‘s requirements explicitly (Samanlioglu and Ayağ 2016). Hence, fuzzy set theory (with interval numbers) incorporated with the ANP helps to reduce the uncertainties and overcoming the ambiguous in the human preferences (Uygun and Dede 2016; Samanlioglu and Ayağ 2016; Choubin et al. 2017). Therefore, in this study weights of criteria in step 1 of the ANP are calculated by using fuzzy extent analysis. Table 2 indicates fuzzy linguistic terms and corresponding triangular fuzzy numbers for relative importance in Fuzzy ANP in the pairwise comparisons.
Stages of fuzzy extent analysis are described below (Chang et al. 2015):

1. Calculation of the fuzzy synthetic extent (S_k) with regards to the k th object:

$$\mathrm{S_k} = \sum_{q=1}^{Q} M_{E_k}^{q} \times \left[\sum_{k=1}^{K}\sum_{q=1}^{Q} M_{E_k}^{q}\right], k = 1, 2, \ldots, K \tag{2}$$

$$\sum_{q=1}^{Q} M_{E_k}^{q} = \sum_{q=1}^{Q}\left(l_q^k, m_q^k, u_q^k\right) = \left(\sum_{q=1}^{Q} l_q^k, \sum_{q=1}^{Q} m_q^k, \sum_{q=1}^{Q} u_q^k\right) \tag{3}$$

$$\begin{aligned}\sum_{k=1}^{K}\sum_{q=1}^{Q} M_{E_k}^{q} &= \sum_{k=1}^{K}\left(\sum_{q=1}^{Q} l_q^k, \sum_{q=1}^{Q} m_q^k, \sum_{q=1}^{Q} u_q^k\right)\\ &= \left(\sum_{k=1}^{K}\sum_{q=1}^{Q} l_q^k, \sum_{k=1}^{K}\sum_{q=1}^{Q} m_q^k, \sum_{k=1}^{K}\sum_{q=1}^{Q} u_q^k\right)\end{aligned} \tag{4}$$

$$\left(\sum_{k=1}^{K}\sum_{q=1}^{Q} M_{E_k}^{q}\right)^{-1} = \left(\frac{1}{\sum_{q=1}^{K}\sum_{q=1}^{Q} u_q^k}, \frac{1}{\sum_{q=1}^{K}\sum_{q=1}^{Q} m_q^k}, \frac{1}{\sum_{q=1}^{K}\sum_{q=1}^{Q} l_q^k}\right) \tag{5}$$

2. Calculating of the degree of possibility of $M_1(l_1, m_1, u_1) \geq M_2(l_2, m_2, u_2)$:

$$V(M_1 \geq M_2) = \begin{Bmatrix} 1, & m_1 \geq m_2 \\ \frac{l_2 - u_1}{(m_1 - u_1) - (m_2 - l_2)} & m_1 < m_2, u_1 \geq l_2 \\ 0, & \text{otherwise} \end{Bmatrix} \quad (6)$$

The degree of possibility of a convex fuzzy number greater than k in the convex fuzzy numbers M_k and k = 1, 2,…, K shown as:

$$V(M \geq M_1, M_2, \ldots, M_k, \ldots, M_K) = \min_{k=1,2,\ldots,K} V(M \geq M_k) \quad (7)$$

3. Computation of the vector of weights

If we assume that

$$d'(A_p) = \min V(Sp \geq Sk), p \in \{1, 2, \ldots k, \ldots, K\} \quad (8)$$

Then, the vector of weights can be defined as:

$$w' = (d'(A_1), d'(A_2), \ldots, d'(A_n))^T \quad (9)$$

where A_i and i = 1, 2,…, n indicate in ith component and n number of components.

4. Finally, computation of the weights of criteria based on the normalization

$$w = (d(A_1), d(A_2), \ldots, d(A_n))^T \quad (10)$$

where w is a nonfuzzy number.

After calculating the weights, for each of the attributes in each table of the pairwise comparison, weights directly transfer to designed network in Super Decision software. Judgments in super decision can be carried in the five ways (Graphical, Verbal, Matrix, Questionnaire, and Direct). Since weights in our study were extracted by fuzzy sets, therefore, our judgments exerted through way into designed network.

Designing the Network Based on the Effective Factors on Gully Erosion

Based on questionnaires and opinions of the academic and soil experts, structure of network with considering interrelationship among criteria was designed. In Fig. 4, the developed network structure for evaluating gully erosion is presented. The interdependence relationships between the variables are observable through the direction of the arrows. One-sided arrows indicate the influence of a criterion on another, whereas two-sided arrows demonstrate mutual effects between the variables. ANP is considered both inner and external dependence between factors. In this study, we have no inner dependencies, so loops that are representative of inner dependence are not shown.

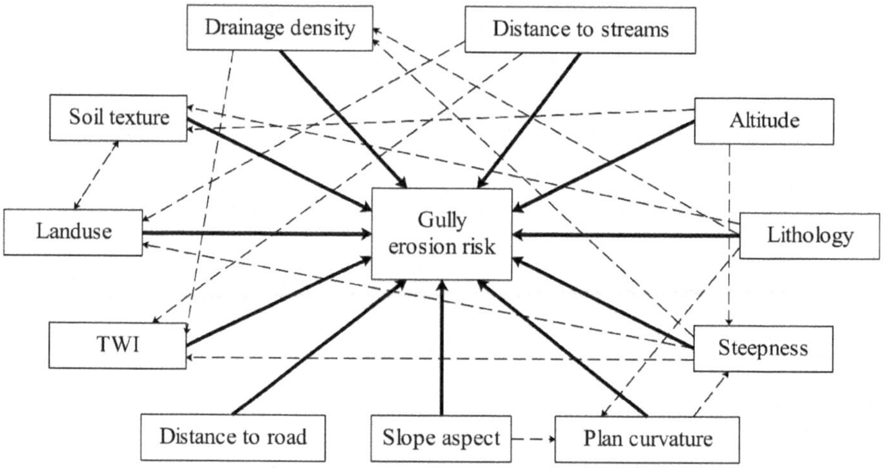

Fig. 4 The interrelationship network structure among criteria for the gully erosion susceptibility evaluation

2.2.3 Validation of the Fuzzy ANP Method

The skill of fuzzy ANP in the extraction of the gully erosion susceptibility map was evaluated by receiver operating characteristics (ROC) curves (Fawcett 2006; Swets 1988). The area under the ROC curve (AUC) measures the overall performance of predictive models (Pereira et al. 2012). The AUC value closer to 1 indicates the better performance in predicting gully erosion susceptibility. A detailed classification of the AUC was presented by Yesilnacar (2005) as follows: poor accuracy (50–60%), moderate accuracy (60–70%), good accuracy (70–80%), very good accuracy (80–90%), and excellent accuracy (90–100%).

3 Results and Discussion

3.1 Fuzzy ANP Results in Estimating the Relative Importance of Factors

Table 3 represents the normalized weights of the effective factors on the gully erosion. As can be seen, the soil texture factor is the most important factor (0.1862), while drainage density is the second most important factor (0.1695). The relative importance of soil texture and drainage density are in agreement with the Rahmati et al. (2017).

Among the all factors plan curvature is the less important factors in the gully erosion (0.0186). Figure 5 illustrates the relative importance of factors. Effective factors are soil texture, drainage density, lithology, distance to stream, steepness, altitude, distance to road, slope aspect, landuse, TWI, and plan curvature, respectively.

Table 3 The normalized weights of the effective factors on the gully erosion

Factors	Weights
Altitude	0.0782
Distance from road	0.0521
Distance from streams	0.1304
Drainage density	0.1695
Landuse	0.0335
Lithology	0.1490
Plan curvature	0.0186
Slope aspect	0.0484
Soil texture	0.1862
Steepness	0.1062
TWI	0.0279

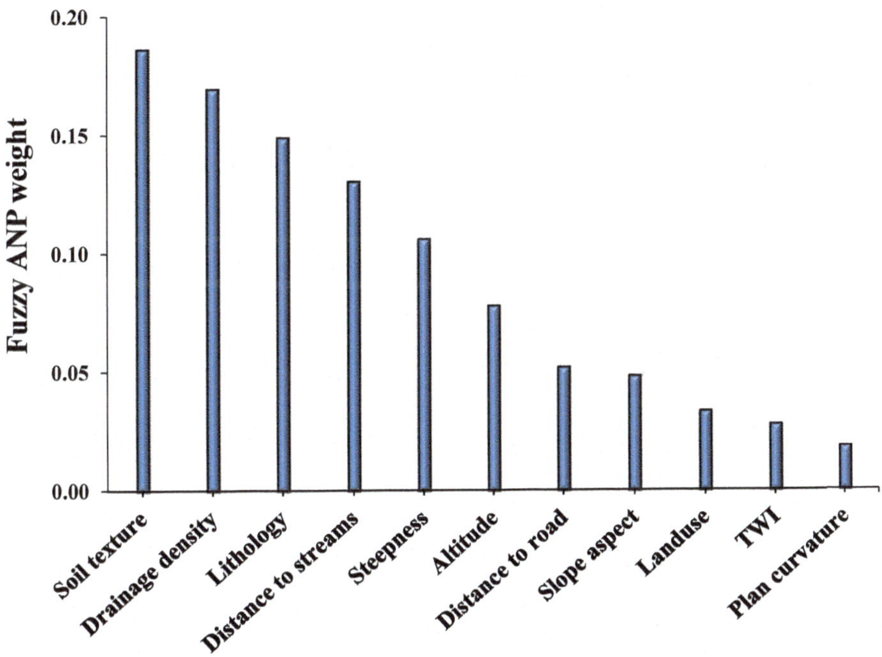

Fig. 5 The relative importance of factors

3.2 *Relative Importance of Classes in Each Factor*

Results of the Fuzzy ANP in estimating the importance of classes in each factor is shown in Fig. 6. The sum of the classes' weight in each factor is equal with 1. Among the soil texture classes, sandy loam and sand are the most important and the

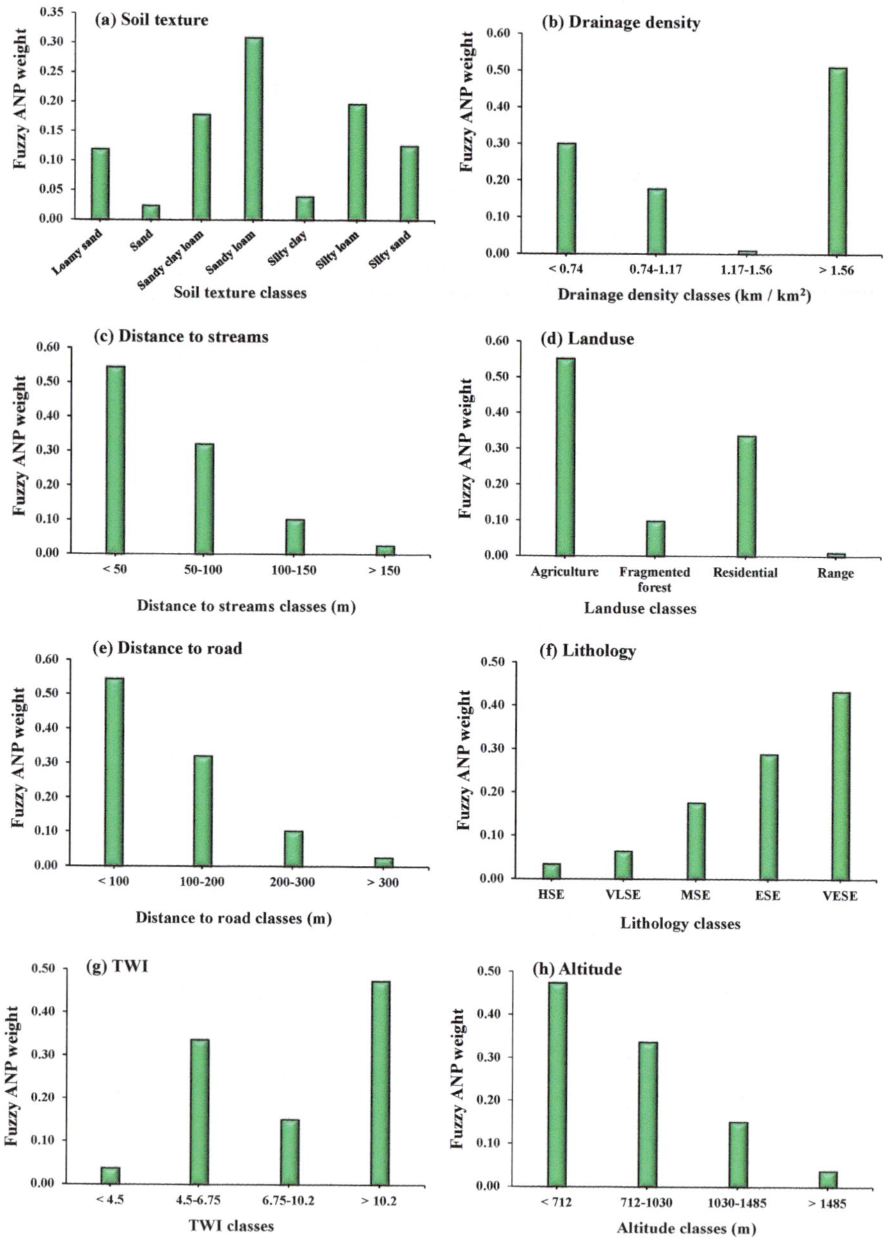

Fig. 6 Relative importance of classes in each factor: **a** soil texture, **b** drainage density, **c** distance to streams, **d** landuse, **e** distance to road, **f** lithology, **g** TWI, **h** altitude, **i** steepness, **j** slope aspect, and **k** plan curvature

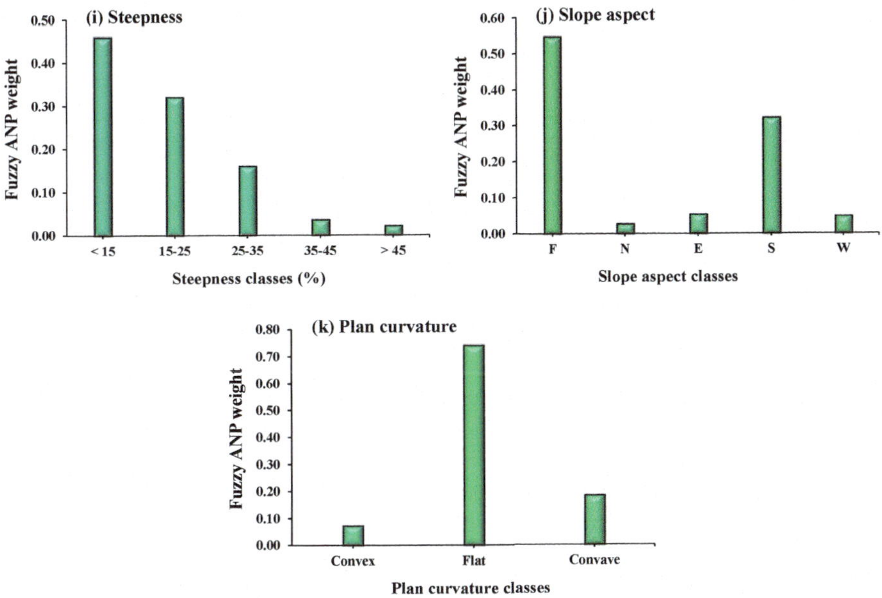

Fig. 6 (continued)

less important classes, respectively, in building gully erosion (Fig. 6a). In drainage density factor, the highest and the lowest classes (more than 1.56 and less than 0.74 km/km^2) are known as the most important classes (Fig. 6b). Shellberg et al. (2016) and Rahmati et al. (2017) mentioned that poor drainage network due to concentrating the runoff is the cause of gully erosion, in addition to the high drainage density. 50 m distance to river have the most weight (about 0.55), while distances more than 150 m have the weight about 0.03 (Fig. 6c). Fuzzy ANP results denoted that the agricultural areas are the most important landuse in the occurrence of the gully erosion (Fig. 6d). This is in agreement with the Serpa et al. (2015) and Rahmati et al. (2016). Results of distance from road indicated that with decreasing the gully erosion, the distance to road increases (Fig. 6e). The weight of the lithology resistance classes was represented in Fig. 6f. Geological materials with the very easy susceptibility to erosion (VESE) have the value about 0.43, while materials with the hardly susceptibility to erosion (HSE) have the value about 0.03 (Fig. 6f). Assessment of TWI confirmed that the highest class is most important in comparison with other classes (weight value of it is about 0.47; Fig. 6g). In the case of altitude, the results indicated that as the altitude increases the gully erosion susceptibility entirely decreases (Fig. 6h). Figure 6i disclosed that gentle slopes have the high gully susceptibility, it is because of surface flow accumulation in these areas, in consequence, is caused the gully initiation (Valentin et al. 2005; Rahmati et al. 2015). Results of the slope aspect demonstrated that slopes facing flat, south, and east have highest probability of gully erosion occurrence, respectively (Fig. 6j). Water flows from the hillside of the mountains and joins in flat areas (i.e. lower terrains) which cause gullying. Since south and east faces have low vegetation cover and high duration of sunlight exposition, these aspects are more susceptible to gully erosion. Investigation of plan curvature indicated that flat curvature is the most important class in occurrence of the gully. This result is similar to the results of other researches such as Conforti et al. (2011) and Rahmati et al. (2017), which mentioned flat, concave, and convex curvatures are most important area to gully erosion development, respectively.

3.3 Developing the Gully Erosion Susceptibility Map

The final weights were calculated based on the designed network in Super decision software. Limit supermatrix (Table 4) indicates the weight of the each class in the designed network, which was computed by Fuzzy ANP method. To extraction of the gully erosion susceptibility map weights of limit supermatrix was employed. Final weights in Table 4 are established based on the clusters weight (Table 3 and Fig. 5) and classes' weight in each cluster (Fig. 6). Means that by applying relative important of the both each cluster and each class limit supermatrix is constructed.

Figure 7 shows the gully erosion susceptibility map in study area. This map was produced in GIS environment by obtained weight through Fuzzy ANP method. After overlaying the layers, the resulted map reclassified into four category using the equal interval classification scheme. The final map was classified in low,

Table 4 Limit supermatrix

Clusters	Class	Weight	Clusters	Class	Weight
Altitude	<712	0.0371	Distance to road	<100	0.0285
	712–1030	0.0263		100–200	0.0168
	1030–1485	0.0119		200–300	0.0054
	>1485	0.0030		>300	0.0014
Distance to streams	<50	0.0712	Drainage density	<0.74	0.0510
	50–100	0.0420		0.74–1.17	0.0303
	100–150	0.0135		1.17–1.56	0.0017
	>150	0.0036		>1.56	0.0865
Landuse	Agriculture	0.0185	TWI	<4.5	0.0011
	Fragmented forest	0.0033		4.5–6.75	0.0094
	Residential	0.0113		6.75–10.2	0.0042
	Range	0.0004		>10.2	0.0132
Lithology	HSE	0.0052	Slope aspect	F	0.0265
	VLSE	0.0097		N	0.0013
	MSE	0.0264		E	0.0026
	ESE	0.0432		S	0.0156
	VESE	0.0645		W	0.0024

(continued)

Table 4 (continued)

Clusters	Class	Weight	Clusters	Class	Weight
Steepness	<15	0.0487	Soil texture	Loamy sand	0.0225
	15–25	0.0341		Sand	0.0046
	25–35	0.0170		Sandy clay loam	0.0334
	35–45	0.0039		Sandy loam	0.0577
	>45	0.0024		Silty clay	0.0075
Plan curvature	Convex	0.0014		Silty loam	0.0369
	Flat	0.0138		Silty sand	0.0236
	Concave	0.0034			

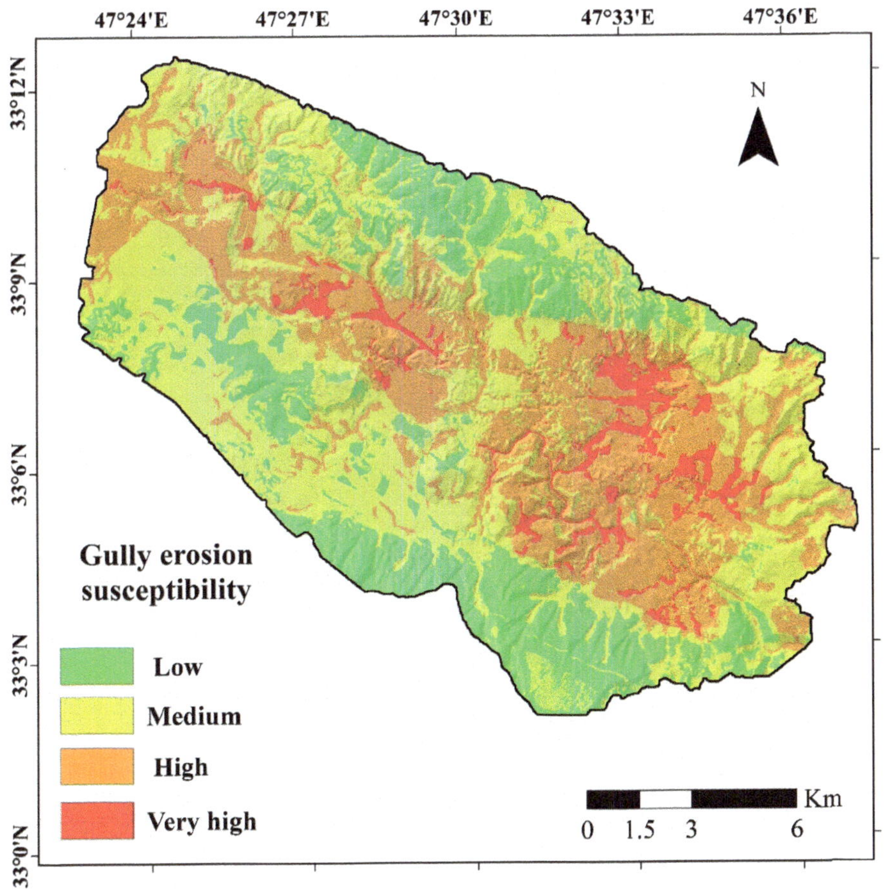

Fig. 7 Gully erosion susceptibility map

medium, high, and very high categories, which are contained 55, 104, 74, and 12 km^2 of the study area, respectively. High and very high susceptibility zones are in the locations with flat curvature, gentle slope, low slope percent, low elevation, and high drainage density, which mostly are in the central area of the watershed.

3.4 Validation of the Fuzzy ANP Method

To investigate the susceptibility model's performance the ROC curve was applied (Rahmati et al. 2016, 2017). The validation of Fuzzy ANP method was investigated considering the total 65 gully erosion features. Figure 8 depicts ROC curve of gully erosion susceptibility model based on the complete gully inventory, which was constructed using pROC-package, an open-source package for R (Robin et al. 2011). The area under the curve (AUC) in ROC exhibits the predicting performance of a model by displaying the capability of the model to simulate the predetermined occurrences or non-occurrences. Results indicated that AUC value for Fuzzy ANP model is 0.904. Thus, this method indicates 90.4% accuracy for study area which it have excellent performance in gully erosion susceptibility mapping according to the Yesilnacar (2005) classification.

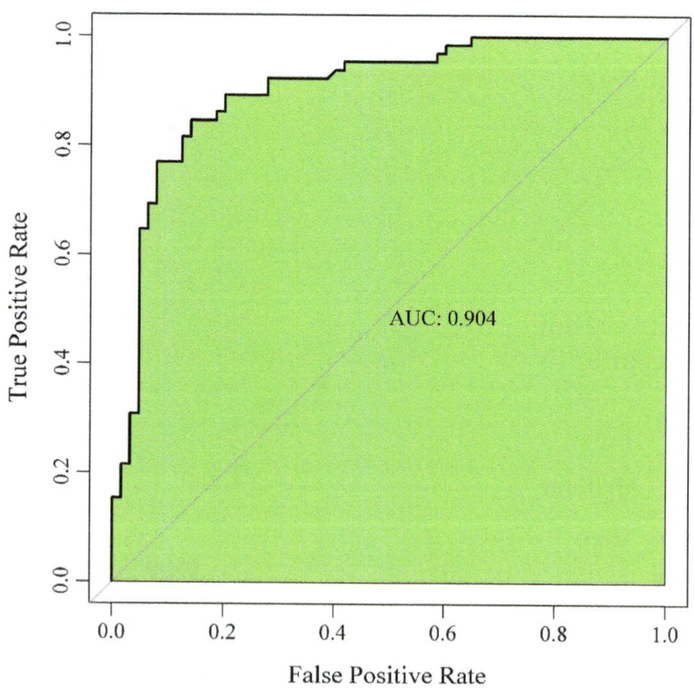

Fig. 8 ROC curve of gully erosion susceptibility model based on the complete gully inventory

4 Conclusion

Despite the hazardous attributes of gullies erosion, many researchers have concentrated on investigating the gully erosion susceptibility. In this study, Fuzzy ANP as a new approach was used to extracting the gully susceptibility map. Primarily, the effective factors in building the gully erosion such as soil texture, drainage density, distance to streams, landuse, distance to road, plan curvature, lithology, TWI, steepness, slope aspect, and altitude were obtained. Then, relative important of the criteria were determined through Fuzzy ANP and gully susceptibility map constructed after overlaying the layers. Then, the location of 65 gully erosion features was recorded through field surveys to evaluate the model performance. Fuzzy ANP method applies knowledge and intuition of the experts to resolve the problems. The major advantage of the ANP method in comparison with the traditional methods (such as AHP) is considering the direct and indirect influences which criteria have on gully erosion (with a network structure). Incorporation of the Fuzzy with ANP method leads to reduce the uncertainties and overcoming the ambiguous in the human preferences according to interval judgments rather than fixed value judgments. Results of the Fuzzy ANP was revealed that soil texture, drainage density, and lithology are most important factors in occurring the gully erosion. After the obtaining the weights through Fuzzy ANP, gully erosion susceptibility map was produced. Skill of the predicting map was confirmed using ROC curve and AUC value. Results of our study demonstrate that Fuzzy ANP method could be an efficient method for producing the gully susceptibility map especially in developing countries such as Iran, which is overly exposed to the erosion hazards. Furthermore, natural resources planners and managers will be able that understand and find vulnerable areas to gully erosion to controlling it and to reducing the damages.

Acknowledgements We thank the Editors, Dr. Pourghasemi and Dr. Rossi, and two anonymous reviewers for their suggestions and comments.

References

Angileri SE, Conoscenti C, Hochschild V, Märker M, Rotigliano E, Agnesi V (2016) Water erosion susceptibility mapping by applying Stochastic gradient treeboost to the imera Meridionale River Basin (Sicily, Italy). Geomorphology 262:61–76

Chang KL, Liao SK, Tseng TW, Liao CY (2015) An ANP based TOPSIS approach for Taiwanese service apartment location selection. Asia Pacific Manag Rev 20(2):49–55

Chaplot V, Coadou le Brozec E, Silvera N, Valentin C (2005) Spatial and temporal assessment of linear erosion in catchments under sloping lands of northern Laos. CATENA 63:167–184

Chen Z, Chen W, Li C, Pu Y, Sun H (2016) Effects of polyacrylamide on soil erosion and nutrient losses from substrate material in steep rocky slope stabilization projects. Sci Total Environ 554:26–33

Choubin B, Darabi H, Rahmati O, Sajedi-Hosseini F, Kløve B (2018) River suspended sediment modelling using the CART model: a comparative study of machine learning techniques. Sci Total Environ 615:272–281

Choubin B, Solaimani K, Roshan MH, Malekian A (2017) Watershed classification by remote sensing indices: a fuzzy c-means clustering approach. J Mountain Sci 14(10):2053–2063

Conforti M, Aucelli PPC, Robustelli G, Scarciglia F (2011) Geomorphology and GIS analysis for mapping gully erosion susceptibility in the Turbolo stream catchment (northern Calabria, Italy). Nat Hazards 56:881–898

Conoscenti C, Agnesi V, Angileri S, Cappadonia C, Rotigliano E, Märker M (2013) A GIS-based approach for gully erosion susceptibility modelling: a test in Sicily, Italy. Environ Earth Sci 70 (3):1179–1195

Conoscenti C, Angileri S, Cappadonia C, Rotigliano E, Agnesi V, Märker M (2014) Gully erosion susceptibility assessment by means of GIS-based logistic regression: a case of Sicily (Italy). Geomorphology 204:399–411

Cui P, Lin YM, Chen C (2012) Destruction of vegetation due to geo-hazards and its environmental impacts in the Wenchuan earthquake areas. Ecol Eng 44:61–69

De Vente J, Poesen J, Govers G, Boix-Fayos C (2009) The implications of data selection for regional erosion and sediment yield modelling. Earth Surf Process Landf 34:1994–2007

Deng Q, Qin F, Zhang B, Wang H, Luo M, Shu C, Liu H, Liu G (2015) Characterizing the morphology of gully cross-sections based on PCA: a case of Yuanmou Dry-Hot Valley. Geomorphology 228:703–713

Dube F, Nhapi I, Murwira A, Gumindoga W, Goldin J, Mashauri DA (2014) Potential of weight of evidence modelling for gully erosion hazard assessment in Mbire District-Zimbabwe. Phys Chem Earth, Parts A/B/C 67:145–152

Fawcett T (2006) An introduction to ROC analysis. Pattern Recogn Lett 27:861–874

Fox GA, Sheshukov A, Cruse R, Kolar RL, Guertault L, Gesch KR, Dutnell RC (2016) Reservoir sedimentation and upstream sediment sources: perspectives and future research needs on streambank and gully erosion. Environ Manag 57(5):945–955

Geissen V, Kampichler C, López-de Llergo-Juárez JJ, Galindo-Acántara A (2007) Superficial and subterranean soil erosion in Tabasco, tropicalMexico: development of a decision tree modeling approach. Geoderma 139:277–287

Gholipour R, Jandaghi G, Rajaei R (2014) Contractor selection in MCDM context using fuzzy AHP. Iranian J Manag Stud 7(1):151–173

Gómez-Gutiérrez Á, Conoscenti C, Angileri SE, Rotigliano E, Schnabel S (2015) Using topographical attributes to evaluate gully erosion proneness (susceptibility) in two mediterranean basins: advantages and limitations. Nat Hazards. https://doi.org/10.1007/s11069-015-1703-0

Ibáñez J, Contador JL, Schnabel S, Valderrama JM (2016) Evaluating the influence of physical, economic and managerial factors on sheet erosion in rangelands of SW Spain by performing a sensitivity analysis on an integrated dynamic model. Sci Total Environ 544:439–449

Kheir RB, Wilson J, Deng Y (2007) Use of terrain variables for mapping gully erosion susceptibility in Lebanon. Earth Surface Process Landforms 32(12):1770–1782

McCloskey GL, Wasson RJ, Boggs GS, Douglas M (2016) Timing and causes of gully erosion in the riparian zone of the semi-arid tropical Victoria River, Australia: management implications. Geomorphology 266:96–104

Moore ID, Grayson RB, Ladson AR (1991) Digital terrain modeling: a review of hydrological, geomorphological and biological applications. Hydrol Process 5:3–30

Mukai S (2017) Gully erosion rates and analysis of determining factors: a case study from the semi-arid main Ethiopian Rift Valley. Land Degradation Dev 28(2):602–615

Nazari Samani A, Ahmadi H, Mohammadi A, Ghoddousi J, Salajegheh A, Boggs G, Pishyar R (2010) Factors controlling gully advancement and models evaluation (Hableh Rood Basin, Iran). Water Resour Manag 24(8):1531–1549

Nekhay O, Arriaza M, Boerboom L (2009) Evaluation of soil erosion risk using Analytic Network Process and GIS: a case study from Spanish mountain olive plantations. J Environ Manage 90:3091–3104

Pereira S, Zêzere JL, Bateira C (2012) Technical note: assessing predictive capacity and conditional independence of landslide predisposing factors for shallow landslide susceptibility models. Nat Hazards Earth Syst Sci 12:979–988

Poesen J, Nachetergaele J, Verstraeten J, Valentin C (2003) Gully erosion and environmental change: importance and research needs. CATENA 50(2–4):91–133

Pourghasemi HR, Kerle N (2016) Random forests and evidential belief function-based landslide susceptibility assessment in Western Mazandaran Province, Iran. Environ Earth Sci. https://doi.org/10.1007/s12665-0154950-1

Rahmati O, Zeinivand H, Besharat M (2015) Flood hazard zoning in Yasooj region, Iran, using GIS and multi-criteria decision analysis. Geomat Nat Hazards Risk. https://doi.org/10.1080/19475705.2015.1045043

Rahmati O, Haghizadeh A, Pourghasemi HR, Noormohamadi F (2016) Gully erosion susceptibility mapping: the role of GIS-based bivariate statistical models and their comparison. Nat Hazards 82(2):1231–1258

Rahmati O, Tahmasebipour N, Haghizadeh A, Pourghasemi HR, Feizizadeh B (2017) Evaluating the influence of geo-environmental factors on gully erosion in a semi-arid region of Iran: an integrated framework. Sci Total Environ 579:913–927

Robin X, Turck N, Hainard A, Tiberti N, Lisacek F, Sanchez JC, Müller M (2011) pROC: an open-source package for R and S+ to analyze and compare ROC curves. BMC Bioinformatics 12(1):77

Saaty RW (2003) Decision making in complex environment: the analytic hierarchy process (AHP) for decision making and the analytic network process (ANP) for decision making with dependence and feedback. Super Decisions, Pittsburgh

Saaty TL (1996) Decision making with dependence and feedback: the analytic network process, vol 4922. RWS publications, Pittsburgh

Samanlioglu F, Ayağ Z (2016) Fuzzy ANP-based PROMETHEE II approach for evaluation of machine tool alternatives. J Intelligent Fuzzy Syst 30(4):2223–2235

Serpa D, Nunes JP, Santos J, Sampaio E, Jacinto R, Veiga S, Lima JC, Moreira M, Corte-Real J, Keizer JJ, Abrantes N (2015) Impacts of climate and land use changes on the hydrological and erosion processes of two contrasting Mediterranean catchments. Science of the Total Environmental 538:64–77

Shellberg JG, Spencer J, Brooks AP, Pietsch TJ (2016) Degradation of the Mitchell River fluvial megafan by alluvial gully erosion increased by post-European land use change, Queensland, Australia. Geomorphology 266:105–120

Svoray T, Michailov E, Cohen A, Rokah L, Sturm A (2012) Predicting gully initiation: comparing data mining techniques, analytical hierarchy processes and the topographic threshold. Earth Surf Process Landforms 37:607–619

Swets JA (1988) Measuring the accuracy of diagnostic systems. Science 240(4857):1285–1293

Uygun Ö, Dede A (2016) Performance evaluation of green supply chain management using integrated fuzzy multi-criteria decision making techniques. Comput Ind Eng 102:502–511

Valentin C, Poesen J, Yong L (2005) Gully erosion: impacts, factors and control. CATENA 63:132–153

Yesilnacar EK (2005) The application of computational intelligence to landslide susceptibility mapping in Turkey. Ph.D Thesis Department of Geomatics the University of Melbourne, p 423

Zakerinejad R, Maerker M (2015) An integrated assessment of soil erosion dynamics with special emphasis on gully erosion in the Mazayjan basin, southwestern Iran. Nat Hazards 79(1):25–50

Zhao Z, Chow TL, Rees HW, Yang Q, Xing Z, Meng FR (2009) Predict soil texture distributions using an artificial neural network model. Comput Electron Agric 65(1):36–48

Zucca C, Canu A, Della Peruta R (2006) Effects of land use and landscape on spatial distribution and morphological features of gullies in an agropastoral area in Sardinia (Italy). CATENA 68 (2):87–95

Landslide Susceptibility Prediction Maps: From Blind-Testing to Uncertainty of Class Membership: A Review of Past and Present Developments

Andrea G. Fabbri and Chang-Jo Chung

Abstract This contribution reviews the spatial characterization originally stimulated by mineral exploration and later by environmental concern. Research network programmes of the European Commission triggered cross-breeding of disciplines and approaches to hazard prediction in particular for the Deba Valley study area in northern Spain. Examples of results of spatial prediction modelling using blind tests to obtain prediction-rate curves and uncertainty patterns allow considerations on the role of such modelling for research, surveying and civil protection.

Keywords Spatial prediction modelling · Cross-validation · Blind tests
Landslide hazard · Empirical likelihood ratio · Prediction-rate curves
Uncertainty of class membership · Prediction patterns

1 Introduction to Spatial Characterization

With all probability the idea of digitizing geological maps came about as a reaction to Allais (1957) report on the statistical distribution of mineral deposits in Africa as directly proportional to the area of study without particular connection to geology. Of course we disagreed about that. The report had a political weight and so did, although we were unaware of it, our first employment at the GSC, the Geological Survey of Canada. For us the Survey had a short name of GSC and was then a well respected federal branch of the Department (Ministry) of Energy, Mines and Resources, EMR. It appeared that a political strategy to stressing the relevance of the GSC was to start a pilot project on statistical methods for mineral exploration. At that time, it was 1969, a well known young scientist, Dr. Frits P. Agterberg, head

A. G. Fabbri (✉)
DISAT, University of Milano-Bicocca, Milan, Italy
e-mail: andrea.fabbri@unimib.it

C.-J. Chung
SpatialModels Inc, Ottawa, Canada

H. R. Pourghasemi and M. Rossi (eds.), *Natural Hazards GIS-based Spatial Modeling Using Data Mining Techniques*, Advances in Natural and Technological Hazards Research 48, https://doi.org/10.1007/978-3-319-73383-8_6

of the Geomathematics Section, was paving the way to such area of application. It fell upon one of us co-authors, AGF, to collect data on several Canadian study areas rich in mineral deposits. It became an opportunity for working at constructing databases of mineral deposits, occurrences and showings (i.e., indicators of likely presence of exploitable mineralization, or just of mineralization, or of possible mineralization). At the time, it seemed like a shot in the dark to construct databases: nobody new how to do that and whether that would be any useful…! From a regional point of view, such features as mineral occurrences appeared point like with some kind of geographical coordinates for location. In addition to that they had to have a list of characteristics such as genetic type of mineralization, amount of ore, reserves proven or suspected, mineral associations and geologic descriptions of host rocks at the location point and at its surroundings.

Beside ensuring the geologic and metallogenic quality of the objects to be described, was the fact that there were literally thousands and thousands of them. Furthermore, many maps were available that used different scales, projections, area cover and legends. Data on mineral prospects existed as separate compilations, public and/or private, using different location accuracies, metric and non-metric units for grade of ore and tonnage. The standardization of such spatial and non-spatial information became a major part of the effort.

Later on, a project was approved entitled "*Quantification of Geological Variables*" that officialised that activity of database construction. It was the end of the 1960s and at that time digitizing was done manually with transparent grids to get density contouring of points and statistics for 10 km square cells (with grids of 20 × 20 points), thousands of them! But in the 1970s, the other co-author, CJC, a mathematician-statistician, joined our Geomathematics Section and the quantification and analyses tasks led to the much referred to 1972 GSC Paper 71–41 on mineral exploration potential (Agterberg et al. 1972), and subsequently to a first paper on databases for mineral potential evaluation (Fabbri 1975). Many more contributions followed and much discussion was to take place on the assessment of mineral potential by quantitative spatial techniques.

Mineral exploration remained an important field for the GSC and the Canadian geological community of prospectors for some years to come, however, by 1986 most of the prospectors and exploration geologists were out of work and became private consultants: mineral exploration was not considered so important any more. By then, progress had been made in the area of digital data capture and picture processing (early version of geographic information systems) with a book on "Image Processing of Geological Data (Fabbri 1984) and on SIMSAG, a software package for mineral resource evaluation (Chung 1983). We were heavily committed to what became later known as geographical information systems or GIS (Burrough 1986; Aronoff 1989; Bonham-Carter 1994). Those experiences became the basis for the spatial prediction modelling work that followed. In 1986 a NATO Advanced Study Institute on "*Quantitative Analysis of Mineral and Energy resources*" was held at *Il Ciocco*, in Castelvecchio Pascoli, Italy and produced a volume with that title (Chung et al. 1988). Nevertheless, the interest on mineral exploration was by then at an historical low.

Of course, as it happens for most attitudes in society, as well as in scientific fields, a new trend brought in environmental and security concerns and more and more advanced technologies, including remote sensing and various types of data interpretations and analyses. That became a critical area of development, training and research when later on one of us co-authors, AGF, joined the International Institute for Aerospace Surveys and Earth Sciences, ITC, in the Netherlands heading the Division of Geological Surveying. Much attention was paid to the development of GIS as an analytical instrument for capturing and managing spatial information, predictive modelling with emphasis on land use planning, environmental security of resources, prediction of natural hazards, risk assessment and the use of remotely sensed data from several new sensors. This became the focus of another NATO Advanced Study Institute on "*Deposits and Geo-Environmental Models for Resource Exploitation and Environmental Security*" held in Mátraháza, Hungary, in 1998, that also produced a volume with that title in 2002 (Fabbri et al. 2002a). Its unusual focus was on the environmental footprints of mineral resources, spatial data integration and prediction modelling.

It was, however, in the beginning of the 1990s, that the other co-author, a former colleague from the GSC, CJC, proposed to set-up a consistent mathematical framework for spatial prediction modelling (Chung and Fabbri 1993). That framework has continued being developed since, under a variety of scientific circumstances and opportunities. Fortunately it also led to much international research and cooperation.

2 Cross Breeding of Persuasions, Curiosities and Strivings

It was also in the 1990s that a number of European Commission's network research projects started to materialize. Structured management and integration of activities of several research teams was a necessary condition for support of those projects. They became vehicles of opportunities for collaboration and exchange of methods and ideas between countries, institutions and disciplines. The focus was on the relationships between environment and natural hazard and coincided with the new emphasis on training and research in environmental geosciences at ITC in the Netherlands. This became even more explicit when one of us, AGF, joined in 2002 the Department of Environmental Sciences and Territory, DISAT, of the University of Milano-Bicocca in Milan, Italy. It is partly due to those projects that many of the developments and application described in Sect. 4 were made possible. In particular the sequential progress the six projects documented was used as an argument to reconsider the field of GIS from mostly management and representation of spatial data into a complete analytical methodology (Fabbri 2007, p. 4).

1. "*Geomorphology and Environmental Impact Assessment: a network of researchers in the European Community*" (1993–1996) HC&M, ERBCXRXCT 930311;
2. "*New Technologies for Landslide Hazard Assessment and Management in Europe*", **NEWTECH** (1996–1998) CEC Environmental Programme, ENV-CT96-024;

3. *"A European Research Network for the Application of Geomorphology and Environmental Impact Assessment to Transportation Systems"*, **GETS** (1998–2001) TMR, ERBFMRXCT970162;
4. *"Quantitative Indicators and Indices of Environmental Quality: a Euro-Latin American Network for Environmental Assessment and Monitoring"*, **ELANEM** (1999–2002) INCO-DC, ERBIC18CT980290;
5. *"Assessment of Landslide Risks and Mitigation in mountain Areas"*, **ALARM** (2001–2004) EVG1-2001-00018.
6. European Commission's Project **Mountain Risks**: *from prediction to management and governance* (MRTN-CT-2006-035978, 2007–2010).

Those projects revealed that geo-information was becoming more and more relevant and that, to represent and comprehend natural and environmental risks, desirable cross-disciplinary applications beside the physical sciences had to involve both economics and sociology. Predicting future events goes beyond documenting the past ones and that was hard to learn. Some of the teams in those networks preferred to limit their research to specific traditional study areas focussing on static detail of geomorphologic cartography. Other teams, however, became definitely committed to a novel use of geo-information for predictive modelling of hazards and risks (Remondo et al. 2003a, b; Zêzere et al. 2004).

Furthermore, those network projects have to be seen as very precious to cross-breeding of experiences. Two examples of the many doctoral theses that originated from those projects are the ones by (Remondo 2001) and by Bonachea (2006). They provided us much material for further training from a study area in northern Spain, in the Deba Valley.

Such cross breeding could take place through those network projects and the joint papers that were produced were a proof of it (Bonachea et al. 2005; Fabbri and Cendrero 1995; Fabbri et al. 2000, 2002b, 2003). The focus, indeed, was the exchange of concepts across disciplines. For instance the INCO project ELANEM that dealt with environmental indicators and indices and the human factor, e.g., the HDI or human development index (Cendrero et al. 2002). Another novel aspect was the human impact on geomorphic processes (Remondo et al. 2005b).

A number of advanced short courses could be offered to the young researchers involved in the EC projects, e.g., in Lisbon and Oviedo (Portugal and Spain) in 1996, in Vechta (Germany) in 1999, in Florence (Italy) in 2004 during the 32nd International Geological Congress, also as part of graduate courses at the DISAT in Milan during 2003–2011, and eventually at WIT, the Wessex Institute of Technology at Ashurst Lodge in the UK in 2012 (WIT 2012). In addition, several visiting fellowships made it possible for those researchers to work at foreign institutions.

3 What Are the Results of Applying Models?

Let us now consider the evolution of spatial modelling during the past two decades. Since the proposal of a basic mathematical framework for data integration in spatial data analysis that was made in 1992 (Chung and Fabbri 1993), experiments via many applications and study areas brought to a number of developments. The single critical reason for applying models of spatial relationships was for us to arrive at the classification of a study area: i.e., a classification in which relatively few classes were to express the clustering of events. Also, under certain conditions and assumptions, the classes were to identify not only areas with known events but also areas with future events, whose characteristics were identical or similar to the ones containing known events.

Examples of conditions were that the events should be of uniform type, thus justifying the similarity of spatial setting and/or phenomena typology, or that the spatial/temporal distribution within the study area be a representative sample of the generating process. Examples of assumptions were that sub-areas of the study area or temporal sub-sets of the events had comparable characteristics so that they could be used to generate comparable classifications.

Initially the events of concern were mineral occurrences of consistent genetic type due to the emphasis on mineral exploration. Later on, the concern on environmental impact and natural hazards led to focusing on hazards and risks caused by floods, avalanches and landslides. Many applications were directed to landslide hazard due to its impact and the availability of cartographies and compilations of mass movements. The spatial representation of the hazardous events was in many cases that of shallow translational mass movements rendered as polygons of their trigger areas. Their spatial distribution had to be related with the polygons of the mapping units in corresponding categorical data layers as well as the values in continuous data layers. These data layers, making up the spatial database of co-registered digital maps, had to express the typical hazardous conditions. The event distribution was considered as direct evidence in support of a proposition, DSP or *direct supporting pattern*. Categorical and continuous field layers were considered instead as indirect evidence, ISP or *indirect supporting patterns*. The ISPs represent the spatial evidence relatively to the DSP. Such numerical spatial support was for a mathematical proposition of the type that "**a point in the study area is affected by a trigger zone of a shallow translational landslide given the presence of the categorical mapping unit and the continuous value of the corresponding data layers**."

Different interpretations of the spatial relationships between DSP and ISP were proposed and compared: for instance, based on Fuzzy Sets, Likelihood Ratio, Logistic and Linear Regression, and Bayesian Probability. These were some of the spatial prediction models considered, for what we named "favourability modelling" as an all compassing term. Their application in the study area was to compute and combine spatial evidence and led to different and comparable relative classifications of all pixels.

The comparability of the different classifications was obtained via the generation of equal-area classes using rank order statistics. The equal-area classes with the higher computed values were, hopefully, to contain higher proportions of events than the subsequent classes. This, however, could not be obtained unless the relative size of the classes was sufficiently large. The aggregation of adjacent ranked classes was done to obtain monotonically decreasing proportions of the events in the ordered classes, representing the relative fitting of the events in the classes.

Such a fitting distribution of known events, however, could not describe the classification power of the database and of its spatial relationships to generate classes with greater or smaller proportion of **future events**. Those relationships were to provide a measure of relative quality of the classes as predictors and that requires particular strategies.

For assessing the prediction quality of the classifications, the set of events needed to be preferably separated into temporal or spatial subsets: e.g., one, possibly older, to obtain a classification and the other, possibly younger, to calculate the proportion of the younger events distributed among the classes generated using only the older events. Such later events were not to be used for the classification. This led to the development of a variety of cross-validation strategies via so called "blind tests", i.e., pretend not to know the existence of the later events for classification. Prediction-rate tables, histograms and cumulative curves were thus used to interpret the different classifications obtained in separate experiments over the study area.

The classification results were termed *prediction patterns* and the associated prediction-rate curves could then be used to assess the relative quality of the classifications obtained by modelling the DSP/ISP spatial relationships and cross-validating such distribution of past events with that of the distribution of "future" events. For instance, the steeper were the prediction-rate curves at the origin, the better could be considered the classes, and conversely the shallower became the curves the less significant the classes would be.

The modelling of prediction-rate curves was also the key to evaluating the impact of the landslide hazard *prediction patterns* on risk assessment. Under given realistic scenarios, the prediction-rate curve could be converted into a probability of occurrence curve to be used in combination with the *prediction pattern* to satisfy a risk equation with the introduction of vulnerability and exposed element patterns.

The term prediction referred to the capability of a classification to generate classes containing high numbers of future events. While the prediction-rate curves could represent the relative "goodness" of a prediction pattern, little was known that far of its reliability, stability or robustness.

For that, iterative procedures were developed in which incremental variations of the DSP in number and/or distribution were used to obtain sets of *prediction patterns* whose prediction-rate curves could be compared, averaged or otherwise combined. The iterative procedures led to assessing the uncertainty of class membership of each future event predicted. They also permitted to generate a new averaged *prediction pattern* as combination of the iterated predictions, termed *Target Pattern*. Of that we could compute statistical measures such as the variance, thus obtaining an *Uncertainty Pattern* and a *Combination Pattern*, using a given threshold of variance to threshold the *Target Pattern*.

Due to a multitude of unsatisfactory applications found in the scientific literature on the subject, strategies of blind testing for the interpretation of *prediction patterns* had to be stressed and demonstrated (Fabbri and Chung 2008). Later on, the development of their application in the area of natural hazard for training decision-makers was proposed (Fabbri and Chung 2009). A recent version of a computer system for spatial prediction modelling, SPM, to be used as a tool for research and training was documented (Fabbri and Chung 2012) for regional and local surveys. A more advanced version of it is the spatial target mapping system, STM, used in training (Fabbri and Chung 2009) and in the experiments discussed in the next section.

4 Exploiting the Deba Valley Database

Our efforts to interpret the results of spatial data integrations concerned study areas in different mountainous regions in the world. Besides Colombia, Canada and Italy, where our first applications of spatial prediction modelling of landslide hazard were made, European Commission network projects, student supervision and scientific collaboration permitted more applications in Portugal, Spain, Austria, Korea, Germany, Belgium and Slovenia. In particular, a study area rich with data and with hundreds of landslides that became available is a database of the Deba Valley, in the Basque Country of northern Spain. Support for its construction was from the EC Network Project NEWTECH (see Sect. 2) and its study became the focus of two doctoral theses: one on hazard prediction (Remondo 2001) and a second one that used it in part to extend research to landslide risk assessment (Bonachea 2006). In addition, the spatial database became an opportunity for international collaboration thus permitting the comparison of different approaches and experiments (Remondo et al. 2003a, b; Fabbri et al. 2002b, 2003; Remondo et al. 2005a, 2008). It eventually became part of the training material for documentation and exercises for advanced courses on spatial prediction modelling for natural hazard prediction and risk assessment (WIT 2012; Fabbri and Chung 2009).

Some critical results from its analysis make the database and the problems it represents still a worthwhile challenge today and probably in the future. Let us consider some blind testing experiments for cross-validation that led to the generation of uncertainty maps that characterize the *Target Pattern* in the Deba Valley study area.

The study area database used by us consists of the following seven data layers to become DSP and ISPs: (1) one-pixel trigger zone distribution of shallow translational landslides and associated flows with their sequential identification and separation into 906 pre-1997 and 217 post-1997, i.e., 1998–2001; (2 and 3) the distribution of 26 and 9 categorical map units, respectively, of lithology and land-use; and (4–7) four continuous digital maps derived from the digital elevation model, elevation, aspect, curvature and slope. The resolution of pixels of all the seven data layers is 10 m x 10 m. The study area occupies 1,393,541 pixels, 906 + 217 pixel corresponding with the distribution of the landslides. Figure 1 shows the locations of the landslides in the Deba Valley. Because the average area occupied by each landslide is relatively small, i.e., about 400 m^2, only one pixel was used to locate its trigger zone.

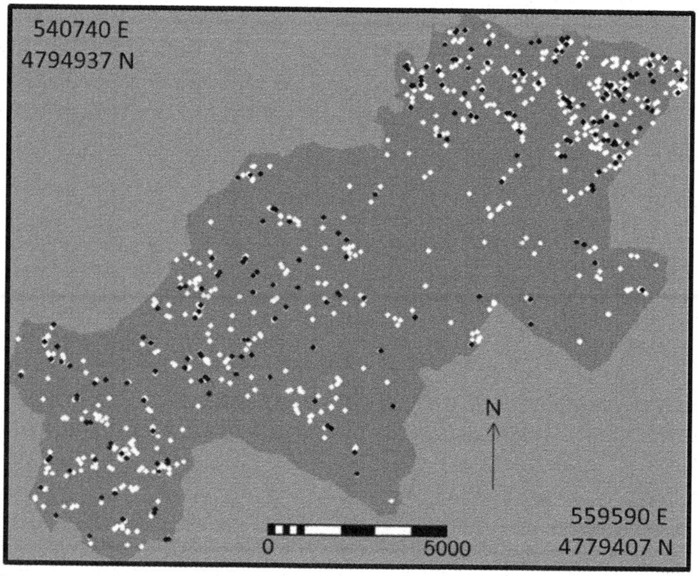

Fig. 1 Locations of the 906 pre-1997 shallow translational landslides, as white dots, and of the 217 post-1997, as black dots. The Deba Valley study area is in dark grey and its outside in light grey. The size of the dots, representing single pixel landslides, has been exaggerated for visibility. UTM northings and eastings of centres of pixels are at top left and bottom right

The empirical likelihood ratio, ELR, is one of several models used for establishing spatial relationships. It is based on the ratio between two density functions of the spatial evidence (i.e., the presence of map units, ISP): that in the presence of the landslide pixels (the DSP) and the one in their absence. Under a number of assumptions, the ELR is computed from the different ISPs, integrated and used to rank the likelihood of occurrence of future hazardous occurrences. The ranks refer to space, and possibly also to time intervals, should information on their time of occurrence be available.

Tables 1 and 2 show the ELR values for the map units of the lithology and the land use data layers, respectively, not shown here. In Table 1, the ELR values for muddy flysch, marly limestone and calcareous flysch are higher than 2 and therefore considered significant. The value 2 is just an arbitrary one that represents a density function in the presence of the trigger areas that is twice that in their absence in the study area. That is so only for land use unit grassland and cultivations in Table 2. It means that those four map units are the most effective ones in supporting the proposition of the presence of landslides. This was because the contribution of the other categorical units resulted either weaker or null. As to the continuous data layers, their contribution is always below 2, except for the curvature ISP slightly above 2, showing a rather marginal contribution to the prediction rates and the *prediction pattern* of Fig. 2. The ELR values provide measures of relevance of the ISPs in predictive capability. Here the integrated ELR values of the six ISPs, ranging between 0 and infinity, are sequenced in descending order so that

Table 1 Two empirical frequency distribution functions and the corresponding empirical likelihood ratio function for the units of lithology from the Deva Valley study area

Lithology unit	Description	Frequency at 906 pre-1997 landslides	Frequency at non-landside areas	Likelihood ratio
1	Silicoclastic-calcareous Flysch	0.000	0.0001	0.0000
2	Muddy Flysch	0.1300	0.0549	**(2.3732)**
3	Stratified limestone	0.0011	0.0017	0.648
4	Marly limestone	0.0265	0.0116	**(2.2934)**
5	Marl	0.2748	0.1988	**1.3824**
6	Sandstone and conglomerate	0.0033	0.0093	0.3547
7	Sandy Flysch	0.1391	0.1842	0.7549
8	Massive limestone	0.0177	0.1508	0.1508
9	Calcareous lutite and sandy marl	0.0331	0.0787	0.4208
10	Marly limestone	0.1998	0.1110	**1.7990**
11	Calcarenite, marl and calcareous breccia	0.0232	0.0220	**1.0524**
12	Calcareous Flysch	0.0408	0.0091	**(4.4765)**
13	Pyroclastics	0.0044	0.0174	0.02539
14	Lavas	0.0563	0.0812	0.6933
15	Polygenic breccias	0.0000	0.0002	0.0000
16	Siliceous breccias	0.0000	0.0002	0.0000
17	Well graded gravel	0.000	0.0007	0.0000
18	Poorly graded gravel	0.0010	0.0007	**1.7441**
19	Clayey gravel	0.0044	0.0189	0.2341
20	Well graded sand	0.0000	0.0002	0.0000
21	Poorly graded sand	0.0000	0.0013	0.0000
22	Silty sand	0.0298	0.0174	**1.7175**
23	Clayey sand	0.0000	0.0014	0.0000
24	Silt and fine sand	0.0033	0.0067	0.4922
25	Residual clay	0.0022	0.0086	0.2576
26	Anthropogenic deposits	0.0000	0.0017	0.0000

Likelihood ratio values higher than 1 are in bold and those higher than 2 are also within brackets

200 ranks of equal-area corresponding to the 0.5% of the area of study, could be generated and grouped for visibility into the eleven classes shown in the legend of the illustration. Maintaining fixed the groupings does facilitate the comparison of patterns. Ranking of equal-area classes is considered a fundamental procedure for evaluating and comparing the *prediction patterns* used in different analyses. The integrated ELR values themselves are not considered an interpretable function beyond their ranks.

Table 2 Two empirical frequency distribution functions and the corresponding empirical likelihood ratio function for the units of land-use from the Deba Valley area

Land use classes	Description	Frequency at 906 pre-1997 landslides	Frequency at non-landslide areas	Likelihood ratio
1	Water bodies	0.0000	0.0003	0.0000
2	Very dense forest	0.0066	0.0476	0.1392
3	Dense deciduous forest	0.0353	0.0986	0.3582
4	Half open deciduous forest	0.0000	0.0261	0.0000
5	Very dense coniferous	0.2804	0.4370	0.6416
6	Scrubs and bushes	0.0408	0.0593	0.6887
7	Grasslands and cultivations	0.6082	0.2444	**(2.4883)**
8	Areas without vegetation	0.0287	0.0868	0.3308

The only likelihood ratio value higher than 2 is in bold within brackets

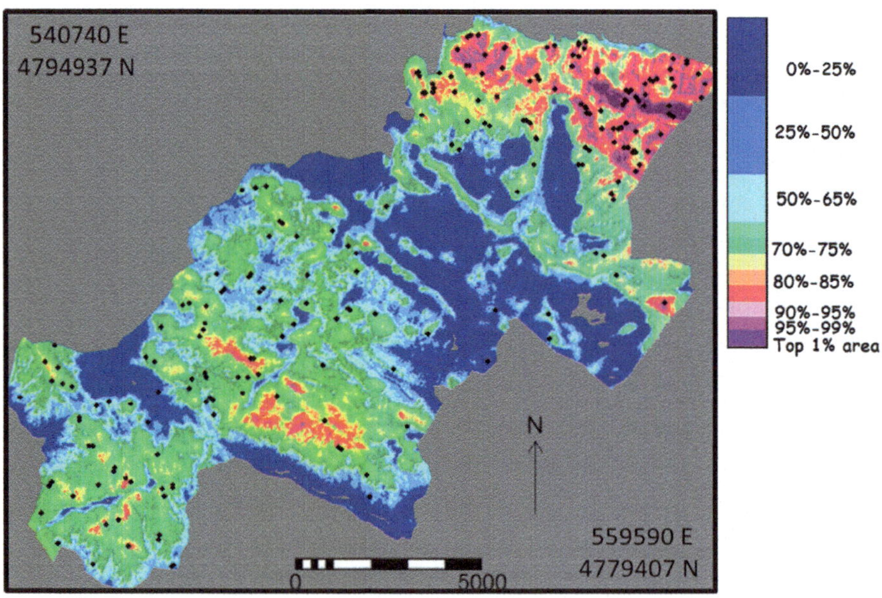

Fig. 2 Empirical likelihood ratio *prediction pattern* of the Deba Valley study area obtained using the distribution of the 906 pre-1997 landslides as DSP, not shown here, and as ISPs the six categorical and continuous data layers. The locations of the 217 post-1997 landslides used for cross-validation are as magnified black dots

The relatively weak support of the continuous ISPs can be interpreted in many ways, however, it is not surprising given the limited spatial support of only 906 landslide pixels used to classify the remainder of (1,393,541–906 =) 1,392,635 pixels of the study area!

The relative quality or "goodness" of the classification as a prediction can be assessed by considering the proportion of future events, here the 217 post-1997 landslides, falling within each equal area class, as described in the prediction-rate histogram of Fig. 3. The 200 classes each consist of approximately 6967 pixels.

In the illustration of Fig. 3 the histogram shows a weak monotonically decreasing trend. It only covers 5 classes with the highest predicted values of 3% each of the study area. The top 15% is the only part of conveniently high values of prediction rates. Histograms of equal-area classes are convenient for eventually setting the levels of hazard as high, intermediate, low, etc. It is implicit in the concept of classification that we have to have a monotonically decreasing set of classes in which the faster is the decrease the sharper is the classification.

Figure 4 shows the entire cumulative prediction-rate curve of the 217 post-1997 landslides. Interpreting the curve in terms of costs-benefits we have that the highest 10% class contains 40% of the validation landslides, the highest 15% contains 50%, and the highest 20% contains close to 60%. Further increments of the study area that could be considered as hazardous do not contain conveniently higher proportions of predicted occurrences. The higher is the curve inclination near the origin the greater becomes the proportion of validation-occurrence distribution through higher classes, i.e., the better are their prediction scores. This type of cumulative curve can be of general use in the assessment of the quality of predictions.

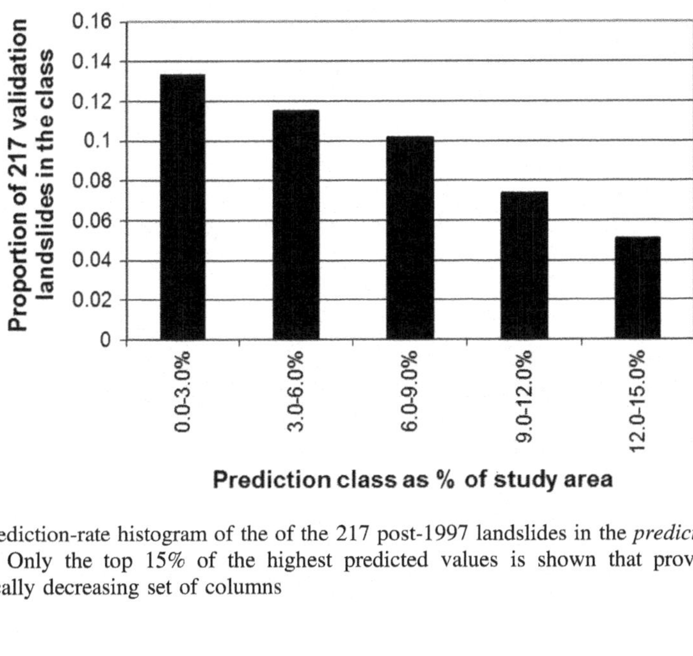

Fig. 3 Prediction-rate histogram of the of the 217 post-1997 landslides in the *prediction pattern* of Fig. 2. Only the top 15% of the highest predicted values is shown that provide a slow monotonically decreasing set of columns

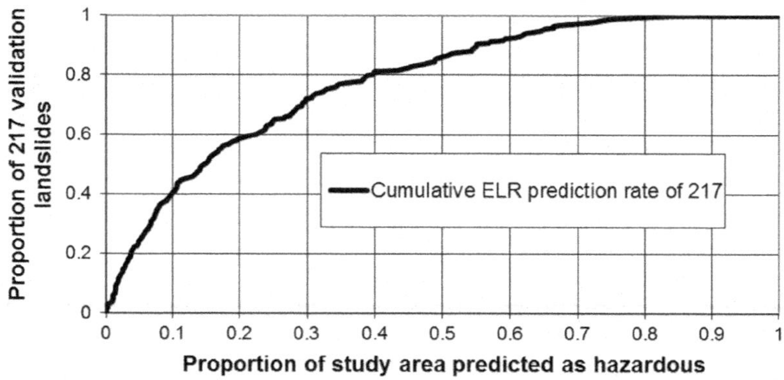

Fig. 4 Cumulative prediction-rate curve for the 217 post-1997 landslides in the *prediction pattern* of Fig. 2

One critical question, given the observed quality of the database for prediction modelling, is the following: "What is the degree of certainty of the 0.5% classes from the 906 pre-1997 landslides used as DSP, and the six ISPs, used for predicting the locations of the 217 post-1997 landslides?"

Let us provide a measure of uncertainty of class membership of the classification. We could proceed in many ways using different strategies. For instance, one strategy that can be used is to repeat the predictions with 50 landslides less than the 906, i.e., 856, and successively iterate that 18 times to predict each time the remaining 50. Such a "50-less" procedure will produce 18 *prediction patterns* and the corresponding 18 prediction-rate curves, shown in Fig. 6. This is termed sequential elimination. Sequential selection or random selection can also be used to obtain further uncertainty measures of the *prediction patterns*. From the spread of cumulative curves shown in Fig. 6 we obtain a visual impression of the variation of prediction scores around the average at the sequential increments of study area considered as hazardous.

A new *prediction pattern* can be generated considering the values for each pixel of the study area (18 in our case) and computing their average with the associated variance, for instance. We have termed the average pattern as the *Target Pattern* and the variance pattern as the *Uncertainty Pattern*. The *Target Pattern*, not shown here, is visually identical to the prediction pattern shown in Fig. 2, however, it is accompanied by the *Uncertainty Pattern*, whose pixels have variance values. This can be used to threshold the *Target Pattern* at a tentative % threshold of variance. The sample average and variance used here are just one example of statistics that can be used to generate *Target* and *UncertaintyPatterns*. Various other statics can also be used to generate and interpret these patterns (Fabbri and Chung 2014).

As an example, the 2% *Combination Pattern* shown in Fig. 5 shows all predicted values of the *Target Pattern* at or below the 2% variance. The light grey areas indicate the distribution of uncertainty of class membership above the 2%. The distribution of the 217 post-1997 landslides over the 2% *Uncertainty Pattern* obtained with the "50-less" iterative procedure separates the predicted (validation) landslides in areas with uncertainty above the variance threshold from the ones in areas with uncertainty below it. For instance, the area occupied by the pixels with variance $\leq 2\%$ consists of 846,268 pixels, i.e., 37.49% of the study area. It contains 70 post-1997 landslides pixels, i.e., 47.62% of the 217. The selection of the 2% as variance threshold is just a tentative one to exemplify a strategy. Only a solid knowledge of the processes represented in the database can guide to a preferable statistics to obtain *Uncertainty Patterns* and their threshold for generating *Combination Patterns*.

Each *prediction pattern* of the 18 generated has provided a prediction-rate curve and an average curve computed for the *Target Pattern*, as shown in Fig. 6. Here we can see another use of the cumulative prediction-rate curve for comparing different *prediction patterns*. In this case the patterns were obtained by sequential elimination of 50 occurrences in each. Also the *prediction patterns* generated with different mathematical models can be compared in a similar manner.

These experiments and their results are now part of the training material in spatial prediction modelling for courses on natural hazards and risk assessment. Besides revealing some of the properties and limitations of the database, they also offer opportunity for further experimentation to refine and change the analytical strategy as a data mining process. This is still a challenge today in spite of the many new mathematical models being proposed.

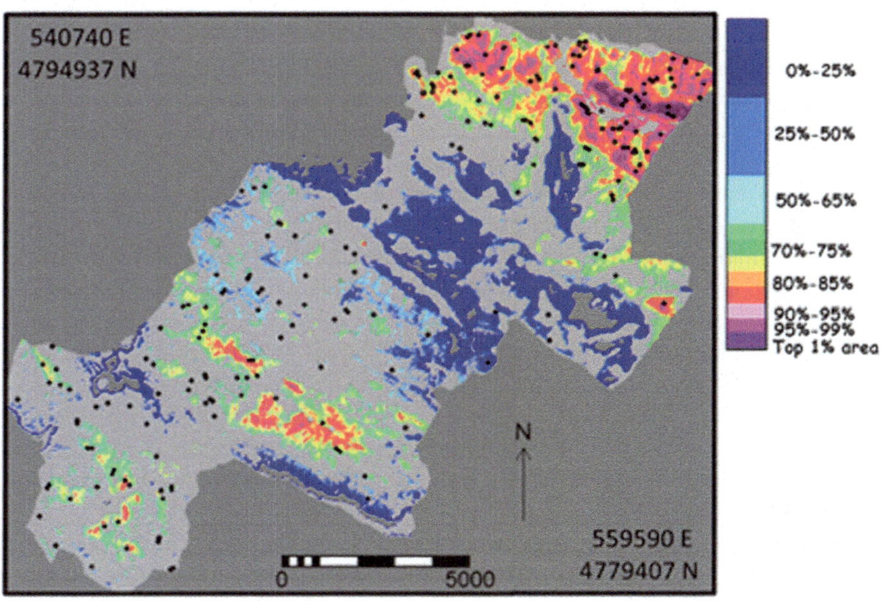

Fig. 5 *Combination Pattern* for uncertainty (variance) values $\leq 2\%$ and the corresponding part of the *Target Pattern* for iterative cross-validation strategy "50 less" (*Target Pattern* not shown here but visually similar to the *prediction pattern* in Fig. 2). The locations of the 217 post-1997 landslides used for cross-validation are as magnified black dots

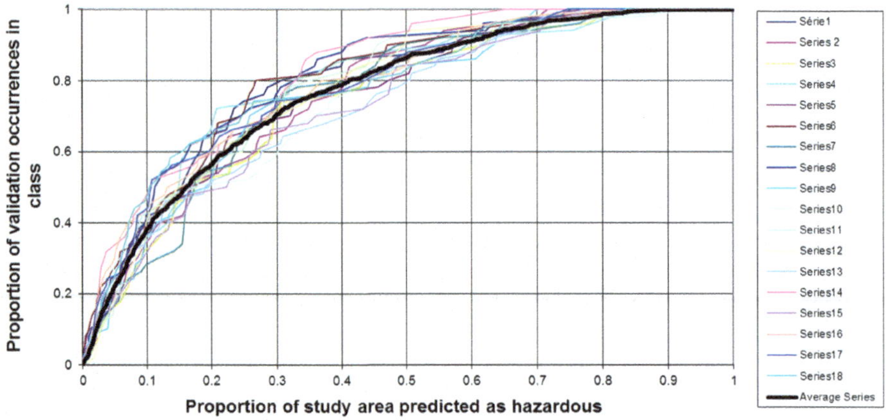

Fig. 6 The 18 cumulative prediction-rate curves from the "50 less" cross-validation analyses performed to generate the corresponding 18 *prediction patterns* to obtain the *Target Pattern*, whose prediction-rate curve is shown as heavier solid black curve

Let us review other studies that made use of the data on the Deba Valley study area. The *Uncertainty Pattern* of class membership has been used to identify the areas with lower uncertainty of hazard class membership when obtaining a risk pattern (Fabbri et al. 2014). Other strategies of analysis were selected according to the known or suspected characteristics of the database and the processes it portrays.

For example, suspecting that the zone of influence of the occurrences of the landslide trigger zones is limited to their immediate surroundings, pixel neighbourhoods of 5, 9 or larger were used for the modelling of spatial relationships and the relative statistics then applied to the remainder of the study area. This generated different *prediction patterns* for interpretation (Fabbri et al. 2008).

Furthermore, assuming the boundaries of categorical mapping units of ISPs to be fuzzy and/or the continuous ISP to be noisy, the ISPs were gradually modified (fuzzyfied) and the analyses repeated to obtain again different *prediction patterns* to evaluate and compare them along with their impacts on risk assessment (Fabbri and Chung 2014). Another strategy applied was the one of selecting sub-regions of the study area for predictions and validations, or different spatial or temporal subsets of the landslides (Fabbri and Chung 2016). Also experiments were made in training sessions substituting the ELR model with other models comparing the respective prediction-rate curves. Only minor differences in the resulting *prediction patterns* were obtained. That revealed that the result of prediction modelling was in our case essentially driven by the data and not by the models.

Furthermore, other collaborative studies were made sharing the Deba Valley database for advanced modelling. Using the same study area and a larger number of ISPs, mostly derived from the digital terrain model, Melchiorre et al. (2006) modelled landslide susceptibility by Artificial Neural Networks. They used likelihood ratio criteria to exclude less significant ISPs and compared *prediction patterns* obtained with all of them with those obtained by a reduced set. The reduced set appeared to moderately improve the resulting prediction-rate curves, particularly when lithology was part of the ISPs. To test the robustness of the patterns, database subdivisions were generated partitioning it 10 times into training, validation and test sets. While it remains a matter of debate whether the database justified the usage of such a sophisticated methodology, it is of interest to observe how the quality of the *prediction patterns* obtained was similar in all of their experiments and how it was considered important to employ iterative processes. Similarly, in another study with the same database, Felicisimo et al. (2013) studied *prediction patterns* obtained by four new mathematical models: multiple logistic regression, multivariate adaptive regression splines, classification and regression trees, and maximum entropy. All models appeared to perform similarly and the inclusion of lithology as ISP provided somewhat better results. The analyses were iteratively replicated using sets of randomized samples.

Such experiments have led to novel results and considerations on the significance of the spatial data for the purpose of prediction. Testing for uncertainty, robustness or reliability is considered a critical part of the modelling. The different *prediction patterns* eventually have to be interpreted and explained in geomorphologic terms for the processes in the study area, a challenging task that has to follow the mathematical modelling. What is remarkable is the richness of the approaches by different teams sharing the database. Such sharing seems a rare opportunity in spatial analysis.

5 Concluding Remarks

A review was made of the development by the authors of spatial characterization originally stimulated by mineral exploration and later by environmental concern. Research network programmes of the European Commission brought together different expertises and backgrounds, encouraging cross-breading of disciplines and approaches, some particularly related to research work in the Deba Valley study area of northern Spain. Its spatial database constructed for landslide hazard prediction and risk assessment has become not only a scientific ground for cooperation and also useful material for training young scientists, graduate students and decision-makers. The experimental results have now a general significance that transcends the study area. In addition, not all the possibilities of experiments have been exhausted. The steps from the application of spatial prediction models to cross-validations by blind testing and the iterative calculation of *Target Patterns* and *Uncertainty Patterns*, or robustness or reliability, are now beyond the stage of research. They have turned into application challenges at institutional levels. It is now up to surveys and civil protection or environmental agencies to construct and analyze spatial databases for hazard and risk assessment under their jurisdiction and produce maps of the likelihood of future hazardous events for current practice. Encouraging is the present conditions of ubiquitous availability of spatial data in easily accessible digital formats that facilitate the construction of spatial databases.

Nevertheless, the hazard situation is worrisome in many parts of the world considering: (1) the progressive invasion of urbanizations into hazardous grounds that aggravate the impact of hazardous processes and exposure to risk; (2) the frequent occurrence of wildfires in forested areas, mostly man-induced, that reduce soil cohesion; and (3) the increasing variability of climatic conditions in the direction of global warming, that increase the intensity of hazardous processes.

The present challenges of spatial prediction modelling are: (1) to redirect scientific activity in spatial prediction modelling from the preference for formulating new sophisticated models to the accurate analysis, improvement and exploitation of spatial databases for data mining and knowledge extraction, (2) to share those databases to compare and contrast different models and study areas, (3) to standardize cross-validation procedures that by many are still interpreted in more restrictive and unsatisfactory ways, and (4) for decision-makers to establish and maintain a tighter connection with the reality of natural hazards and the spatiotemporal expression of hazard representations.

The authors are committed to experiments of spatial prediction modelling for measuring uncertainty, leading to the separation of occurrences into more cohesive groupings, to the extension of predictions from one area to another with similar geomorphologic characteristics and to a more generalized procedure for evaluating and comparing modelling results.

Acknowledgements We are grateful to an anonymous reviewer and RAC Garcia who helped in correcting and improving this manuscript.

References

Agterberg, FP, Chung CF, Fabbri AG, Kelley AM, Springer, JA (1972) Geomathematical evaluation of copper and zinc potential of the Abitibi Area, Ontario and Quebec. Geological Survey of Canada Paper, pp 71–41, 55 p

Allais M (1957) Method of appraising economic prospects of mining exploration over large territories. Manage Sci 3:285–347

Aronoff S (1989) Geographic information systems: a managements perspective. WDL Publications, Ottawa, p 294

Bonachea J (2006) Desarrollo, aplicación y validación de procedimientos y modelos para la evaluación de amenazas, vulnerabilidad y riesgo debidos a procesos geomorfológicos. Ph.D. thesis, Universidad de Cantabria, Santander, Spain, 356 p, July 2006 (http://www.tesisenred.net/TDR-1124106-134112/index_cs.html)

Bonachea J, Bruschi VM, Remondo J, Alberto Gonzalez-Diez A, Salas L, Bertens J, Cendrero A, Otero C, Giusti C, Fabbri A, Gonzalez-Lastra JR, Aramburu JR (2005) An approach for quantifying geomorphological impacts for EIA of transportation infrastructures: a case study in northern Spain. Geomorphology 66:95–117

Bonham-Carter GG (1994) Geographic Information Systems for Geoscientists. Modelling with GIS. Pergamon (Elsevier Science Ltd.), Tarrytown, New York, 398 p

Burrough PA (1986) Principles of geographical information systems for land resources assessment. Clarendon Press, Oxford, p 194

Cendrero A, Frances E, Latrubesse EM, Prado R, Fabbri A, Panizza M., Cantu MP, Hurtado M, Gimenez, JE, Martinez O, Cabral M, Tecchi RA, Hamity V, Ferman JL, Quintana C, Ceccioni A, Recatalá L, Bayer M, Aquino S (2002) Projecto Relesa-Elanem: uma Nova Proposta Metodológica de Índices e Indicadores para Avaliação da Qualidade Ambiental. Revista Brasileira de Geomorfologia, Ano 3(1):33–47

Chung CF (1983) SIMSAG: Integrated computer system for use in evaluation of mineral and energy resources. Math Geol 15:47–58

Chung CF, Fabbri AG (1993) Representation of geo-science data for information integration. J Non-Renew Res 2(2):122–139

Chung CF, Fabbri AG, Sinding-Larsen R (eds) (1988) Quantitative Analysis of Mineral and Energy resources. D. Reidel, Dordrecht, 738 p

Fabbri AG (1975) Design and structure of geological data banks for regional mineral potential evaluation. The Can Min Metall Bull 69:91–98

Fabbri AG (1984) Image Processing of Geological Data. Van Nostrand-Reinhold, New York, p 244

Fabbri AG (2007) On spatial prediction models or the unbearable lightness of GIS. Inaugurele Rede Vrije Universiteit Amsterdam, presented on 17 Apr 2007, 20 p, www.feweb.vu.nl; see also https://www.researchgate.net/publication/241927019_On_Spatial_Prediction_Models_or_the_Unbearable_Lightness_of_GIS

Fabbri AG, Cendrero A (1995) Changes and the environment. ITC Journal 4:354–357

Fabbri AG, Chung CF (2008) On blind tests and spatial prediction models. Nat Res Res 17 (2):107–118. Also in, Bonham-Carter G, Cheng Q (eds) Progress in geomathematics. Springer, Berlin, Heidelberg, pp 315–332

Fabbri AG, Chung CJ (2009) Training decision-makers in hazard spatial prediction and risk assessment: ideas, tools, strategies and challenges. In: Duncan K, Brebbia CA (eds) Disaster management and human health risk. WIT Press, Southampton, pp 285–296, or WIT Transactions on the Built Environment, www.witpress.com, ISSN 1743-3509 (online) doi:10_2495/DMAN09025

Fabbri AG, Chung CJ (2012) A spatial prediction modeling system for mineral potential and natural hazard mapping. Proceedings 7th EUREGEO, Bologna, Italy, 12–15 June 2012, vol. II, pp 756–757

Fabbri AG, Chung C-J (2014) On spatial uncertainty in hazard and risk assessment. In: Brebbia CA (ed) Risk Analysis IX. WIT Press, Southampton, Boston, pp 3–15

Fabbri AG, Chung C-J (2016) Blind-testing experiments for interpreting spatial-prediction patterns of landslide hazard. Int J Saf Secur Eng 6(2):193–208

Fabbri AG, Chung CF, Cendrero A, Bauer B (2000) A strategy for sustainable development of natural resources based on prediction models. Proc. ISPRS 2000, Amsterdam, July 2000

Fabbri AG., Gaal G, McCammon RB (eds) (2002a) Deposit and geoenvironmental models for resource exploitation and environmental security. Kluwer Academic Publishers, Dordrecht, 532 p, and a CD

Fabbri AG, Chung CF, Napolitano P, Remondo J, Zezere JL (2002b) Prediction rate functions of landslide susceptibility applied in the Iberian Peninsula. In: Brebbia CA (ed) Risk Analysis III. Southampton, Boston, WIT Press, pp 703–718

Fabbri AG, Chung CF, Cendrero A, Remondo J (2003) Is prediction of future landslides possible with a GIS? Nat Hazards 30:487–499

Fabbri AG, Remondo J, Ballabio C, Poli S, Chung CF, Scholten HJ (2008) Occurrence neighborhoods and risk assessment from landslide hazard in northern Spain. In: Brebbia CA, Beriatos E (eds) Risk analysis VI. Simulation and hazard mitigation. WIT Press, Southampton, Boston, pp 29–41. Also available as WIT Transactions on Information and Communication, v. 39, www.witpress.com, ISSN 1743-3517 (online), https://doi.org/10.2495/risk080041

Fabbri AG, Remondo J, Chung CJ (2014) Landslide risk assessment with uncertainty of hazard class membership. An application of favourability modeling in the Deba Valley area, northern Spain. Proceedings of IAEG XII Congress, International Association for Engineering Geology, Springer, Turin, Italy, 15–18 Sep 2014. Springer, Lollino G, Giordan D, Crosta GB,

Corominas J, Azzam R, Wasowski J, Sciarra N (eds.), Engineering Geology for Society and Territory—Volume 2, Ch. 021, 4 p

Felicisimo Á, Cuartero A, Remondo J, Quiros E (2013) Mapping landslide susceptibility with logistic regression, multiple adaptive regression splines, classification and regression trees, and maximum entropy methods: a comparative study. Landslides 10(2):175–189

Melchiorre C, Matteucci M, Remondo J (2006) Artificial neural networks and robustness analysis in landslide hazard zonation. In: The 2006 IEEE international joint conference on Neural Networks Proceedings; 4375–4381, Vancouver, BC, Canada, 16–21 July 2006

Remondo J (2001) Elaboración y validación de mapas de susceptibilidad de deslizeamientos mediante técnicas de análysis espacial. Ph.D. thesis, Universidad de Oviedo, Spain, 404 p, Annexos 95 p, and CD

Remondo J, González-Diéz A, Diaz de Teran JR, Cendrero A (2003a) Landslide susceptibility models utilising spatial data analysis techniques. a case study from the lower Deba Valley, Guipúzcoa (Spain). Nat Hazards 30:267–279

Remondo J, González- A, Diaz de Teran JR, Cendrero A, Fabbri AG, Chung CF (2003b) Validation of landslide susceptibility maps: examples and applications from a case study in Northern Spain. Nat Hazards 30:437–449

Remondo J, Bonachea J, Cendrero A (2005a) A statistical approach to landslide risk modeling at basin scale: from landslide susceptibility to quantitative risk assessment. Landslides 2(4):321–328

Remondo J, Soto J, González-Díez A, Díaz de Terín JR, Cendrero A (2005b) Human impact on geomorphic processes and hazards in mountain areas in northern Spain. Geomorphology 66(1–4):69–84

Remondo J, Bonachea J, Cendrero A (2008) Quantitative landslide risk assessment and mapping on the basis of recent occurrences. Geomorphology 94(3–4):496–507

WIT (2012) http://www.wessex.ac.uk./news/courses-and-seminars/772-hazard-and-risk-assessment

Zêzere JL, Reis E, Garcia R, Oliveira S, Rodriques ML, Vieira G, Ferreira AB (2004) Integration of spatial and temporal data for the definition of different landslide hazard scenarios in the area north of Lisbon (Portugal). Natural Hazards Earth Sys Sci 4:133–146

Earthquake Events Modeling Using Multi-criteria Decision Analysis in Iran

Marzieh Mokarram and Hamid Reza Pourghasemi

Abstract Kerman Province in Iran is known as an earthquake prone area, with different serious damages. In this study, GIS-based ordered weight averaging (OWA) with fuzzy quantifier algorithm is used to model earthquake events in north of Kerman Province, Iran. For this aim, at first using attraction model was tried to increase DEM resolution from 30 to 10 m. Then, using the mentioned DEM, three layers such as aspect, slope, and elevation was prepared. Also, different layers including lithology, land use, river, road, fault, and earthquake occurrences were prepared in ArcGIS software. Subsequently, the importance of each factor in earthquake events was defined using trapezoidal membership function. Finally, the earthquake events map with different risk level (six levels) was prepared using OWA method. The results showed that with decreasing risk (no trade-off), many parts of the study area had not earthquake events hazard. While, with increasing risk (no trade-off), all of the study area had earthquake events hazard. Low level of risk and no trade-off had the highest area in the very low class (98%), while high level of risk and average trade-off had the highest area in the very low class (15.62%). So, for the study where has high earthquake should use low level risk maps in order to better management and damage decreasing.

Keywords Earthquake events modeling · Ordered weighted averaging (OWA) Fuzzy quantifiers · GIS · Iran

---

M. Mokarram (✉)
Department of Range and Watershed Management, College of Agriculture and Natural Resources of Darab, Shiraz University, Shiraz, Iran
e-mail: m.mokarram@shirazu.ac.ir

H. R. Pourghasemi
Department of Natural Resources and Environmental Engineering, College of Agriculture, Shiraz University, Shiraz, Iran
e-mail: hr.pourghasemi@shirazu.ac.ir

H. R. Pourghasemi and M. Rossi (eds.), *Natural Hazards GIS-based Spatial Modeling Using Data Mining Techniques*, Advances in Natural and Technological Hazards Research 48, https://doi.org/10.1007/978-3-319-73383-8_7

1 Introduction

Iran is one of 10 countries which are determined as earthquake prone areas.

Many earthquakes in Iran are related to the 21st century. 14 earthquakes of magnitude ~7.0 have killed more than 14,600 people. The earthquakes have occurred mostly in urban and rural lands. Unfortunately, there are no detailed statistics from the casualties of earthquakes in Iran (Iranian Studies Group). By increasing population, the risk of earthquake is increasing; so, it is important to study this natural hazard and to decrease its mortal effects. In the literature review, here are different methods for studying of earthquake. Such as Yagoub (2015) used remote sensing and GIS application for preparing earthquake map in UAE (1984–2012). To preparing hazard mapping of earthquake was used seven parameters such as geology, soil, slope, land use, historical earthquake events, fault line, and roads. The results showed that map made could helpful in proper use of land for planning and population reduction. Champatiray et al. 2005, Lillesand et al. 2008, Roustaei et al. 2005 using Remote Sensing (RS) images with different resolution investigated changes of land surface before and after earthquake. Theilen-Willige et al. (2012) used RS and Geographic Information Systems (GIS) for determination of prone areas to earthquakes. The results showed that the use of geographic data such as satellite images and topographic information was useful for determination of prone areas and reduce the risk of damage.

GIS and RS had been widely used to identify and predict risk areas (Balaji et al. 2002; Laefer et al. 2006; Roy et al. 2000). GIS and RS using a variety of models have a great ability to analyze data spatially (Henning 2011; Miles and Ho 1999). So in the study used ordered weighted averaging (OWA) in GIS for earthquake events modeling.

One of the methods for preparing earthquake mapping with different risk levels is OWA algorithm. OWA is a type of multi-criteria evaluation that is often used for environmental engineering analysis (van Westen et al. 2000; Malczewski et al. 2003; Komac 2006a; Malczewski 2006; Gorsevski and Jankowski 2008, 2010; Mokarram and Aminzadeh 2010; Mokarram and Hojati 2016). Kerman Province one of the most cites in Iran where is prone to earthquakes. Kerman province is not exception in this case and by having huge and active fault, every year will be observed a lot of earthquakes which some of them cause a lot of damages. So in the study aim is Earthquake events modeling using multi-criteria decision analysis in north of Kerman region. Kerman region over 2000 years of history, has witnessed many moderate to strong earthquakes. So investigation of earthquake in the study is very importance that using OWA method was studied it. In general, OWA is a relatively novel technique that has been used less in natural sciences (e.g., earthquake events). In the study is prepared six maps with different multi-criteria evaluation for earthquake that considering the state budget and different management can use each one of them. Thus, the region was selected OWA method to investigate earthquake events susceptibility. In fact by changing the parameters, OWA can made a wide range of effective decisions to earthquake managements (Fig. 1).

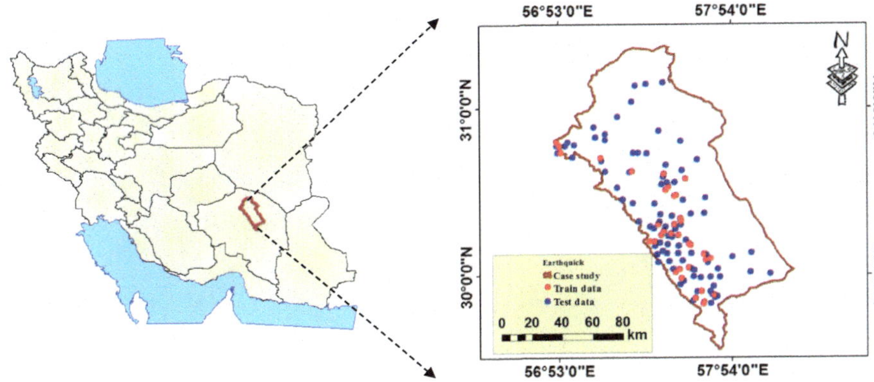

Fig. 1 Location of the study area

2 Materials and Methods

2.1 Case Study

The study area was located in north of Kerman Province, Iran. The area of the study area is about 8854 km^2, and is located at latitudes of 29° 40′ to 31° 18′ N and longitude of 56° 54′ to 58° 11′ E (Fig. 2). The elevation of the study area ranges from 219 to 3539 m a.s.l. The principal agricultural crops consisted of wheat, beans, barley, and rice. In term of geology, the case study consists of cretaceous limestone, young conglomerate (Neogene–Quaternary) and quaternary sediments (IV period).

Data Preparation

Earthquake points in the study are was prepared from Natural Resources and Watershed management of Fars Province, Iran. For earthquake events mapping in the study area different layers including aspect, altitude, lithology, slope degree, land use, distance from rivers, distance from roads, and distance from faults were used. Lithology and fault maps derived from geological maps in 1:100,000-scale. Roads and rivers extracted from a topographical map in a scale of 1:25,000. For creating Digital Elevation Model (DEM) was used SRTM DEM and slope degree and aspect layers are extracted of it, respectively. Land use/land cover maps were derived from Landsat 7 ETM+ satellite images with spatial resolution of 15 m.

At first, using attraction model was tried to enhance resolution of the built DEM from 30 to 10 m (http://earthexplorer.usgs.gov). Subsequently, using the built DEM, slope degree and aspect maps were prepared in ArcGIS v.10.2 software (Fig. 2). According to Fig. 2, many parts of south and southeast of the study area have the elevation more than 2500 m. Also, the slope value is between 0 and 84 degrees that the most slope value is in southeast and east directions (green and brown color). The aspect value is between –1 (flat) –360 (north), so that aspects of south and east are sensitive to earthquake (Yagoub 2015) (Table 1).

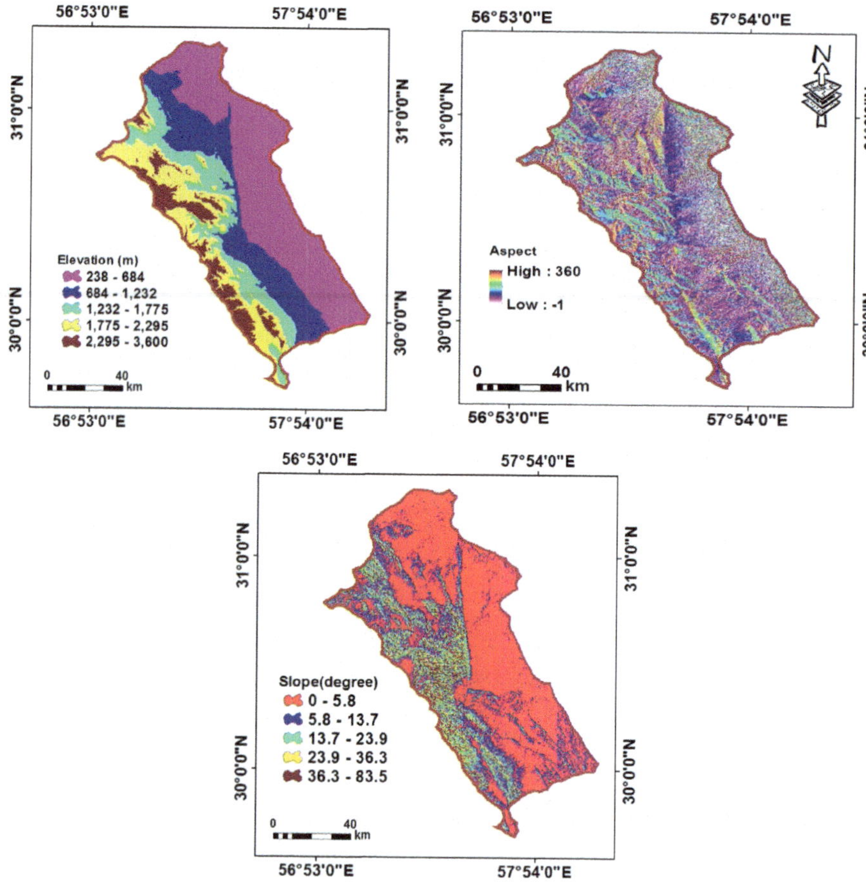

Fig. 2 Slope degree, aspect, and DEM maps for the study area

Table 1 Impact of elevation, slope degree, and aspect

Parameters	Class	Description
Elevation (m)	>2500	Low sensitive to earthquake
	<2500	Very sensitive to earthquake
Slope (°)	0–30	Low sensitive to earthquake
	30–60	Medium sensitive to earthquake
	>60	Very sensitive to earthquake
Aspect	Flat, North, Northeast	Low sensitive to earthquake
	East, Southeast, South	Medium sensitive to earthquake
	Southwest, West, Northwest	Very sensitive to earthquake

Table 2 Distance from faults, rivers and roads

Distance from faults (km)			
Class 1 (<2)	Class 2 (2–5)	Class 3 (5–10)	Class 4 (10–30)
Distance from rivers (m)			
Class 1 (0–50)	Class 2 (50–100)	Class 3 (100–150)	Class 4 (>150)
Distance from roads (m)			
Class 1 (0–500)	Class 2 (500–1000)	Class 3 (1000–1500)	Class 4 (>1500)

The earthquake points in the study area consisted of 104 points where are shown in Fig. 1. According to Fig. 1, earthquake points were divided to training and validation data (70% of the database for training and 30% for testing). For preparing raster maps for distance of road, fault and river, using buffer tools in ArcGIS were made the buffer maps for them (Yagoub 2015). According to Table 2 were prepared buffer maps using distance from features where show in Fig. 3.

The land use map for study area was prepared from Organization of Agriculture Jahad Fars that consist of as Forest, wood land, garden, agriculture, salt land, bare land, range, urban, and sand dune (Fig. 4). So, there is low the possibility of earthquake in the green areas (Yagoub 2015).

Finally, for preparing earthquake events map was used sensitivity map of geological formations to earthquake. So, for determination of formations sensitivity, geological formations map in six classes was produced by Iranian Geological Organization and is shown in Fig. 5. The description of each class is given in Table 3.

2.2 Methods

In the current study after preparing thematic layers with 10 m spatial resolution (Mertens et al. 2004), were applied fuzzy method and Analytical Hierarchy Process (AHP) techniques for overlaying of layers and preparing final earthquake events map.

2.2.1 Fuzzy Inference

Zadeh (1965) defined a fuzzy set by trapezoidal membership functions from properties of objects. According to membership function (MF) was prepared fuzzy map for slope, elevation, sensitive, land use and aspect. The MF for these parameters was shown in the following (Yagoub 2015):

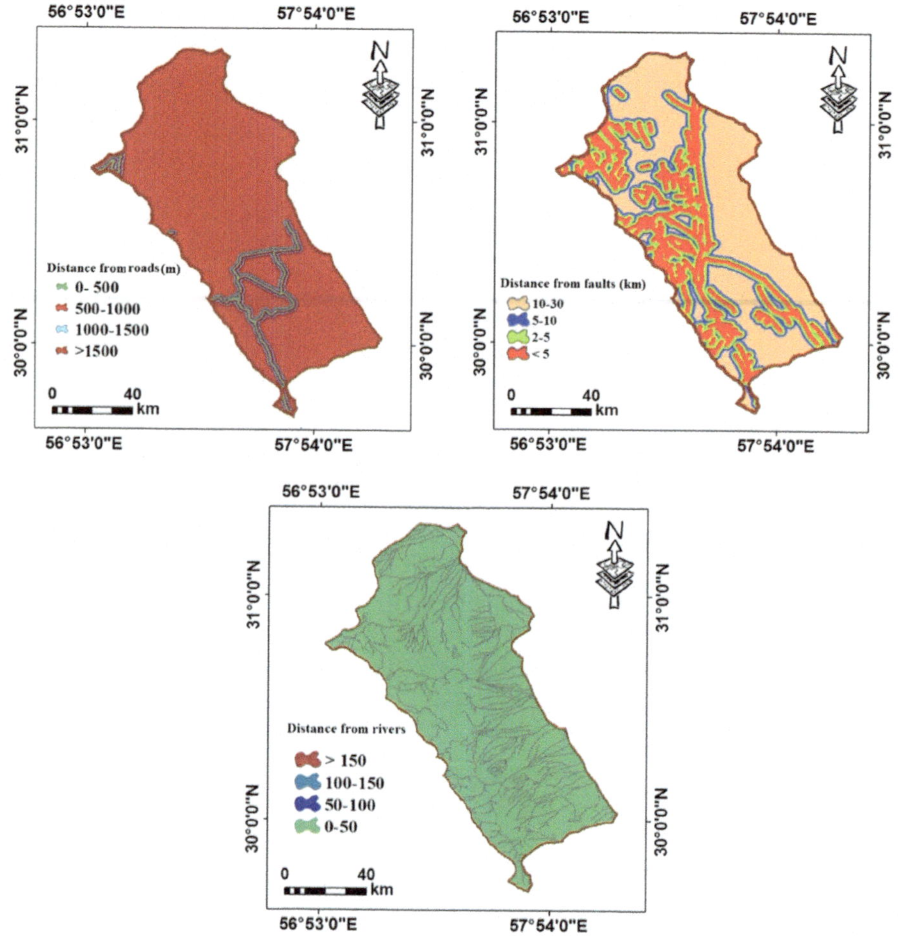

Fig. 3 Distance from Faults, rivers, and road maps in the study area

$$\mu_A(x) = f(x) = \left\{ \begin{array}{cc} 0 & x \leq m \\ x - m/n - m & m \prec x \prec m \\ 1 & x \geq n \end{array} \right\} \quad (1)$$

where x is the input data and m, n are the limit values.

For distance of river, distance of fault, and distance of road the following MF was used (Yagoub 2015):

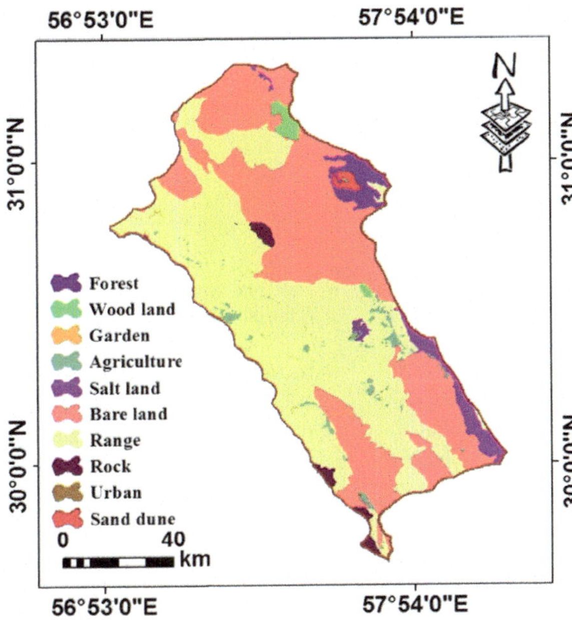

Fig. 4 Land use map for the study area

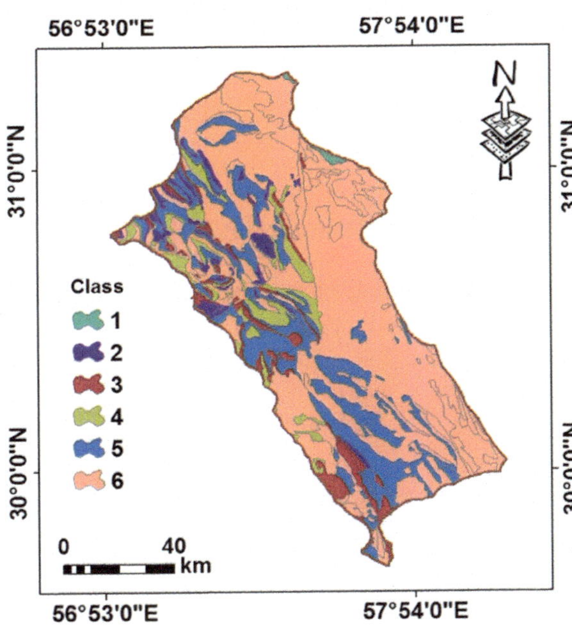

Fig. 5 Geological formations map of the study area

Table 3 Description of geological formations sensibility classes of lithology to erosion

Classes	Sensitivity classes	Description	Age
1	Very low sensitive	Limestone rock	Quaternary
2	Very low sensitive	Bedded to massive fossiliferous limestone	Cretaceous, Cenozoic, Early-Middle. Triassic
3	Low sensitive	Hale and chert, bedded to massive orbitolina limestone	Cretaceous, Early-Middle. Triassic, Cambrian
4	Low sensitive	Bedded argillaceous limestone and calcareous shale, bedded sandstone	Jurassic, Pleeocene
5	Very sensitive	Piedmont conglomerate and sandstone, shelly limestone	Carboniferous, Devonian, Pliocene
6	Very sensitive	Bedded argillaceous—limestone, low level piedmont fan and valley terrace deposits	Early-Middle. Jurassic, Quaternary

$$\mu_A(x) = f(x) = \left\{ \begin{array}{cc} 1 & x \leq m \\ n - x/n - m & m \prec x \prec n \\ 0 & x \geq n \end{array} \right\} \quad (2)$$

where, x is the input data and m, n are the limit values.

2.2.2 Analytical Hierarchy Process (AHP)

AHP (Saaty 1980) as MCDA (Multi-criteria decision analysis) procedure is applied for the elicitation of criteria weights. Using pairwise comparison in AHP can apply the quantitative and qualitative data in various studies (Malczewski 1999). This method uses a matrix of pairwise comparison of each of the parameters that according to the Table 4, the parameters are valued between 1 and 9.

Table 4 Scales for pairwise comparisons (Saaty and Vargas 1991)

Intensity of importance	Definition
1	Equal importance
3	Moderate importance of one over another
5	Essential importance
7	Demonstrated importance
9	Absolute importance
2, 4, 6, and 8	Intermediate values between the two adjacent judgments

2.3 *Ordered Weighted Averaging (OWA)*

One of the methods of multi-criteria evaluation is OWA. based on criterion weights and criterion map layers, the OWA combination operator associates with the i-th location, a set of order weights w = $w_1, w_2, \ldots, w_n$ such that $w_j \in [0, 1]$, j = 1,2,…,n, $\sum_{j=1}^{N} w_j = 1$, so on are defined as follows (Yager 1988; Malczewski et al. 2003):

$$OWA_t = \sum_{j=1}^{N} \left(\frac{u_j w_j}{\sum_{j=1}^{n} u_j w_j}\right) m_{tf} \tag{3}$$

where $m_{i1} \geq m_{i2} \geq \ldots \geq m_{in}$ is the sequence obtained by reordering the attribute values $a_{i1}, a_{i2}, \ldots, a_{in}$, and u_j are the criteria weights reordered according to the attribute value m_{ij}. OWA operator from smallest (OR) to largest (AND) showed in Table 5.

In general the different process of OWA method is as following (Fig. 6).

3 Results and Discussion

3.1 *Fuzzy Method*

As regards changes of each parameters effective on earthquake is linearly, trapezoidal membership function (MF) in order to determine fuzzy map for each parameter was used in ArcGIS software (Hadji et al. 2016; Mahalingam and Olsen 2016). The maximum and minimum values for membership functions are determined in Table 6. For example, the MF value for DEM >3000 m is 1, whereas the

Table 5 The OWA weight of each criteria (Malczewski 2006)

α	Quantifier	Order Weights (w_{ik})	GIS combination procedure	ORness	Trade-off
$\alpha \rightarrow = 0$	At least one	$w_{i1} = 1$; $w_{ik} = 0$, $(1 < k \leq n)$	OWA (OR)	1.0	0
$\alpha = 0.1$	At least a few	a	OWA	a	a
$\alpha = 0.5$	A few	a	OWA	a	a
$\alpha = 1$	Half (identity)	$w_{ik} = 1/n$, $1 \leq k \leq n$	OWA (WLC)	0.5	1
$\alpha = 2$	Most	a	OWA	a	a
$\alpha \rightarrow \infty$	All	$w_{in} = 1$; $w_{ik} = 0$, $(1 \leq k < n)$	OWA (AND)	0	0.0

[a]The set of order weights depends on values of sorted criterion weights and parameter

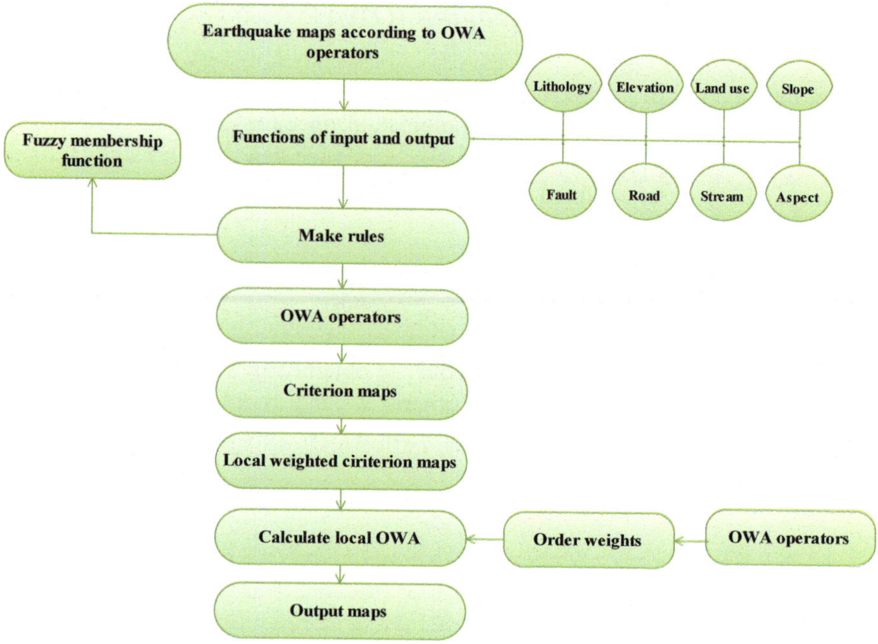

Fig. 6 The stage of determine the earthquake events map using OWA method in the study area

Table 6 Maximum and minimum values of criteria (Yagoub 2015)

Parameters	Minimum	Maximum
Land use	Forest, wood land, garden, agriculture, and range	Rock bodies, bare land, salt land, sand dune, and urban
Distance from roads (m)	>1500	<500
Distance from faults (km)	0–1000	>4000
Distance from rivers (m)	>10	<2
Geological formations sensitivity	5–6	<2
Aspect	Flat	South
Elevation (m)	>3000	<1200
Slope (°)	>60	<30

value of smaller than 1200 m has a MF = 0; in contrast, MF is between 0 and 1 for the value of 1200–3000 m. This law was applied for all of the factors in order to determine membership functions (MFs) (Table 6).

The fuzzy maps prepared for the earthquake events parameters are shown in Fig. 7, where MF is closer to 0 with decreasing earthquake, while MF is closer to 1 with increasing earthquake.

According to Fig. 7 geology and land use map had MF of closer to 1 in east and southeast of the study area. While for other parameters MF was closer to 0 in east and southeast of the study area. The MF of closer to 1 indicated the area with the high earthquake for the study area (Fig. 8).

3.2 Computation of Criterion Weight

In the current research, the AHP method was used as the pairwise comparison method for comparing two criteria (Yagoub 2015). According to Fig. 7, distance from faults and distance from rivers had the highest and lowest weight, respectively.

Finally to overly the input data and prepare the earthquake event, the OWA method was used. In the present study, eight order weights were used for each parameter. According to Table 8, gives six typical sets of order weights for eight factors:

According to the standardized criterion maps and the corresponding criterion weights, authors of the present study applied the OWA operator using Eq. (3) for the selected values of fuzzy quantifiers: at least one, at least a few, a few, identity, most, almost all, and all. Each quantifier is associated with a set of order weights that are calculated according to Eq. (3). The $\left(\sum_{k=1}^{j} u_k\right)^{\propto}$ and $\left(\sum_{k=1}^{j} u_k\right)^{\propto} - \left(\sum_{k=1}^{j-1} u_k\right)^{\propto}$ values for each Quantifier show in Table 8 for the i-th location and eight criterion values and for six linguistic quantifiers: from at least one ($\alpha = 0$) to all ($\alpha = \infty$) (Table 7).

Finally, OWA was used to overlay each parameter and prepare the earthquake events map. As it mentioned, six ordered weights were used to the eight parameters that were rank-ordered for each parameter. Figure 9 shows the six alternative earthquake events patterns. According to Fig. 9, with an average risk (Fig. 9d) all effective parameters of the earthquake events received some weights (0.11). According to Fig. 9, (d) some parts of the study area had high values (drake blue color), a medium value (yellow color) and a low value.

According to Fig. 9a, with decreasing the risk (no trade-off), the area with a high earthquake events was determined. Thus, almost all of the study area had low value that show in the area had low events of the earthquake.

Also, with increasing the risk (no trade-off) (Fig. 9f), all the study area had the high value that show that in the area had high earthquake event. According to Fig. 9f, almost all parts of the study area had a high earthquake value.

Figure 9b, showed a low risk with an average trade-off that show the east and southeast of the study area had more value than the other parts.

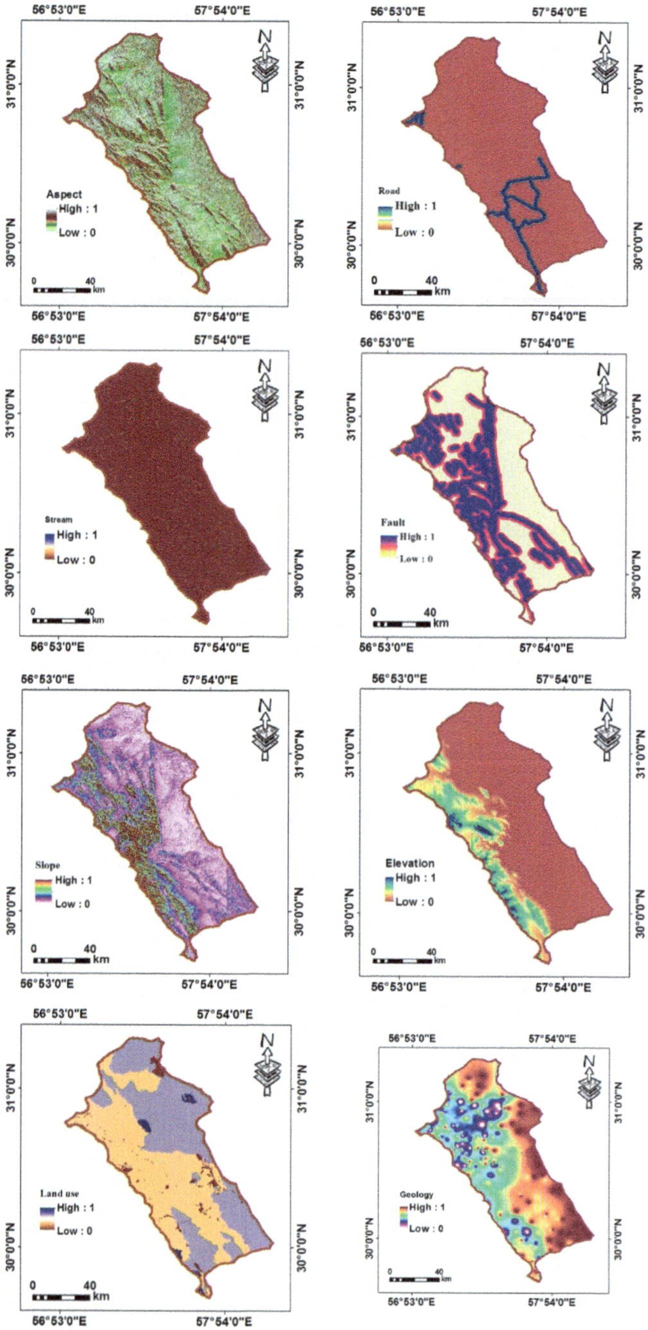

Fig. 7 Fuzzy map of studied area for each earthquake events factors

Fig. 8 Factor weights using pairwise comparison matrix for the input data

Figure 9c, showed a high risk with an average trade-off that in comparison with Fig. 9b, had lower risk for determination of earthquake events. Figure 9e, showed an average risk with no trade-off that had more risk.

According to Table 8, the OWA maps were classified into four classes as shown in Fig. 10 is true. It was determined that high risk areas (Table 8 of class 4 (0.75–1)) were more than the low risk areas (Table 8 of class 1 (0–0.25)). According to Fig. 10 and Table 8, low level of risk and no trade-off (2) had the highest area in the very low class (between 0 and 0.25) while high level of risk and no trade-off (3) had the highest area in the very high class (between 0.75 and 1).

With decreasing risk (no trade-off), almost all of parts of the study area were found to has not earthquake event. On the other hand, with increasing risk (no trade-off), all of the study had earthquake event.

This paper applied OWA operators to prepare different earthquake events maps with different risk levels. From these maps, the appropriate map can be chosen by the used different financial situations and appropriate risk levels for decisions. For example, with low risk, only the some parts of the study has high earthquake events, and for high risk conversely. There is not any study on mapping earthquake events through the use of OWA while there are some studies using OWA method for preparing other hazard map such as landslide susceptibility (Gorsevski et al. 2006; Ayalew et al. 2004, Ayalew and Yamagishi 2005; Guzzetti et al. 2005; Komac 2006b; Feizizadeh and Blaschke 2013). In fact using OWA method can be select any degree of trade-off among criteria, ranging from no trade-off to full trade-off, depending on the decision-making strategy.

4 Conclusions

This research evaluated the spatial distribution of earthquake events with different risk levels evaluation using OWA with fuzzy quantifier approach. The results showed that with decreasing risk (no trade-off), almost all of the study area had not earthquake. In addition, with increasing risk (no trade-off), all of the study area had good earthquake. Based on the importance of public health and level of finance for different regions, the findings of this study can be used to determine the appropriate risk levels for earthquake event. Based on the different conditions of the study area, such as the financial condition of the people and government, age distribution of the population, etc., the earthquake events map with the appropriate risk level can be used.

Table 7 Computing $\left(\sum_{k=1}^{j} u_k\right)^{\propto}$ and $\left(\sum_{k=1}^{j} u_k\right)^{\propto} - \left(\sum_{k=1}^{j-1} u_k\right)^{\propto}$ for the i-th location and eight criterion values for the linguistic quantifiers

j	Quantifier	Criterion weights u^k	$\left(\sum_{k=1}^{j} u_k\right)^{\propto}$	$\left(\sum_{k=1}^{j} u_k\right)^{\propto} - \left(\sum_{k=1}^{j-1} u_k\right)^{\propto}$
Distance from faults	(a) At least one (α → = 0)	0.341	1	1
Lithology		0.154	1	0
Slope		0.119	1	0
Aspect		0.093	1	0
DEM		0.087	1	0
Land use		0.073	1	0
Distance from roads		0.073	1	0
Distance from rivers		0.059	1 1	0 0
Distance from faults	(b) At least a few (α → = 0.1)	0.341	0.898	0.898
Lithology		0.154	0.932	0.0348
Slope		0.119	0.952	0.020
Aspect		0.093	0.966	0.014
DEM		0.087	0.977	0.011
Land use		0.073	0.986	0.009
Distance from roads		0.073	0.994	0.009
Distance from rivers		0.059	1	0
Distance from faults	(c) A few (α → = 0.5)	0.341	0.584	0.584
Lithology		0.154	0.704	0.120
Slope Aspect		0.119 0.093	0.784 0.841	0.080 0.057
DEM		0.087	0.891	0.050
Land use		0.073	0.931	0.040
Distance from roads		0.073	0.970	0.038
Distance from rivers		0.059	1	0.030
Distance from faults	(d) Half (identity) (α → = 1)	0.341	0.125	0.125
Lithology		0.154	0.25	0.125
Slope		0.119	0.375	0.125
Aspect		0.093	0.5	0.125
DEM		0.087	0.625	0.125
Land use		0.073	0.75	0.125

(continued)

Table 7 (continued)

j	Quantifier	Criterion weights u^k	$\left(\sum_{k=1}^{j} u_k\right)^{\propto}$	$\left(\sum_{k=1}^{j} u_k\right)^{\propto} - \left(\sum_{k=1}^{j-1} u_k\right)^{\propto}$
Distance from roads		0.073	0.875	0.125
Distance from rivers		0.059	1	0.125
Distance from faults	(e) Most (α → = 2)	0.341	0.116	0.116
Lithology		0.154	0.245	0.129
Slope		0.119	0.377	0.132
Aspect		0.093	0.500	0.123
DEM		0.087	0.630	0.131
Land use		0.073	0.752	0.121
Distance from roads		0.073	0.884	0.132
Distance from rivers		0.059	1	0.116
Distance from faults	(f) All (α → ∞)	0.341	0	0
Lithology		0.154	0	0
Slope		0.119	0	0
Aspect		0.093	0	0
DEM		0.087	0	0
Land use		0.073	0	0
Distance from roads		0.073	0	0
Distance from rivers		0.059	1	1

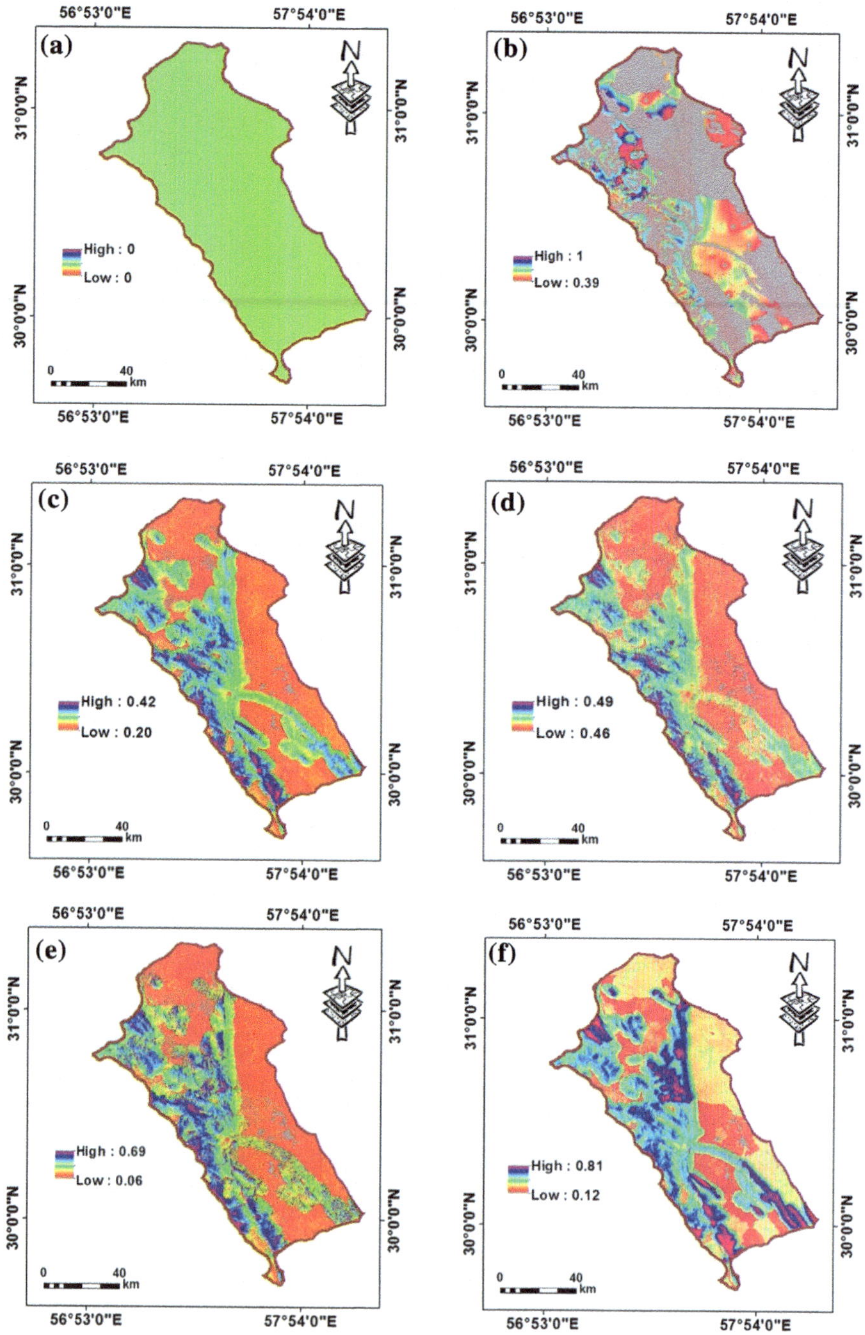

Fig. 9 Earthquake events maps of OWA results for selected fuzzy linguistic quantifiers (description of a to f is in the Table 7)

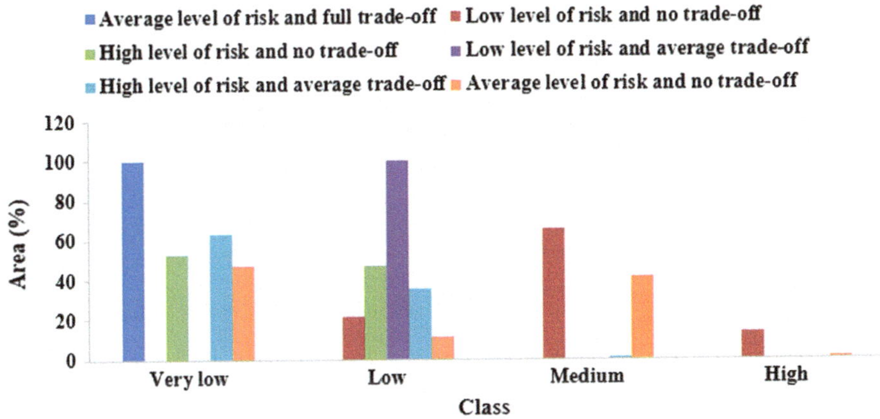

Fig. 10 Area of each class using the OWA method

Table 8 Description of each class

Range	Description
0–0.25	Very low
0.25–0.5	Low
0.5–0.75	Medium
0.75–1	High

Acknowledgements The authors would like to thanks to all personnel of Agricultural Jihad of Fars province for their kind help.

References

Ayalew L, Yamagishi H, Ugawa N (2004) Landslide susceptibility mapping using GIS-based weighted linear combination, the case in Tsugawa area of Agano River, Niigata Prefecture, Japan. Landslides 1:73–81

Ayalew L, Yamagishi H (2005) The application of GIS-based logistic regression for landslide susceptibility mapping in the Kakuda-Yahiko Mountains, Central Japan. Geomorphology 65 (1–2):15–31. https://doi.org/10.1016/J.GEOMORPH.2004.06.010

Balaji D, Sankar R, Karthi S (2002) GIS Approach for disaster management through awareness-an overview. Paper presented at the proceedings of the 5th annual international conference-map India, New Delhi, 6–8 Feb 2002

Champatiray PK, Perumal RJ, Thakur VC, Bhat MI, Malik MA, Singh VK et al (2005) A quick appraisal of ground deformation in Indian region due to the October 8, 2005 earthquake, Muzaffarabad, Pakistan. J Indian Soc Remote Sens 33(4):465–473

Feizizadeh B, Blaschke T (2013) GIS-multicriteria decision analysis for landslide susceptibility mapping: comparing three methods for the Urmia lake basin, Iran. Nat Hazards 65(3):2105–2128

Gorsevski PV, Jankowski P (2008) Discerning earthquake events using rough sets. Comput Environ Urban Syst 32:53–65

Gorsevski PV, Jankowski P (2010) An optimized solution of multi-criteria evaluation analysis of earthquake events using fuzzy sets and Kalman filter. Comput Geosci 36:1005–1020
Gorsevski PV, Jankowski P, Gessler PE (2006) An heuristic approach for mapping landslide hazard by integrating fuzzy logic with analytic hierarchy process. Control Cybern 35:21–141
Guzzetti F, Reichenbach P, Cardinali M, Galli M, Ardizzone F (2005) Probabilistic landslide hazard assessment at the basin scale. Geomorphology 72:272–299
Hadji R, Chouabi A, Gadri L, Raïs K, Hamed Y, Boumazbeur A (2016) Application of linear indexing model and GIS techniques for the slope movement susceptibility modeling in Bousselam upstream basin. Northeast Algeria Arab J Geosci 9(3):1–18
Henning BD (2011) Gridded cartograms as a method for visualising earthquake risk at the global scale. J Maps. https://doi.org/10.1080/17445647.2013.806229
Komac M (2006a) A earthquake events model using the analytical hierarchy process method and multivariate statistics in perialpine Slovenia. Geomorphology 74(1–4):17–28
Komac M (2006b) A landslide susceptibility model using the analytical hierarchy process method and multivariate statistics in perialpine Slovenia. Geomorphology 74(1–4):17–28
Laefer DF, Alison K, Pradhan A (2006) The need for baseline data characteristics for GIS-based disaster management systems. J Urban Plan Dev 132(3):115–119
Lillesand TM, Kiefer RW, Jonthan WC (2008) Remote sensing and image interpretation, 6th edn. Wiley, New York
Mahalingam R, Olsen MJ (2016) Evaluation of the influence of source and spatial resolution of DEMs on derivative products used in landslide mapping. Geomat Nat Hazards Risk 7(6):1835–1855
Malczewski J (1999) GIS and multicriteria decision analysis. Wiley, New York
Malczewski J (2006) Ordered weighted averaging with fuzzy quantifiers: GIS-based multicriteria evaluation for land-use suitability analysis. Int J Appl Earth Obs Geoinf 8:270–277
Malczewski J, Chapman T, Flegel C, Walters D, Shrubsole D, Healy MA (2003) GIS-multicriteria evaluation with ordered weighted averaging (OWA): case study of developing watershed management strategies. Environ Plann A 35(10):1769–1784
Mertens KC, Verbeke LPC, Westra T, De Wulf RR (2004) Sub-pixel mapping and sub-pixel sharpening using neural network predicted wavelet coefficients. Remote Sens Environ 91 (2):225–236.https://doi.org/10.1016/J.RSE.2004.03.003
Miles SB, Ho CL (1999) Applications and issues of GIS as tool for civil engineering modeling. J Comput Civil Eng ASCE 13(3):144–161
Mokarram M, Aminzadeh F (2010) GIS-based multicriteria land suitability evaluation using ordered weight averaging with fuzzy quantifier: a case study in Shavur Plain, Iran. The Int Arch Photogram Remote Sens Spat Inf Sci 38(2):508–512
Mokarrama M, Hojati M (2016) Landform classification using a sub-pixel spatial attraction model to increase spatial resolution of digital elevation model (DEM). The Egypt J Remote Sens Space Sci
Roustaei M, Nazi H, Amirmotallebi N (2005) The seismotectonic and zonation map of Salmas Vastness by GIS modeling according to Landsat satellite images (ETM+) and aeromagnetic data. Paper presented at the Map Middle East, 23–25 Apr 2005, Dubai, UAE
Roy PS, WestenCJ VVK, Lackhera RC, Chapari ray PK (2000) Natural disasters and their mitigation-Remote Sensing and Geographical Information System Perspectives. Indian Institute of Remote Sensing Publication, Dehradun
Saaty TL (1980) The Analytic Hierarchy Process: Planning, Priority Setting, Resource Allocation. McGraw-Hill International Book Co.
Saaty TL, Vargas LG (1991) Prediction, Projection and Forecasting. Dordrecht: Springer Netherlands. https://doi.org/10.1007/978-94-015-7952-0
Theilen-Willige B, Savvaidis P, Tziavos IN, Papadopoulou I (2012) Remote sensing and geographic information systems (GIS) contribution to the inventory of infrastructure susceptible to earthquake and flooding hazards in North-Eastern Greece. Geosciences 2(4):203–220

Van Westen CJ, Soeters R, Sijmons K (2000) Digital geomorphological earthquake events hazard mapping of the Alpago area, Italy. Int J Appl Earth Obs Geoinf 2:51–60
Yager RR (1988) On ordered weighted averaging aggregation operators in multi-criteria decision making. IEEE Trans Syst Man Cybern 18(1):183–190
Yagoub MM (2015) Spatio-temporal and hazard mapping of earthquake in UAE (1984–2012): remote sensing and GIS application. Geoenviron Disasters 2(1):1
Zadeh LA (1965) Fuzzy sets. Inf Control 8(3):338–353. https://doi.org/10.1016/S0019-9958(65)90241-Xdoi:10.1016/S0019-9958(65)90241-X

Prediction of Rainfall as One of the Main Variables in Several Natural Disasters

Vahid Moosavi

Abstract Rainfall is one of the main variables in several natural disasters such as, floods, drought, groundwater depletion and landslides. Therefore, development of robust models for rainfall forecasting is essential in environmental studies. The chief goal of this research is to use Group Method of Data Handling (GMDH) besides signal processing approaches to forecast rainfall in monthly time steps. To that end, three different signal processing approaches i.e. Ensemble empirical mode decomposition (EEMD), wavelet transform (WT) and wavelet packet transform (WPT) were used combined with GMDH model. Four stations were used to apply aforementioned modeling techniques. Results of this research showed that all three abovementioned signal processing approaches can enhance the ability of the GMDH model. The ability of EEMD-GMDH and wavelet packet-GMDH were relatively close to each other. However, wavelet packet-GMDH outperformed EEMD-GMDH model to some extent. The other important note was the effect of exogenous data on the ability of all models. It was shown that forecasting rainfall without using exogenous data does not produce acceptable results.

Keywords Polynomial neural network · GMDH · Rainfall forecasting
Wavelet · Wavelet packet · Ensemble empirical mode decomposition (EEMD)

1 Introduction

Rainfall is one of the main variables contributing in several natural disasters. The shortage of rainfall can cause drought while its excessive amounts can lead to floods, landslides etc. Rainfall links the atmosphere and land processes. Great amounts of rainwater can cause inundation problems, floods, and consequently significant damage over large areas. River floods happen when water levels rise

V. Moosavi (✉)
Department of Watershed Management Engineering, Faculty of Natural Resources, Yazd University, Yazd, Iran
e-mail: moosavi_v66@yahoo.com

H. R. Pourghasemi and M. Rossi (eds.), *Natural Hazards GIS-based Spatial Modeling Using Data Mining Techniques*, Advances in Natural and Technological Hazards Research 48, https://doi.org/10.1007/978-3-319-73383-8_8

over the highest part of the river bank because of heavy rains and thunderstorms. Coastal floods which can be defined as the inundation of lands along the coasts is mainly generated by heavy rainfalls besides onshore winds. Flooding happens once exhaustive precipitation falls during a short period of time or moderate precipitation gathers over several days.

Rainfall can also affect the ecological processes. It can affect the plant, animal and human populations. Shortage of rainfall causes tree mortality mainly in forest boundaries, decreases tree growth and increases leaf shedding. This procedure cause an increase in the canopy openness and consequently the insolation of understory vegetation which itself leads to drying of the accumulated litter. It significantly increases the risk of deforestation and forest fires. Plant die off caused by drought is a significant effect of rainfall scarcity. Lack of water is a main environmental factor which limits the plant productivity. Deteriorating the crop yield caused by climate, undoubtedly is more than losses from all other reasons, because both the severity and duration of the water stress induced by climate conditions are dangerous.

Furthermore, rainfall is one of the most important triggering factors which can induce Landslides. Landslides are of the main natural disasters in many regions. They can impose serious threats to life and property in target areas. Rainfall is the main trigger of landslides particularly in heavy or prolonged rainfalls. Mainly, this effect is due to increasing the pore water pressures in the soil. While the soil fills with water, the resistance to movement is reduced significantly because of the buoyancy. Moreover, fluid can exert a pressure downward and deliver a hydraulic push to the landslide. It can also decrease the slope stability.

The rainfall is also a chief component of the water cycle and is the major source of groundwater recharge. Groundwater is an invaluable resource for agricultural, industrial, and civic purposes. Results of unsustainable groundwater management and consumption have been a severe problem in global scale, particularly in developing countries (Zhan 2005; Mackay et al. 2014). Rainfall on the plain can be the source of direct recharge. The accumulated water in rivers, behind dams etc., which is the result of rainfall is of the main sources of groundwater recharge. Rainwater scarcity can affect all these sources and extremely affect the groundwater resources as the most important water resource in the word.

As mentioned, rainfall as the first component in the hydrologic cycle can affect several different processes. Therefore, forecasting rainfall in different time steps can help mangers to deal with these problems. Physics based numerical models are of important models used in rainfall forecasting. The mentioned models inaugurate a main equation which simplifies the physical process of precipitation and solve it using appropriate initial conditions and boundary states with numerical approaches. These models are very robust and useful, however, a great deal of data are mandatory for modeling, calibration and simulation processes (Yoon et al. 2011). Obtaining sufficient data for model development are usually expensive, time consuming and labor intensive. Therefore, when adequate data is not in hand and when getting precise estimates is more essential than regarding the physical properties of the phenomena, empirical models can be a good alternative to make suitable results with an easier calibration process (Daliakopoulos et al. 2005).

Artificial intelligence approaches, such as Artificial Neural Network (ANN), Adaptive Neuro-Fuzzy Inference System (ANFIS), Support vector regression (SVR) and Linear Genetic Programming (LGP), have been applied in different hydrological studies (Sànchez-Marrè et al. 2004; Daliakopoulos et al. 2005; Wieland et al. 2010; Talei et al. 2010; Alves et al. 2011; Alvisi and Franchini 2011; Yoon et al. 2011; Young et al. 2011; Millie et al. 2012; Fallah-Mehdipour et al. 2013; He et al. 2014; Moosavi et al. 2014; Li et al. 2015; Moosavi et al. 2015; Si et al. 2015; Yoon et al. 2016). One of the most important types of artificial intelligence models is Group Method of Data Handling (GMDH). GMDH proposed by Ivakhnenko (1966) is a method which works by sorting the progressively intricate models and evaluating them based on predefined standards (Ravisankar and Ravi 2010). The chief improvement of this model is to make logical functions using feed forward network according to quadratic polynomial. In this method regression technique is used to calculate weights (Kalantary et al. 2009). Generally, GMDH networks outperform common regression techniques and other artificial intelligence methods (Najafzadeh 2015). Several researchers have used GMDH in different fields e.g. energy, manufacturing, system identification, advertising, economic, geology and hydrology (Witczak et al. 2006; Amanifard et al. 2008; Mehrara et al. 2009; Kalantary et al. 2009; Najafzadeh et al. 2013; Najafzadeh and Lim 2015). Artificial intelligence techniques such as GMDH have a notable flexibility and ability in modeling hydrologic processes. However, these techniques are not capable to cope with the problem of non-stationarity in data in their single form (Cannas et al. 2006; Moosavi et al. 2013). As existing non-stationary data handling methods are not acceptably advanced (Adamowski and Chan 2011), further investigations to solve this problem is inevitable. Specific signal processing approaches for instance wavelet transform and Ensemble Empirical Mode Decomposition (EEMD) may be used to cope with the non-stationarity of natural data. Wavelets can be defined as mathematical functions which provide a time-scale illustration of the time series and their relationships. These functions can be used to assess non-stationary hydrological time series. These approaches can make suitable decompositions of the original time series to sub-series in order to improve the performance of AI models by taking suitable information on different resolutions (Adamowski 2007; Kisi 2009; Nourani et al. 2009; Adamowski and Sun 2010; Nourani et al. 2011; Quiroz et al. 2011; Kim et al. 2014; Moosavi et al. 2013, 2015). As decomposition in ordinary wavelet transform is only performed on the approximation component, the results of decomposition in higher levels are not preferred. This decomposition process cannot be used to obtain required information. Wavelet packet transform can deal with this problem. In fact, the necessary frequency resolution can be attained applying wavelet packet transform.

The other useful signal processing method is Empirical Mode Decomposition (EMD). EMD is an automatic decomposition approach that produce an effective analysis technique for signals which are non-stationary and non-linear (Yu et al. 2015). Despite various benefits, conventional EMD has restrictions such as end-point effect and mode mixing problem. End-point effect can be removed simply and efficiently using end-points continuation (Zhao and Huang 2001; Han et al. 2014). The mode mixing which is typically affected by signal intermittency disparate frequencies exist in a single IMF. Ensemble empirical mode decomposition (EEMD) is the advanced version of EMD method. The key benefit of this method is the decomposition of a signal into a set of wholly adaptive basis functions named Intrinsic Mode Functions (IMF) (Liu et al. 2012). EEMD can overcome the mode mixing problem. Altogether, EEMD is a more effective method than EMD with high reliability and can effectively reduce the noise in the signal (Feng et al. 2012). Wavelet transform and EEMD methods have been widely used in different fields (Adamowski 2007; Kisi 2009; Nourani et al. 2009; Adamowski and Sun 2010; Breaker and Ruzmaikin 2011; Nourani et al. 2011; Guo and Tse 2013; Moosavi et al. 2013, 2014, 2015; Mariyappa et al. 2014; Moosavi and Niazi 2015). Nonetheless, to the best of our knowledge, there is no published research about coupling wavelet, wavelet packet analysis and ensemble empirical mode decomposition (EEMD) with GMDH model to forecast rainfall in different time steps. The main goal of this study is to assess the ability of GMDH for rainfall prediction and the effect of wavelet and wavelet packet transforms as well as EEMD as signal processing approaches on its performance.

2 Materials and Methods

2.1 Study Area and Data

This study was performed on four different stations in four different provinces of Iran i.e. Khorasan Razavi (Mashhad), Alborz (Karaj), East Azerbaijan (Maraghe) and Khozestan (Ramhormoz). Figure 1 shows the four studied stations and their location in Iran. The climate in these regions are different. Average long term (30 years) temperature of Maraghe, Karaj, Mashhad and Khozestan are 12, 16, 15 and 25, respectively. Average long term (30 years) precipitation of Maraghe, Karaj, Mashhad and Khozestan are 300, 255, 245 and 320, respectively.

Two different data sets were used in all modeling approaches. The first dataset includes rainfall data without exogenous data to predict rainfall in different time steps. The second dataset includes exogenous data in addition to rainfall data. The exogenous data were evaporation (E), maximum and minimum temperature (T) and humidity (H). All data are monthly. The lengths of the used data are different in different stations because of the difference in the availability of data. The length of dataset is more than 15 years. The data were divided into two categories i.e. training set (70%) and testing set (30%). In this study, three signal processing approaches i.e. EEMD, wavelet and wavelet packet transforms were used combined with GMDH modeling approach to forecast one, two, three and four-month ahead rainfalls. Figure 2 shows the flowchart of the study. Below are some explanations about the proposed methods.

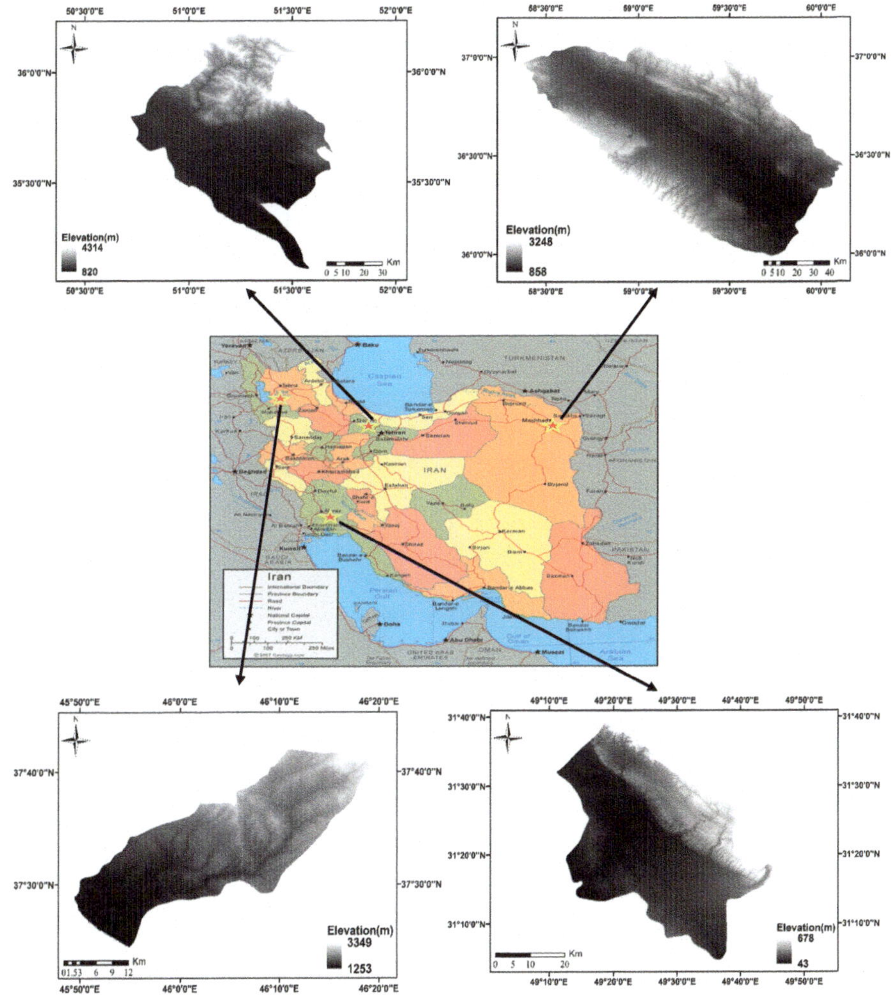

Fig. 1 Iran map and study areas

2.2 *Group Method of Data Handling (GMDH)*

The Group Method of Data Handling (GMDH) algorithm as an inductive self-organizing black-box modeling approach was proposed by Ivakhnenko (1971) based on Darwin's theory of natural selection. This method is similar to back propagation artificial neural network. However, it has several differences in the internal structure of processing element. In the common multilayer perceptron (MLP) neural network, neurons are linked to all nodes of the preceding layer of the network. Nevertheless, each node contains a portion of information obtained from

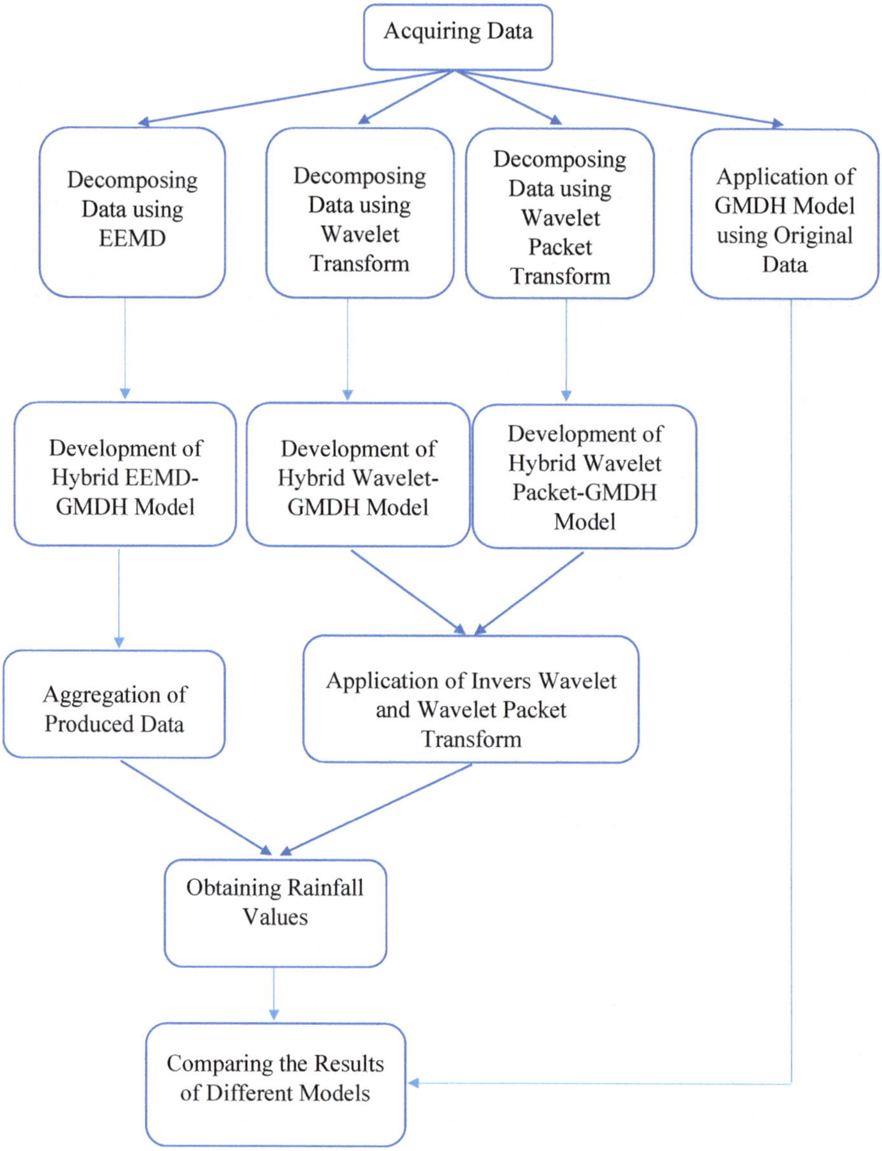

Fig. 2 Flow chart of the study

two nodes of the preceding layer. This process can explain a complicated system without tracking the whole route of input-output. Commonly, the linking between inputs and output is performed in a nonlinear form. This process can be demonstrated using the Volterra series (Volterra 1959):

$$y(t) = \int_0^t h_1(\tau)x(t-\tau)d\tau + \iint_0^t h_2(\tau_1\tau_2)x(t-\tau_1)x(t-\tau_2)d\tau_1 d\tau_2 + \iiint_0^t h_3(\tau_1\tau_2\tau_3)x(t-\tau_1)x(t-\tau_2)x(t-\tau_3)d\tau_1 d\tau_2 d\tau_3 + \cdots \tag{1}$$

The discrete form of this series can be described as Kolmogorov-Gabor polynomial (Ivakhnenko 1971; Mehra 1977):

$$y = a_0 + \sum_i^N a_i x_i + \sum_i^N \sum_i^N a_{ij} x_i x_j + \sum_i^N \sum_i^N \sum_i^N a_{ijk} x_i x_j x_k + \cdots \tag{2}$$

in which x is the input matrix, N shows the number of inputs, and a is the coefficients matrix (Farlow 1984).

A least-squares approach were used to calculate weights. Therefore, Eq. 3 were used as objective function in the optimization process (Najafzadeh et al. 2013):

$$E = \frac{\sum_{i=1}^{M} (y_i - G_i)^2}{M} \rightarrow \min \tag{3}$$

2.3 Hybrid Wavelet-GMDH Model

Wavelet transform (WT) as a good alternative for Fourier transform uses wavelet functions instead of sines and cosines for filtering and decomposing data (Aghajani et al. 2016). WT can be defined as a spectral analysis in time domain which decomposes data in a time-frequency space to produce a timescale explanation of procedures and their relations (Daubechies 1990). This method can reveal the information within the signal in both the time and scale (frequency) domains (Nourani et al. 2009). Therefore, it can deal with the basic disadvantage of Fourier analysis, which is that the Fourier spectrum provides a widespread explanation of the properties of the non-stationary processes providing a mapping procedure that is localized in frequency but global in time (Pal and Devara 2012). Another benefit of this method is the ability to choose the mother wavelet based on the features of the studied time series.

There are two kinds of wavelet transform namely continuous wavelet transform (CWT) and discrete wavelet transform (DWT). CWT of a signal can be defined as follows:

$$CWT_x^{\psi}(\tau, s) = \frac{1}{\sqrt{|s|}} \int_{-\infty}^{+\infty} x(t)\psi^*\left(\frac{t-\tau}{s}\right)dt \tag{4}$$

where s is the scale parameter, τ is the translation parameter and * symbolizes the complex conjugate (Cannas et al. 2006). The mother wavelet ψ is the transforming function. However, it needs a huge quantity of calculation time and resources. Discrete wavelet transform (DWT) needs less computational time and is simpler than CWT. DWT scales and positions are usually based on powers of two (dyadic scales and positions). DWT can estimate general behaviors (low frequency) and local behaviors (high frequency) of data using low pass and high pass filters. The use of these features usually leads to better performance of AI models (Christopoulou et al. 2002; Moosavi et al. 2013). DWT can be represented as:

$$\psi_{j,k}(t) = \frac{1}{\sqrt{|s_0^j|}} \psi\left(\frac{t - k\tau_0 s_0^j}{s_0^j}\right) \tag{5}$$

where j and k are integers that control the scale and translation respectively, s_0 is a fixed dilation step (Cannas et al. 2006) and τ_0 is a translation factor that depends on the abovementioned dilation step. The selection of the appropriate wavelet transform for an application is a very important step in this case. Therefore, mother wavelet and decomposition level should be optimized. In this way, different mother wavelets such as "Haar", "db", "rbio" and "bior" were used to decompose original datasets in 1, 2, 3, 4 and 5 levels. After decomposing the time series, the obtained sub-series were used as an input for the GMDH model. In this step, the wavelet based coefficients of rainfall is forecasted. Afterwards, inverse wavelet transform was performed on the resulted coefficients to obtain the actual rainfalls. Figure 3 shows the steps of hybrid wavelet-GMDH modeling process.

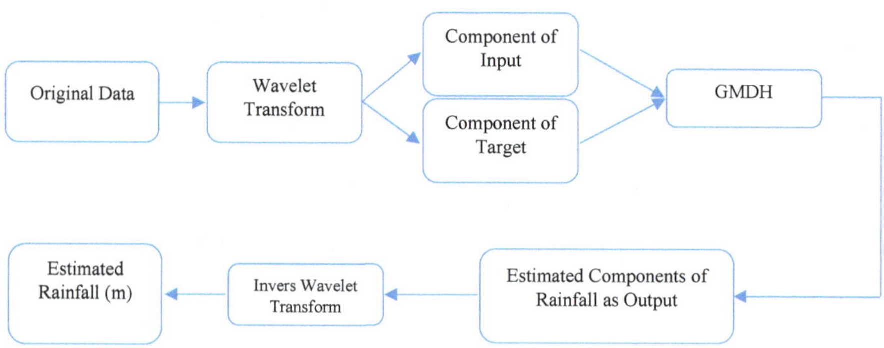

Fig. 3 Wavelet-GMDH modeling process

2.4 *Hybrid Wavelet Packet-GMDH Model*

The wavelet packet decomposition as a generalized form of the ordinary wavelet transform produces a richer signal analysis rather than conventional wavelet transform (Garcia et al. 2000). This method provided finer resolution decomposition. Figures 4 and 5 shows the wavelet and wavelet packet transformation processes. As shown in this figure, in ordinary wavelet transform, the original signal is divided into an approximation and a detail. Subsequently, the approximation is decomposed to approximation and detail components and the process is reiterated. Therefore, in an n-level decomposition process, there are n + 1 probable ways to decompose a specific signal. Nevertheless, in wavelet packet transform, the details can be split in addition to approximations. In fact, the original signal is divided into

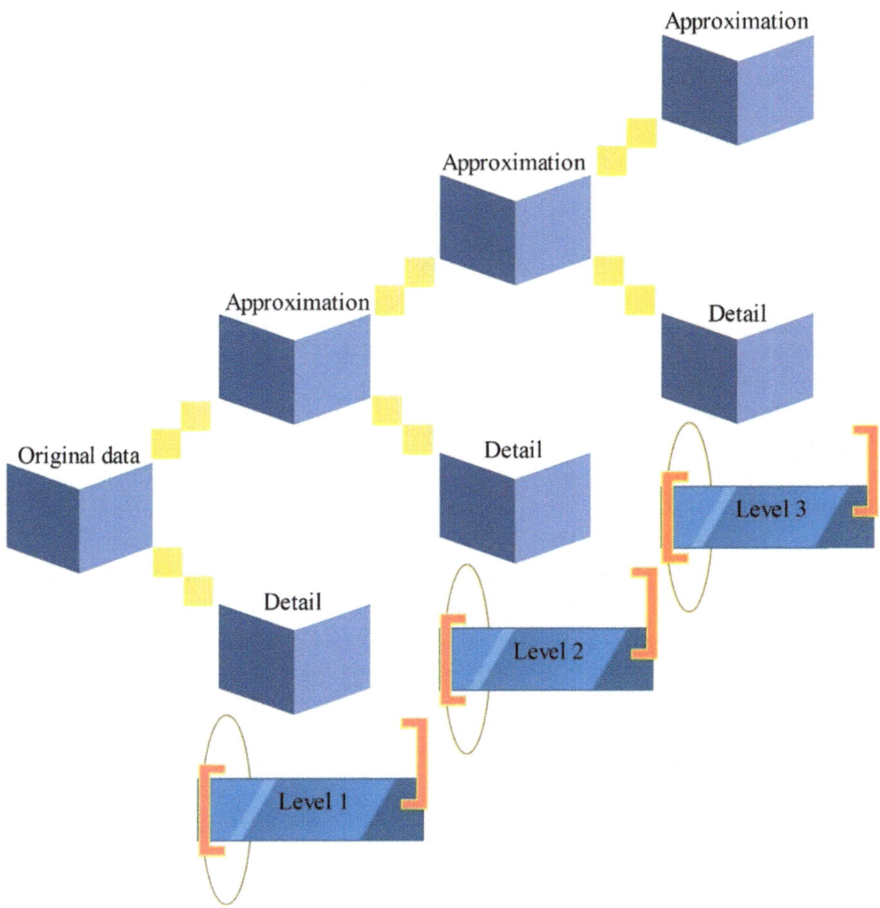

Fig. 4 Schematic demonstration of the wavelet based decomposition process in 3 levels

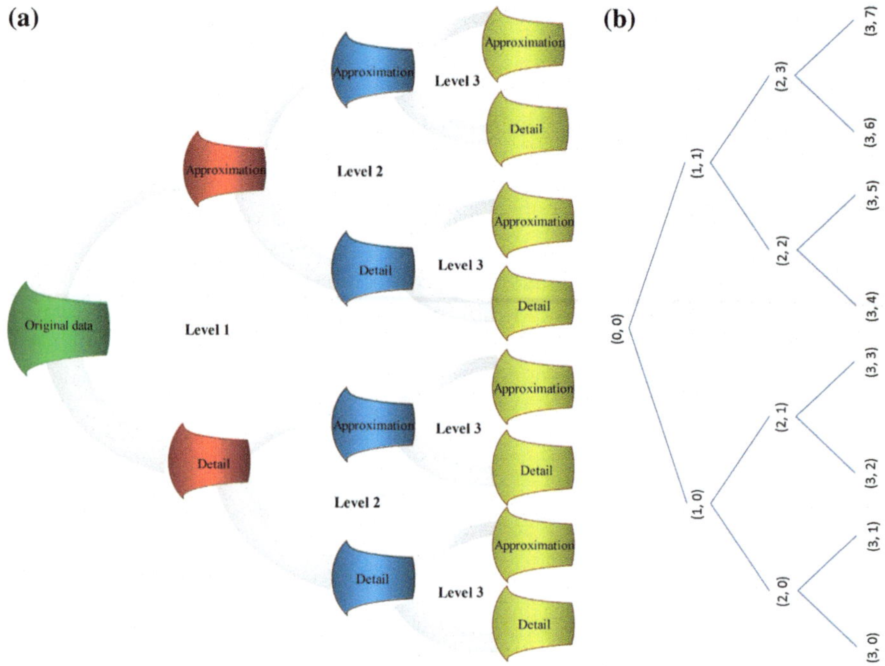

Fig. 5 Schematic demonstration of the wavelet packet based decomposition process in 3 levels

an approximation and a detail in the first step. Then, both approximation and detail components are decomposed to new approximation and details in the next step (Moosavi and Niazi 2015).

With the intention of developing combined wavelet packet-GMDH model, the original data were decomposed into approximation coefficients (low frequency) and detail coefficients (high frequency). This procedure was iteratively performed using different mother wavelets in different levels. It finally resulted in a wavelet decomposition tree as shown in Fig. 5. In this study, different mother wavelets such as "Haar", "db", "rbio" and "bior" were used to decompose original datasets in 1, 2, 3, 4 and 5 levels using Matlab software. Afterwards, the decomposed data were used in the GMDH model. In this step, the wavelet packet based coefficients of rainfall is predicted. Afterwards, the inverse process of wavelet packet based transformation was applied on the resulted coefficients to obtain the actual rainfalls. Figure 6 shows the steps of hybrid wavelet packet-GMDH modeling process.

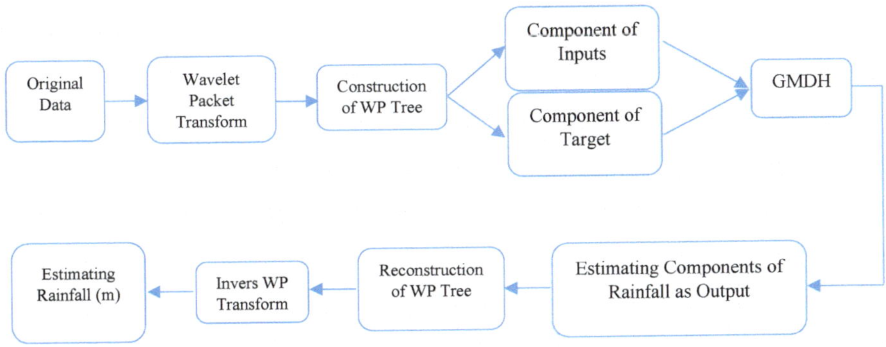

Fig. 6 Wavelet Packet-GMDH modeling process

2.5 *Hybrid EEMD-GMDH Models*

The EMD method like wavelet transform can decompose a complex dataset into a definite number of intrinsic mode functions (IMFs) which can fulfil two conditions. First, the number of extrema and the number of zero crossings cannot be different or the difference should not be more than one. Second, the average of the envelope defined by local maxima and that defined by the local minima should be zero for each data point (Manjulaa and Sarma 2012). As mentioned before, mode mixing is a major problem of the EMD. It can be defined as a particular IMF that encompasses signals with extremely different scales or a signal of the same scale appears in changed IMF components (Wu and Huang 2009). In order to deal with this problem, a novel approach suggested by Wu and Huang (2009) was used in this study. They proposed the ensemble empirical mode decomposition (EEMD) with the average of an ensemble of trails as the true IMF components. All trails contain the decomposition results of the signal in addition to a white noise of finite amplitude.

In order to apply this method on the original data sets, at first, a white noise series were added to the original data. In the next step, the new signals were decomposed to IMFs using EMD. These steps were reiterated by diverse white noises to get the equivalent IMF components. The number of iterations is named the ensemble number. The decomposed data using EEMD was imported to GMDH model to develop combined EEMD-GMDH model. In fact in this step, the IMFs and the residue of the rainfall signal is estimated. In the final step, the mentioned component were accumulated to calculate the final rainfall forecasting results. Figure 7 shows the steps of hybrid EEMD-GMDH modeling process.

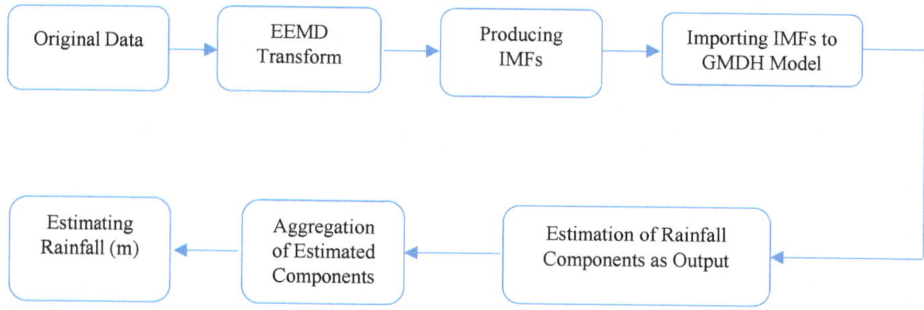

Fig. 7 EEMD-GMDH modeling process

2.6 *Evaluation Criteria*

Coefficient of determination (R^2), root mean squared error (RMSE), normalized RMSE (NRMSE) and the range of errors were calculated and used for comparing the performance and ability of the mentioned models (Sreekanth et al. 2009).

$$r^2 = \left(\frac{\sum_{i=1}^{n} (o_i - \bar{o})(e_i - \bar{e})}{\sqrt{\sum_{i=1}^{n} (o_i - \bar{o})^2} \sqrt{\sum_{i=1}^{n} (e_i - \bar{e})^2}} \right)^2 \tag{6}$$

$$RMSE = \sqrt{\frac{\sum_{i=1}^{n} (o_i - e_i)^2}{n}} \tag{7}$$

$$NRMSE = \frac{RMSE}{range\ of\ observed\ data} \tag{8}$$

where o, e and n, are observed rainfall values, predicted rainfall values and number of data, respectively.

3 Results and Discussion

As mentioned before, GMDH based modeling as an iterative process was performed in order to predict one, two, three and four-month ahead rainfall. In this approach, the polynomial coefficients were calculated by conventional least square method. Figures 8 and 9 demonstrate the results of GMDH modeling for the four studied stations for one month ahead rainfall forecasting for the two aforementioned data sets. These figures show the results for test datasets. Because the results of test

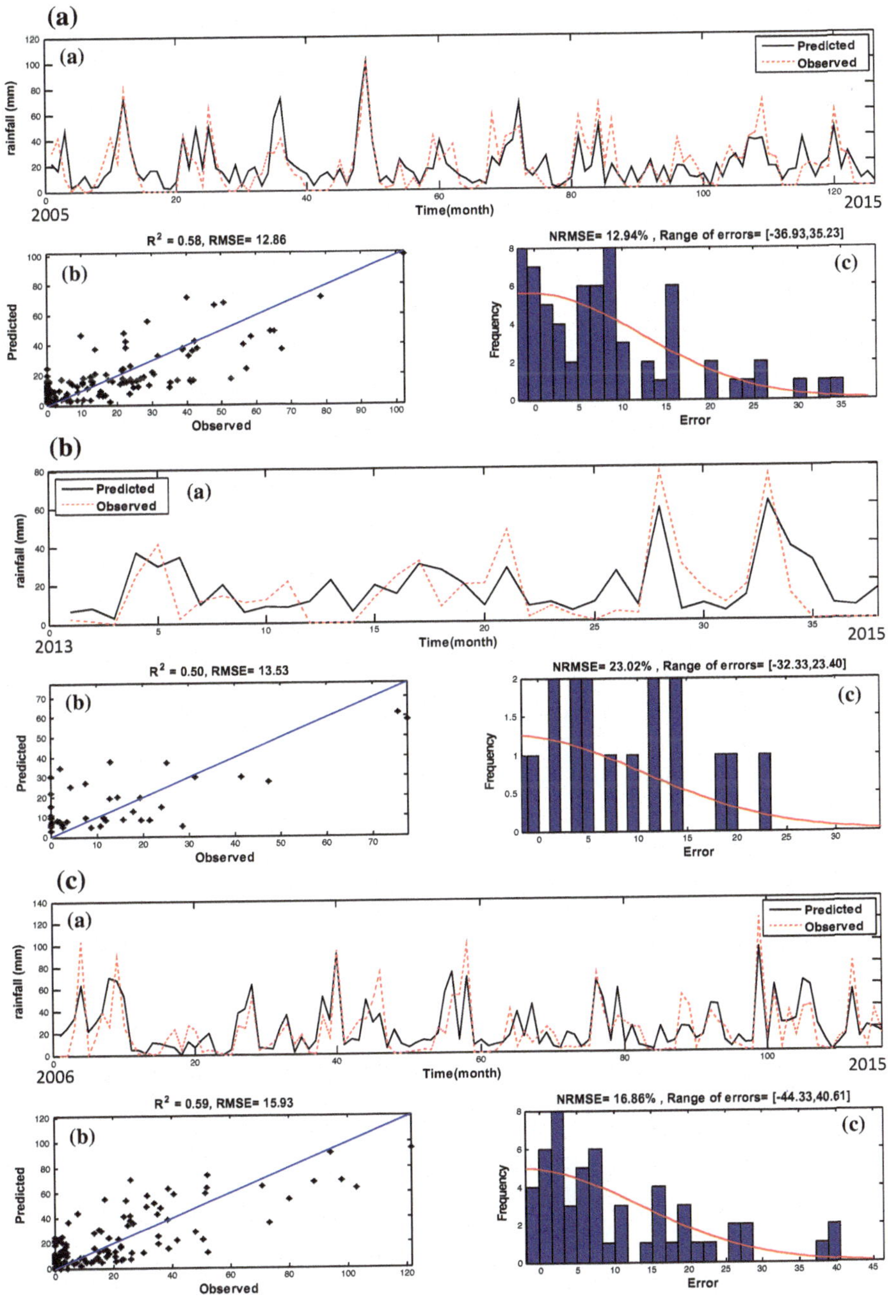

Fig. 8 Results of GMDH modeling using rainfall data. **a** Mashhad, **b** Karaj, **c** Maragheh and **d** Ramhormoz

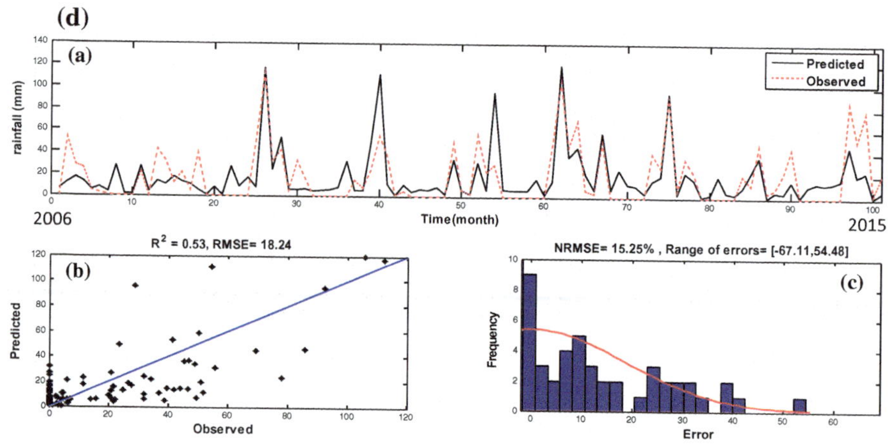

Fig. 8 (continued)

datasets are more useful to evaluate and compare the performance of different models, as these datasets were not used in the training process and are new for the models. Using test datasets for evaluation can help determining and omitting over fitted models. In these figures, Part "*a*" shows the comparison between the observed and predicted rainfall. Scatter plot of observed and predicted rainfall is shown in Part "*b*". Part "*c*" also demonstrates the histogram of error. The RMSE, R^2, Normalized RMSE (NRMSE) and the range of errors are also reported for all models. Figure 8 shows that when the rainfall is the only variable used in the modeling process, the results are not satisfactory. However, Fig. 9 shows that the GMDH has a moderate performance in rainfall modeling for all stations using the exogenous data. The trend and behavior of rainfall are determined relatively well, but the values are not estimated precisely. The first parts of Table 1 show the results of one, two, three and four-month ahead rainfall forecasting for the GMDH model using the rainfall data without exogenous data. As it is shown in this table, the performance of the model is not good. This performance is exacerbated when the time step increased. The first parts of Table 2 show the results of one, two, three and four-month ahead rainfall forecasting for the GMDH model using the rainfall data in addition to exogenous data. As it is obvious in this table, the performance of the model is not very good but is better than just using the rainfall data.

For developing the combined wavelet-GMDH model, the main signals were split into different sub-series using various mother wavelets and levels. Figure 10 demonstrates the main and decomposed rainfall values for all stations, using db4 mother wavelet in level 4, as an example. This figure shows the approximation and detail components obtained from the decomposition of the rainfall data for all stations. As decomposition process were done in four levels (for instance), one

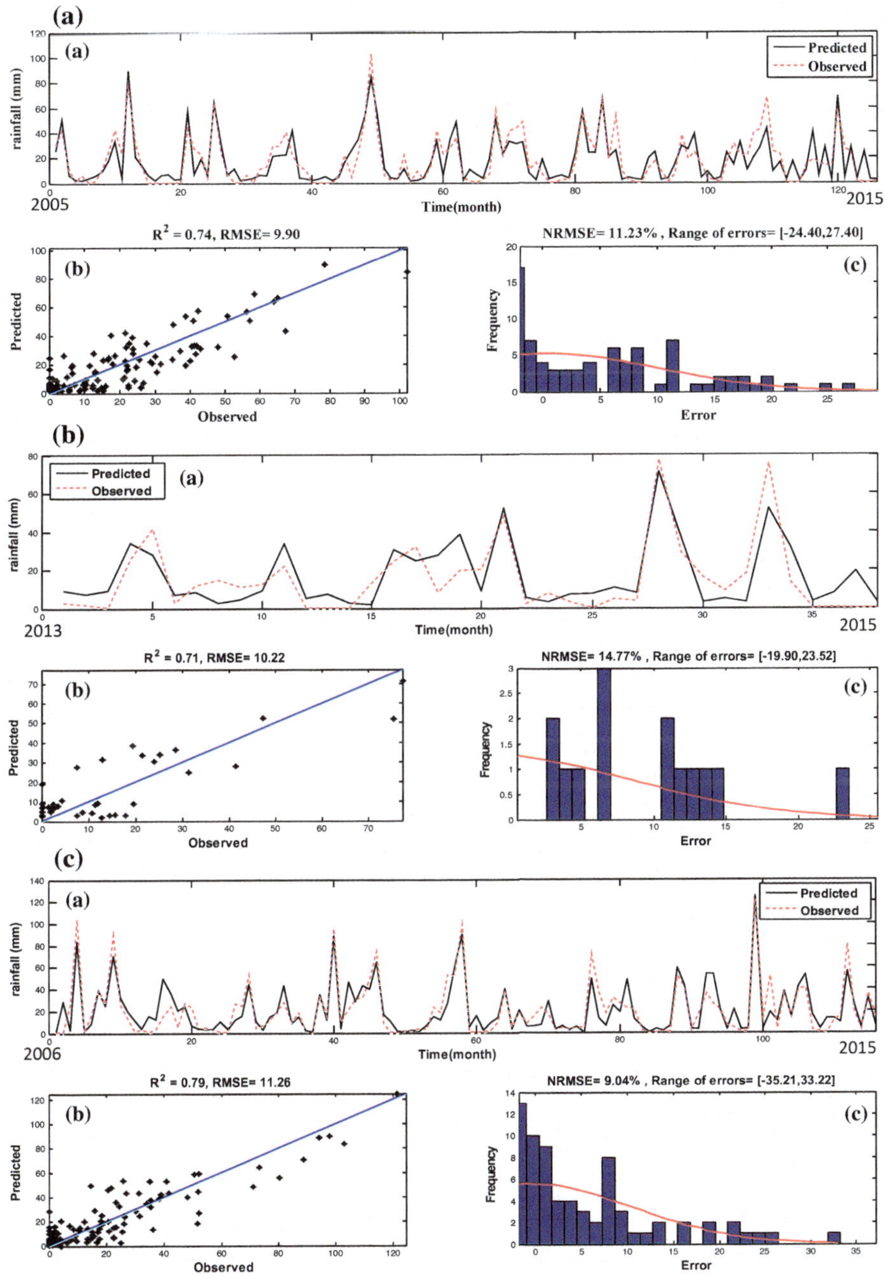

Fig. 9 Results of GMDH modeling using rainfall data in addition to exogenous data. **a** Mashhad, **b** Karaj, **c** Maragheh and **d** Ramhormoz

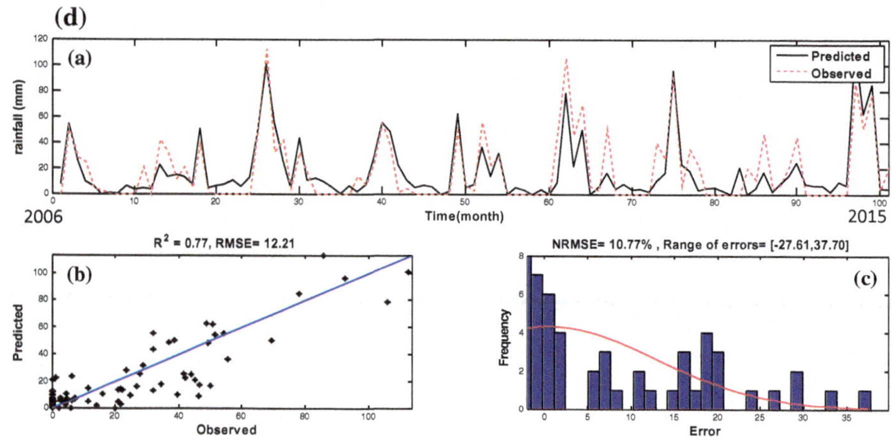

Fig. 9 (continued)

approximation and four detail components are produced. The approximation component demonstrates the low frequency fluctuation of the main signal and the detail components show the high frequency oscillations. In the next step, the combined wavelet-GMDH model was produced using wavelet based decomposed data. Then, the invers process of wavelet based decomposition were performed on the predicted components of target to determine the actual rainfall data. The mother wavelet "db4" and level "2" was selected as the optimum mother wavelet and decomposition level, respectively (according to the results). Figure 11 shows the results of the hybrid wavelet-GMDH model using rainfall data. This shows that although wavelet-GMDH based modeling with rainfall data can enhance the performance of single GMDH model to some extent but the overall performance is still unacceptable.

Figure 12 shows the results of the hybrid wavelet-GMDH model using rainfall data in addition to exogenous data. The performance is better than previous modeling condition. It means that using exogenous data can improve the performance of the modeling. Results indicated that wavelet transform can moderately enhance the ability of the GMDH model. Wavelet improved the R^2 and RMSE to some extent, however, it could not significantly decrease the range of errors. The second parts of Table 1 show the results of one, two, three and four month-ahead rainfall forecasting for the wavelet-GMDH model using the rainfall data without exogenous data. The second parts of Table 2 show the results of one, two, three and four month-ahead rainfall forecasting for the wavelet-GMDH model using the rainfall data as well as exogenous data. As it is shown in this table, the prediction ability of the model is exacerbated when the time step increased. The model can

Table 1 Results of one, two, three and four-month ahead rainfall forecasting using the rainfall data without exogenous data

A	GMDH				Wavelet-GMDH				Wavelet Packet-GMDH				EEMD-GMDH			
	R_{t+1}	R_{t+2}	R_{t+3}	R_{t+4}	R_{t+1}	R_{t+2}	R_{t+3}	R_{t+4}	R_{t+1}	R_{t+2}	R_{t+3}	R_{t+4}	R_{t+1}	R_{t+2}	R_{t+3}	R_{t+4}
R^2	0.58	0.57	0.55	0.51	0.71	0.69	0.65	0.62	0.85	0.82	0.81	0.77	0.82	0.80	0.79	0.77
RMSE	12.86	13.5	14.2	18.36	10.79	11.51	11.98	13.91	7.81	8.24	9.58	11.29	8.34	9.47	11.80	12.96
Max. Abs. Erro	36.93	37.85	39.42	40.53	30.24	32.72	34.62	36.21	17.67	18.25	19.69	23.69	19.54	21.60	22.99	24.69
B	GMDH				Wavelet-GMDH				Wavelet Packet-GMDH				EEMD-GMDH			
	R_{t+1}	R_{t+2}	R_{t+3}	R_{t+4}	R_{t+1}	R_{t+2}	R_{t+3}	R_{t+4}	R_{t+1}	R_{t+2}	R_{t+3}	R_{t+4}	R_{t+1}	R_{t+2}	R_{t+3}	R_{t+4}
R^2	0.50	0.49	0.46	0.41	0.70	0.68	0.66	0.61	0.79	0.76	0.75	0.68	0.80	0.72	0.69	0.62
RMSE	13.53	13.99	15.37	16.89	10.85	11.64	12.81	13.81	9.23	10.6	11.1	13.9	8.46	9.26	10.6	13.5
Max. Abs. Erro	32.33	34.65	36.50	39.81	30.38	32.11	33.95	35.92	30.97	32.44	33.91	36.61	20.83	21.92	23.15	25.73
C	GMDH				Wavelet-GMDH				Wavelet Packet-GMDH				EEMD-GMDH			
	R_{t+1}	R_{t+2}	R_{t+3}	R_{t+4}	R_{t+1}	R_{t+2}	R_{t+3}	R_{t+4}	R_{t+1}	R_{t+2}	R_{t+3}	R_{t+4}	R_{t+1}	R_{t+2}	R_{t+3}	R_{t+4}
R^2	0.59	0.57	0.55	0.51	0.74	0.73	0.71	0.68	0.85	0.84	0.82	0.79	0.84	0.82	0.79	0.75
RMSE	15.93	16.25	16.91	18.96	12.67	13.67	14.92	16.33	9.38	10.45	11.51	13.48	9.76	10.78	11.31	13.90
Max. Abs. Erro	44.33	45.36	46.40	49.67	42.53	44.35	46.38	47.89	25.29	26.32	27.42	29.22	25.23	26.90	28.40	29.81
D	GMDH				Wavelet-GMDH				Wavelet Packet-GMDH				EEMD-GMDH			
	R_{t+1}	R_{t+2}	R_{t+3}	R_{t+4}	R_{t+1}	R_{t+2}	R_{t+3}	R_{t+4}	R_{t+1}	R_{t+2}	R_{t+3}	R_{t+4}	R_{t+1}	R_{t+2}	R_{t+3}	R_{t+4}
R^2	0.53	0.52	0.50	0.47	0.72	0.70	0.69	0.64	0.85	0.83	0.80	0.78	0.84	0.83	0.82	0.80
RMSE	18.24	19.14	20.51	22.59	14.00	15.15	16.93	18.62	9.79	10.81	11.73	13.88	10.00	11.25	12.76	14.52
Max. Abs. Erro	67.11	68.21	69.20	71.69	41.64	43.58	45.68	47.43	24.75	25.10	26.92	29.14	27.22	28.60	30.44	33.90

Table 2 Results of one, two, three and four-month ahead rainfall forecasting using the rainfall data with exogenous data

A	GMDH				Wavelet-GMDH				Wavelet Packet-GMDH				EEMD-GMDH			
	R_{t+1}	R_{t+2}	R_{t+3}	R_{t+4}	R_{t+1}	R_{t+2}	R_{t+3}	R_{t+4}	R_{t+1}	R_{t+2}	R_{t+3}	R_{t+4}	R_{t+1}	R_{t+2}	R_{t+3}	R_{t+4}
R^2	0.74	0.71	0.69	0.6	0.82	0.81	0.78	0.72	0.95	0.93	0.92	0.89	0.94	0.93	0.91	0.87
RMSE	9.90	10.30	10.90	14.00	8.30	9.25	10.95	13.10	3.86	4.10	5.36	6.95	4.58	6.25	7.99	9.14
Max. Abs. Erro	27.41	28.28	30.11	38.96	24.93	25.81	27.52	29.80	9.58	11.57	12.90	14.96	9.96	12.14	13.69	15.43
B	GMDH				Wavelet-GMDH				Wavelet Packet-GMDH				EEMD-GMDH			
	R_{t+1}	R_{t+2}	R_{t+3}	R_{t+4}	R_{t+1}	R_{t+2}	R_{t+3}	R_{t+4}	R_{t+1}	R_{t+2}	R_{t+3}	R_{t+4}	R_{t+1}	R_{t+2}	R_{t+3}	R_{t+4}
R^2	0.71	0.69	0.66	0.60	0.81	0.80	0.76	0.70	0.89	0.88	0.86	0.83	0.90	0.88	0.87	0.82
RMSE	10.22	11.20	12.50	15.60	8.18	8.96	9.64	11.68	5.89	6.24	7.39	9.12	5.93	6.12	7.65	9.92
Max. Abs. Erro	23.52	25.10	26.9	28.9	24.96	25.6	26.9	29.81	9.92	10.85	11.69	13.97	10.68	11.60	12.30	14.27
C	GMDH				Wavelet-GMDH				Wavelet Packet-GMDH				EEMD-GMDH			
	R_{t+1}	R_{t+2}	R_{t+3}	R_{t+4}	R_{t+1}	R_{t+2}	R_{t+3}	R_{t+4}	R_{t+1}	R_{t+2}	R_{t+3}	R_{t+4}	R_{t+1}	R_{t+2}	R_{t+3}	R_{t+4}
R^2	0.79	0.77	0.75	0.72	0.84	0.83	0.81	0.78	0.93	0.92	0.90	0.86	0.93	0.92	0.90	0.85
RMSE	11.26	12.53	13.42	15.21	9.86	10.68	11.25	13.96	6.46	7.25	8.21	10.95	6.42	7.22	8.19	10.58
Max. Abs. Erro	35.21	36.28	37.96	39.17	39.64	41.25	42.36	44.58	13.56	14.32	15.81	18.96	14.94	15.34	17.65	19.89
D	GMDH				Wavelet-GMDH				Wavelet Packet-GMDH				EEMD-GMDH			
	R_{t+1}	R_{t+2}	R_{t+3}	R_{t+4}	R_{t+1}	R_{t+2}	R_{t+3}	R_{t+4}	R_{t+1}	R_{t+2}	R_{t+3}	R_{t+4}	R_{t+1}	R_{t+2}	R_{t+3}	R_{t+4}
R^2	0.77	0.76	0.74	0.70	0.83	0.81	0.79	0.76	0.94	0.92	0.89	0.86	0.94	0.93	0.89	0.85
RMSE	12.21	13.66	14.52	16.33	10.27	11.28	12.39	14.58	6.33	7.02	8.21	10.74	6.36	7.91	8.69	10.97
Max. Abs. Erro	37.70	39.66	39.90	43.27	30.17	32.54	33.90	35.78	16.04	17.14	18.52	20.00	17.60	18.09	19.27	21.30

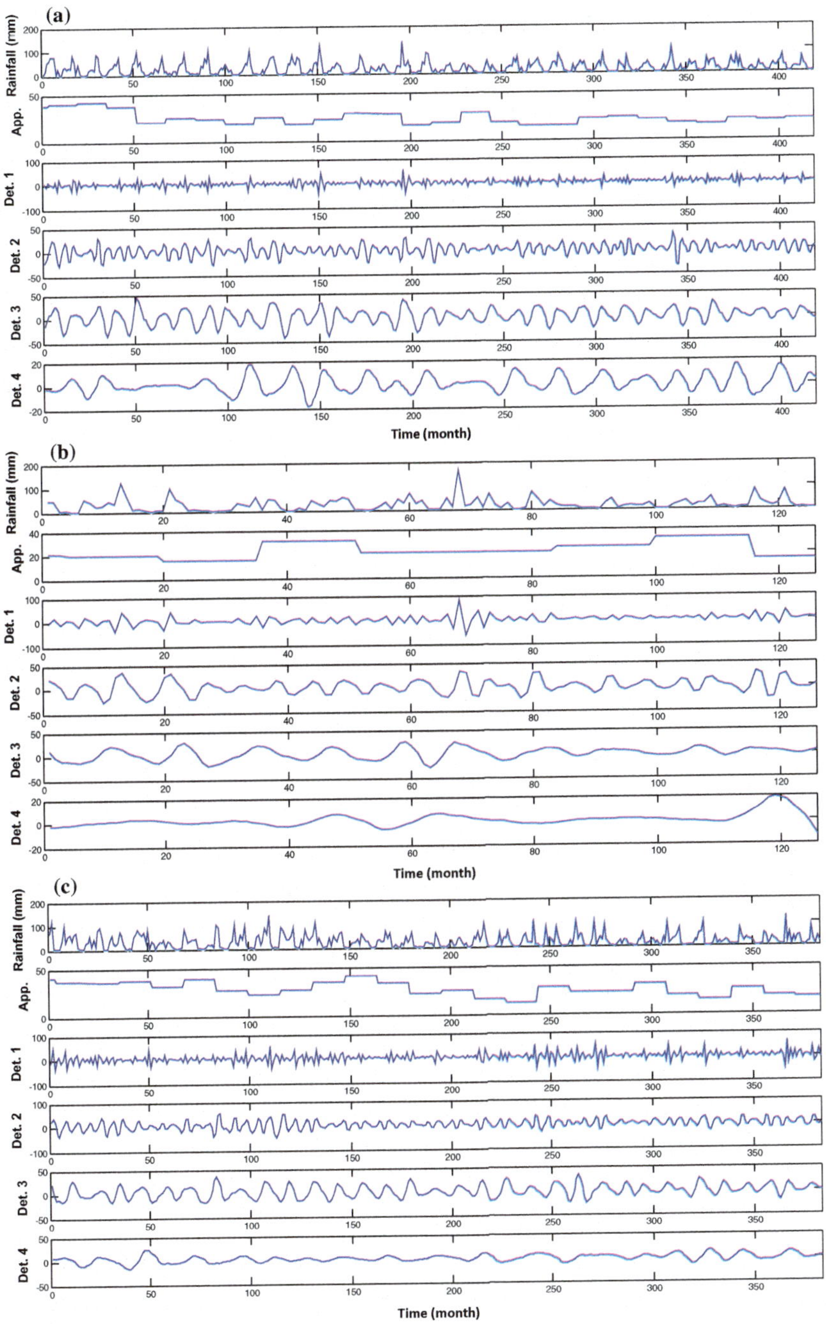

Fig. 10 Results of wavelet decomposition on rainfall data. **a** Mashhad, **b** Karaj, **c** Maragheh and **d** Ramhormoz

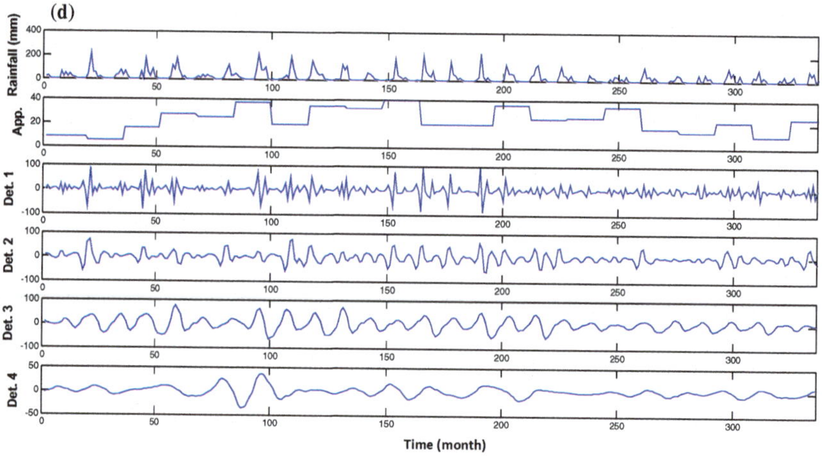

Fig. 10 (continued)

forecast rainfall for the first months moderately, however, it does not appropriately worked for the three last time steps.

With the intention of producing combined wavelet packet-GMDH model, the wavelet packet analysis were applied on original data and the wavelet packet tree were produced. Figure 5 (section B) displays a schematic diagram of a wavelet packet tree obtained from applying a db4 mother wavelet in level 3, as an example. The mentioned tree demonstrates that details are split to their sub-components as well as approximations. As this figure shows, there are several nodes (*j*, *k*) in the wavelet packet tree.

The node *j* shows the depth inside the transformation tree and *k* shows the position of the node in the tree. For example, node (0, 0) shows the original data. Node (1, 0) is produced applying a Low-pass filter and Node (1, 1) is produced applying a high-pass filter on the original data. These nodes demonstrate the approximation and detail components in level 1, respectively. Thereafter, nodes (2, 0) and (2, 1) are created from Node (1, 0) and this procedure continues to the predefined level. The coefficients obtained from the last level were then imported to the hybrid wavelet packet-GMDH models. Figure 13 shows the original and the results of wavelet packet decomposition for both studied stations, using db4 mother wavelet in level 3, as an example. After using the coefficients obtained from wavelet packet transform as inputs for the GMDH model, the wavelet packet tree were reconstructed. In this case the estimated target components were used. Eventually, the inverse process of wavelet packet transformation were applied on the reconstructed wavelet packet tree. This results the actual rainfall values. Figures 14 and 15 show the outcomes of the best wavelet packet-GMDH model for all stations using rainfall data without and with exogenous data, respectively. As

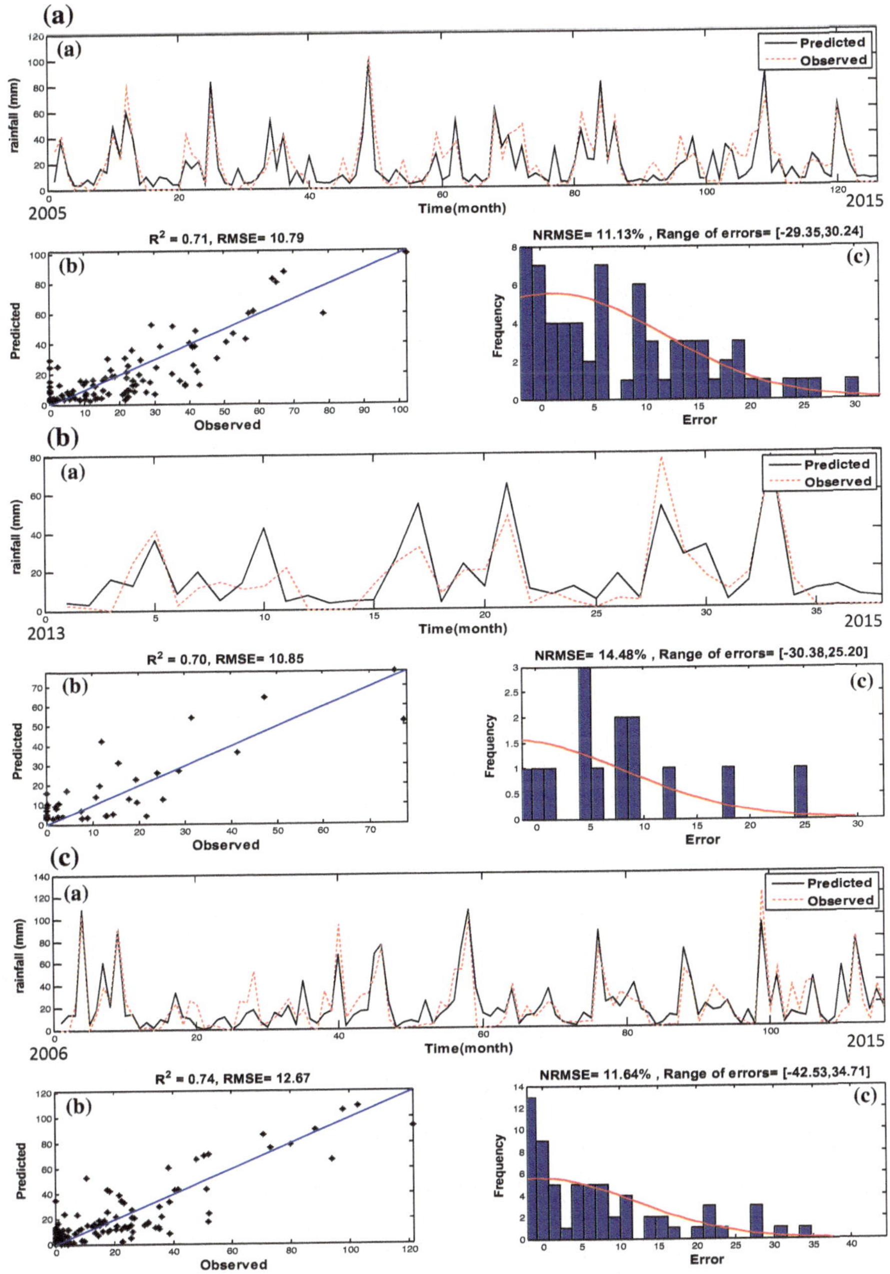

Fig. 11 Results of wavelet-GMDH modeling using rainfall data. **a** Mashhad, **b** Karaj, **c** Maragheh and **d** Ramhormoz

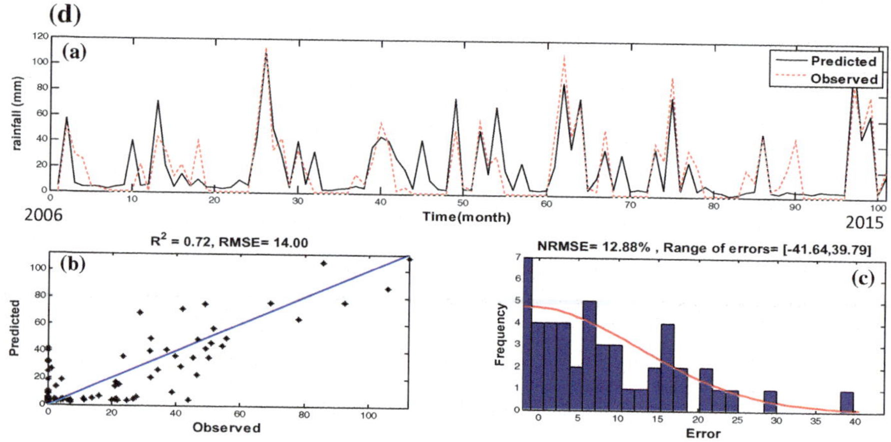

Fig. 11 (continued)

this figure shows, wavelet packet analysis improved the performance of the GMDH model significantly specially when using exogenous data.

In the next step, the original data were decomposed using EEMD method to combine EEMD signal processing approach with GMDH model. Figure 16 shows the original and EEMD based split rainfall values for all stations, as an example. This figure shows different IMFs each of which represent a distinct feature. The first IMFs have higher frequencies and can show the random information of the initial rainfall signal. The next IMFs show the periodic trends of the original signal. IMF6, IMF 7 and "r" can be considered as trend components. After decomposing data by EEMD, the rainfall prediction is transformed into the estimation of each IMF and the residue "r". In fact, GMDH modeling approach was used to forecast each IMF and the residue "r". Figures 17 and 18 show the results of hybrid EEMD-GMDH model for rainfall forecasting using rainfall data without and with exogenous data, respectively. It shows that the performance is enhanced in comparison with GMDH in its single form and wavelet-GMDH model. The performance of EEMD-GMDH model is close to the wavelet packet-GMDH model. However, wavelet packet-GMDH model slightly outperforms EEMD-GMDH in most of the stations.

The third and fourth parts of Table 1 show the results of one, two, three and four-month ahead rainfall forecasting for the wavelet packet-GMDH and EEMD-GMDH models using the rainfall data without exogenous data, respectively. As it is shown in this table, the performance of the model has been improved in comparison with single GMDH and hybrid wavelet-GMDH models. However, the performance is not satisfactory yet. This performance is exacerbated when the time step increased. The third and fourth parts of Table 2 show the results of one, two, three and four-month ahead rainfall forecasting for the wavelet packet-GMDH and

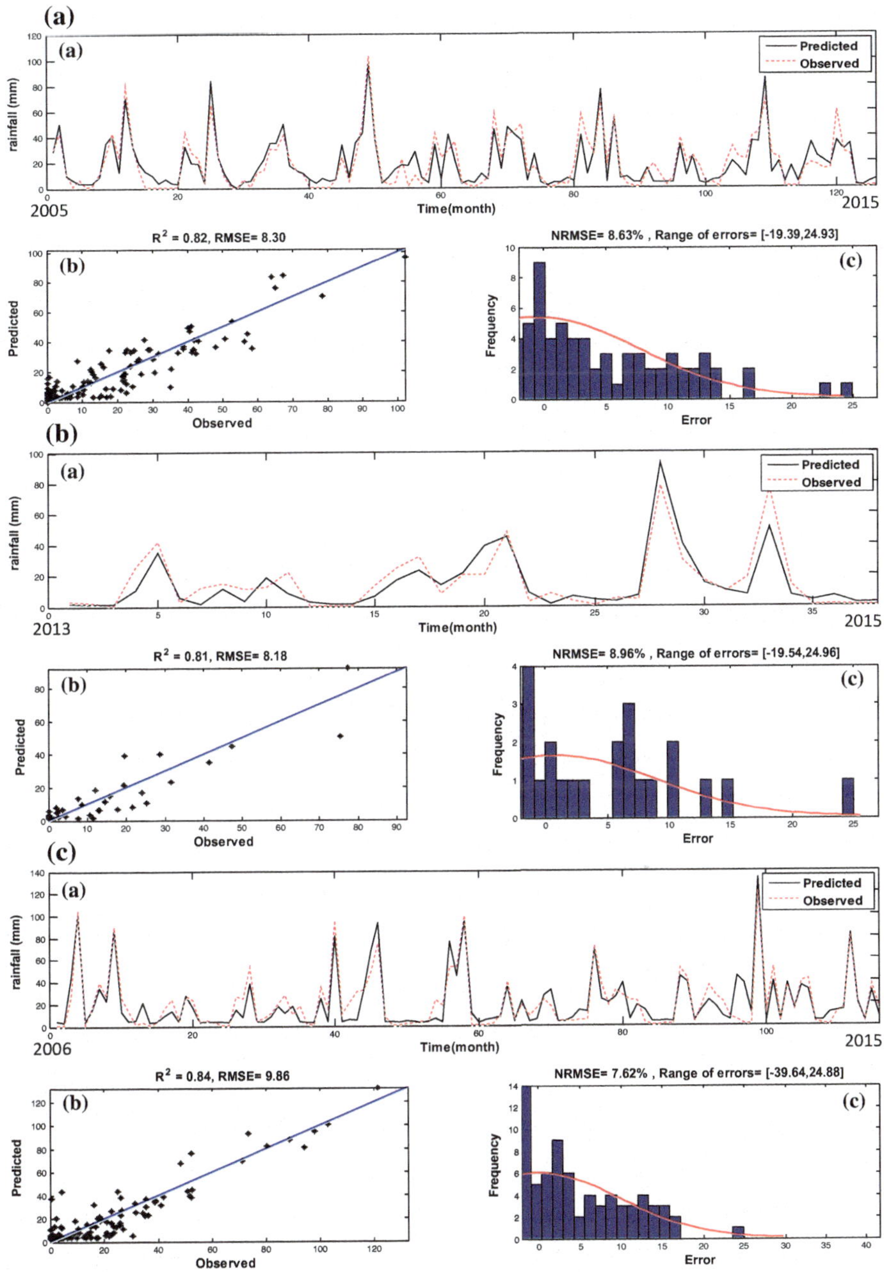

Fig. 12 Results of wavelet-GMDH modeling using rainfall data in addition to exogenous data. **a** Mashhad, **b** Karaj, **c** Maragheh and **d** Ramhormoz

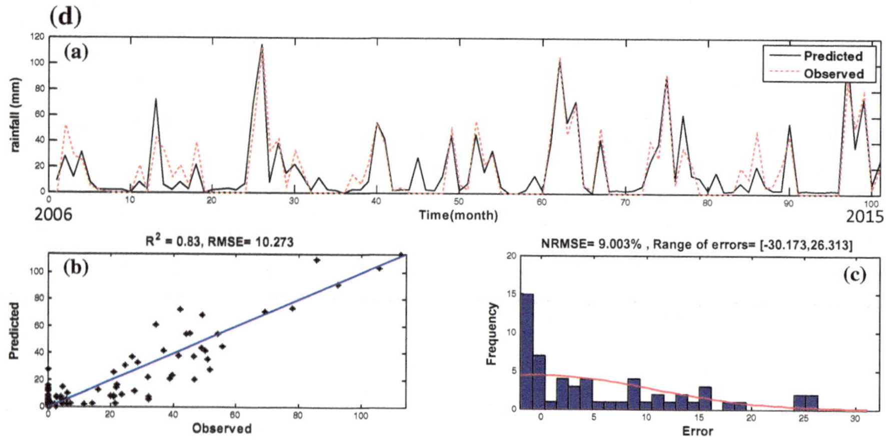

Fig. 12 (continued)

EEMD-GMDH models using the rainfall data in addition to exogenous data, respectively. As it is shown in this table, the prediction ability of the model is significantly enhanced. The maximum absolute error is also considerably decreased.

EMD analysis can be used when the signal does not provide satisfactory information in its global form. As this method provides a time-frequency analysis, it can extract local information in time series. Dividing the specific signal into different frequency bands, it can be analyzed in interested frequency ranges. The produced IMFs in the EEMD method, show various frequencies (high to low). In fact, the first IMF is related to high frequency component and the last IMF corresponds to low frequency component. The problem of mode mixing that indicates a single IMF including signals of intensely different scales or a signal with the same scale performing in different IMF components and regularly producing intermittency of the mentioned signal is solved in EEMD method. The EEMD removes the mode mixing problem and provide a proper distribution of time frequency of the analyzed time series (Wu and Huang 2009). One of the main difficulties of wavelet analysis is to find a mother wavelet which is as close as possible to analyzed signal. Empirical mode decomposition has no such deficiency since it has no basic functions and is fully adaptive to the signal itself. One of the key advantages of empirical mode decomposition over wavelet decomposition is the ability to estimate subtle fluctuations in frequency. However, wavelet transform can enhance the GMDH modeling ability to some extent. In both wavelet and wavelet packet transforms, db4 was the best mother wavelet. Daubechies (db) wavelet family is one of the most widespread wavelet family used in wavelet based signal processing studies, due to orthogonal and compact support abilities. The names of this family wavelets are written dbN, where N shows the order. The superior performance of

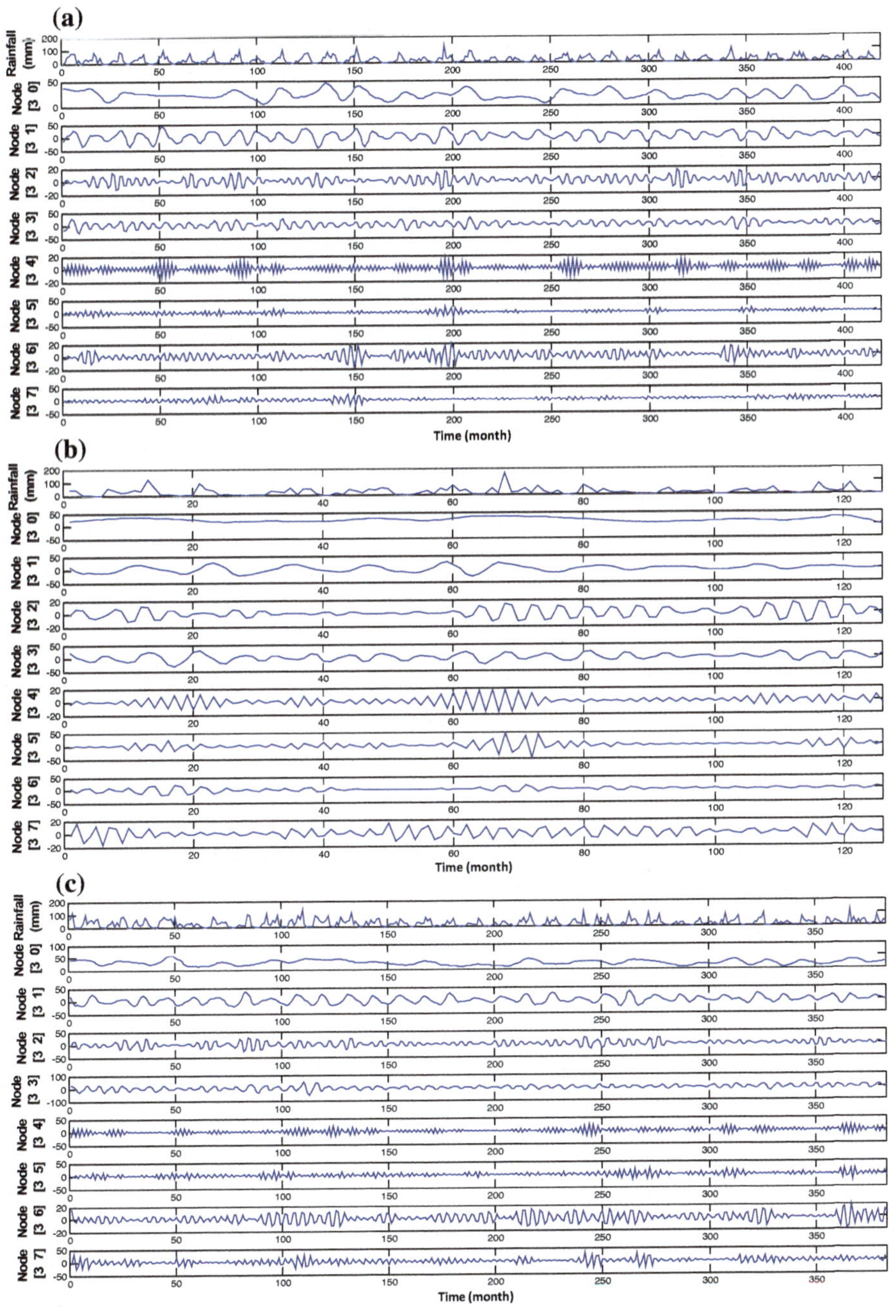

Fig. 13 Results of wavelet packet decomposition on rainfall data. **a** Mashhad, **b** Karaj, **c** Maragheh and **d** Ramhormoz

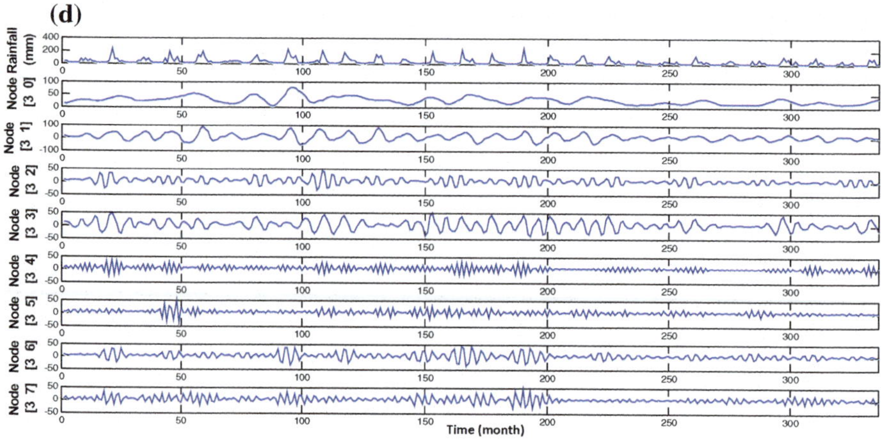

Fig. 13 (continued)

the wavelet packet transform may be related to its higher ability to deal with the problem associated with nonstationary signals. Wavelet packet transform offers a level by level decomposition and transformation process from the time to the frequency domain. In contrast to wavelet transform, the frequency domains are of equal width because the wavelet packet transform decomposes high frequency sub-bands as well as low frequency sub-bands. This transformation provides more detailed decomposition of a signal. Decomposition of high frequency components provides several base function at a specific scale. WPT provides more than $2^{2^{n-1}}$ different ways to translate a specific signal. It may be the reason of the better performance of combined wavelet packet-GMDH model in comparison with ordinary and combined wavelet-GMDH modeling methods.

4 Conclusion

This study examined the ability of three hybrid models for rainfall forecasting. The single form GMDH model was also performed as an initial test. It was demonstrated that GMDH model in its single form had a moderate performance in rainfall modeling. All three signal processing models improved the performance of GMDH. The performance of EEMD-GMDH and wavelet packet-GMDH models were close to each other, however, the ability of wavelet packet was relatively better than EEMD-GMDH transform. EEMD can extract different aspects and properties of the signal. This details can be then imported to the artificial intelligence models. EEMD is particularly suitable for studying periodic signals that includes both high and low

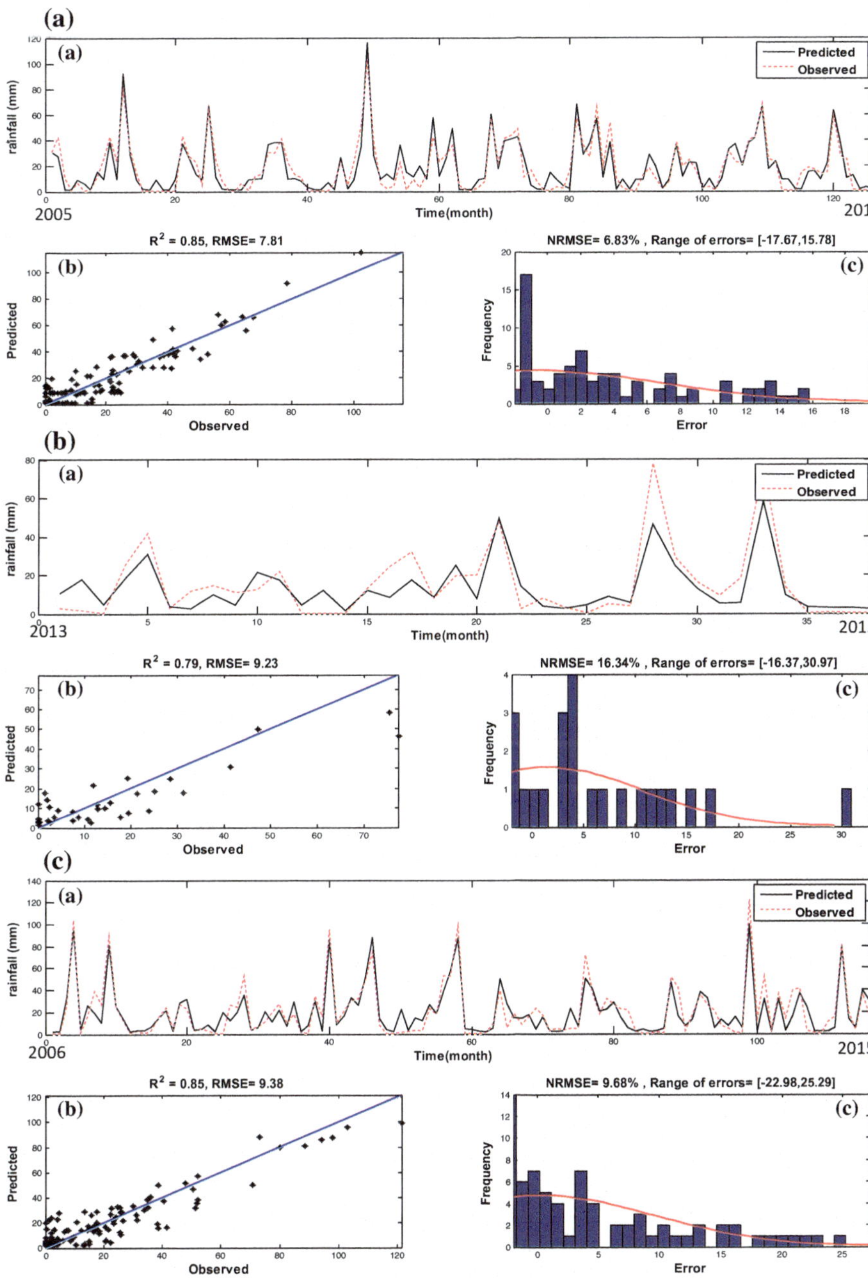

Fig. 14 Results of wavelet packet-GMDH modeling using rainfall data. **a** Mashahd, **b** Karaj, **c** Maragheh and **d** Ramhormoz

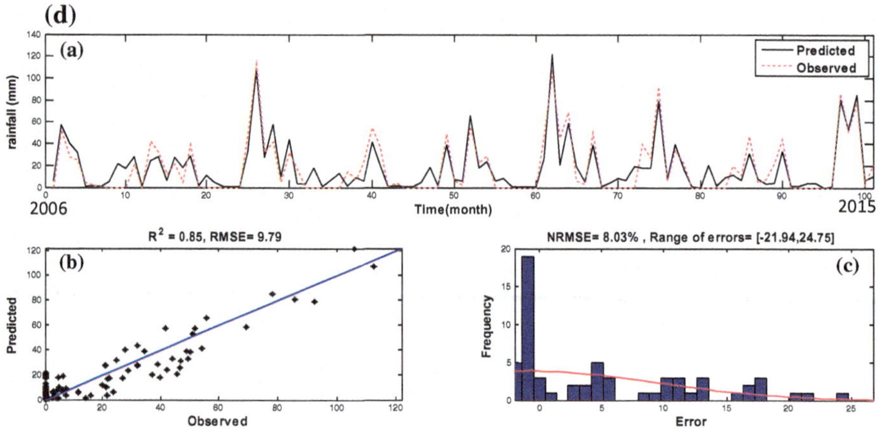

Fig. 14 (continued)

frequency components. The performance of signal processing wavelet packet approach is significant in comparison with the wavelet transform and EEMD signal processing approach since wavelet packet can produce a more detailed information about the frequency resolution. These signal analysis methods can manage the nonstationarity in the time series. Stationarity can be defined as preserving the properties throughout the path of the signals. It is extensively documented that in real and natural applications data are seldom stationary and isotropic. As artificial intelligence approaches have point based performance, the original time series cannot be considered as a single structure. Actually, these models consider the data independently without considering the adjacent data. Therefore, signal processing approaches can help these models to better recognize the properties of the data and the relationship between dependent and independent variables. Thus, using powerful signal processing methods e.g. wavelet packet transform and EEMD as preprocessing techniques can help modelers to cope with the difficulty of simulating the natural events. The other important note is the effect of exogenous data on the performance of the models. It was shown that using appropriate variables can significantly improve the ability of all modeling approaches. Application of other AI models such as multilayer perceptron (MLP), radial basis function (RBF) and Support vector regression (SVR), and other signal processing approaches such as bi-dimensional empirical mode decomposition (BEMD) and multi-dimensional ensemble empirical mode decomposition (MEEMD), in addition to combining EEMD and wavelet transform can be suggested as future works.

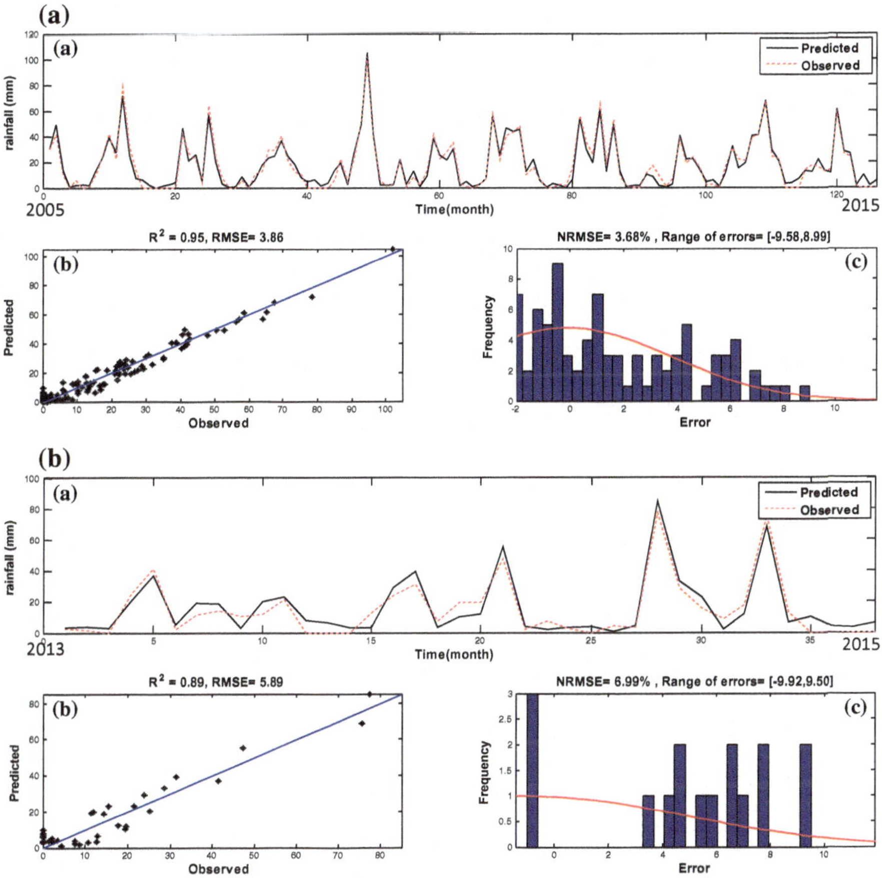

Fig. 15 Results of wavelet packet-GMDH modeling using rainfall data in addition to exogenous data. **a** Mashahd, **b** Karaj, **c** Maragheh and **d** Ramhormoz

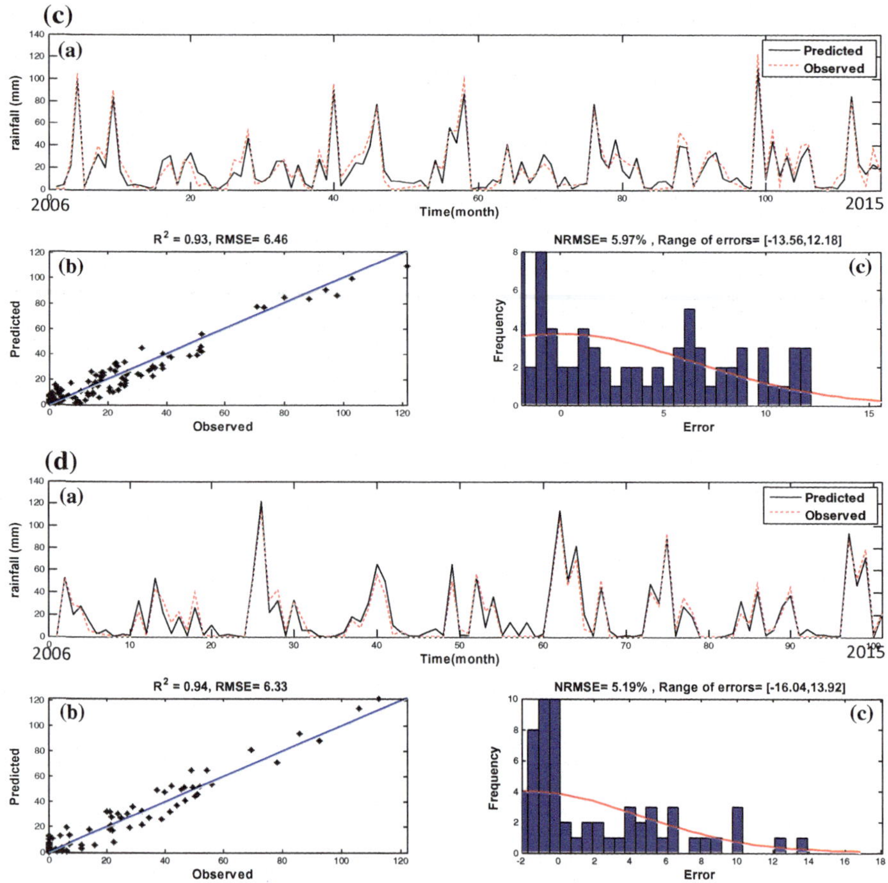

Fig. 15 (continued)

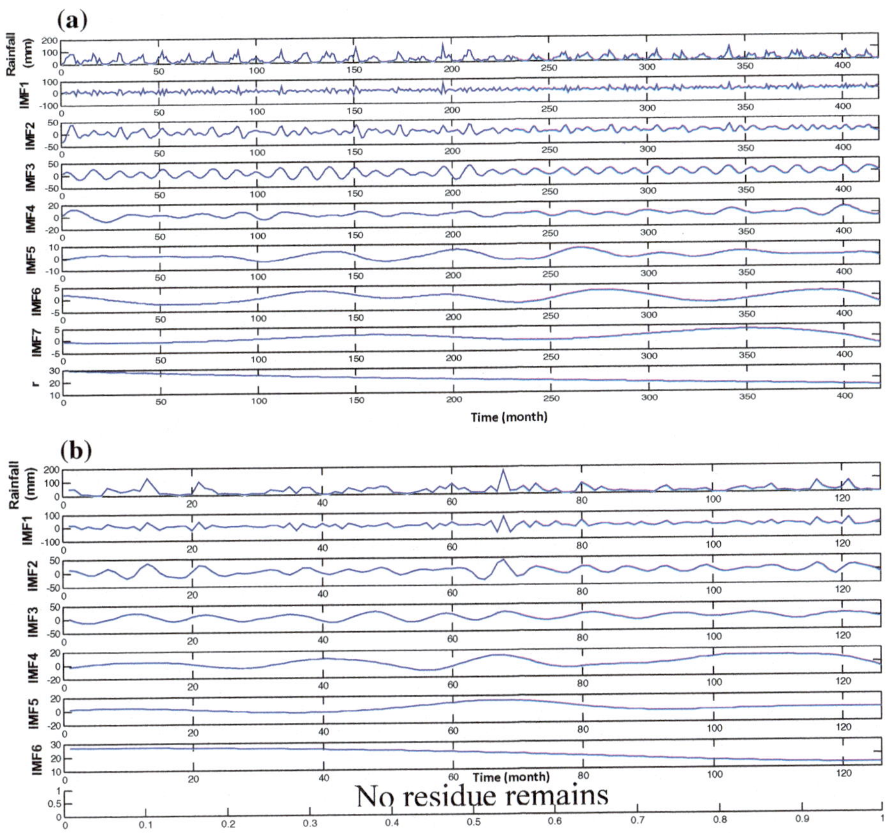

Fig. 16 Results of EEMD decomposition on rainfall data. **a** Mashahd, **b** Karaj, **c** Maragheh and **d** Ramhormoz

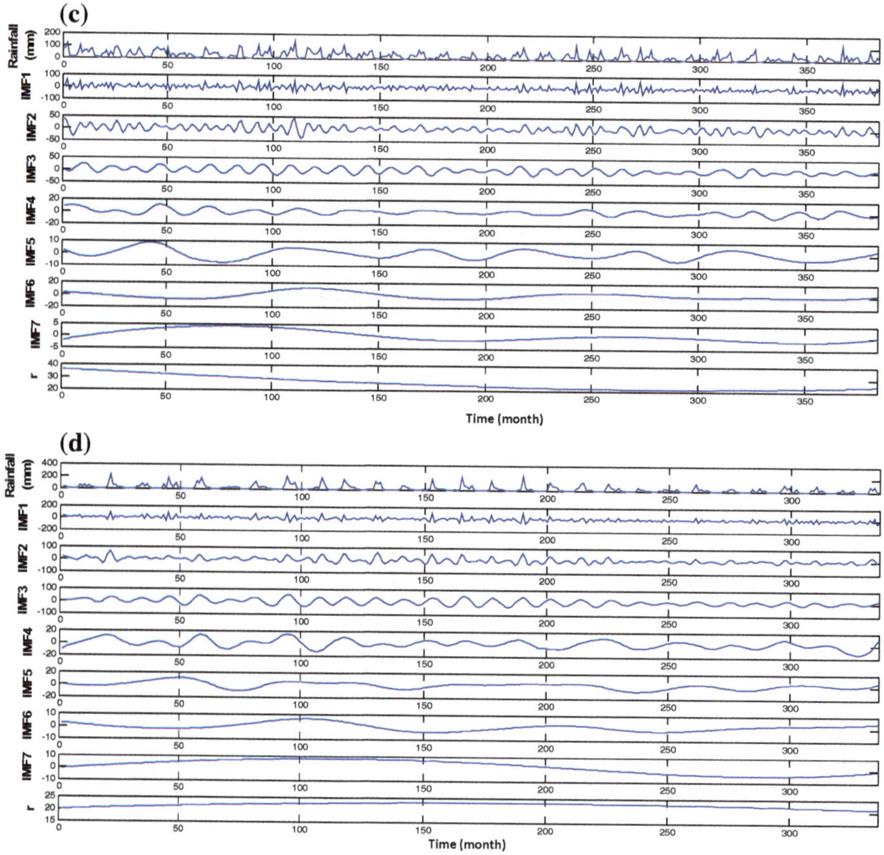

Fig. 16 (continued)

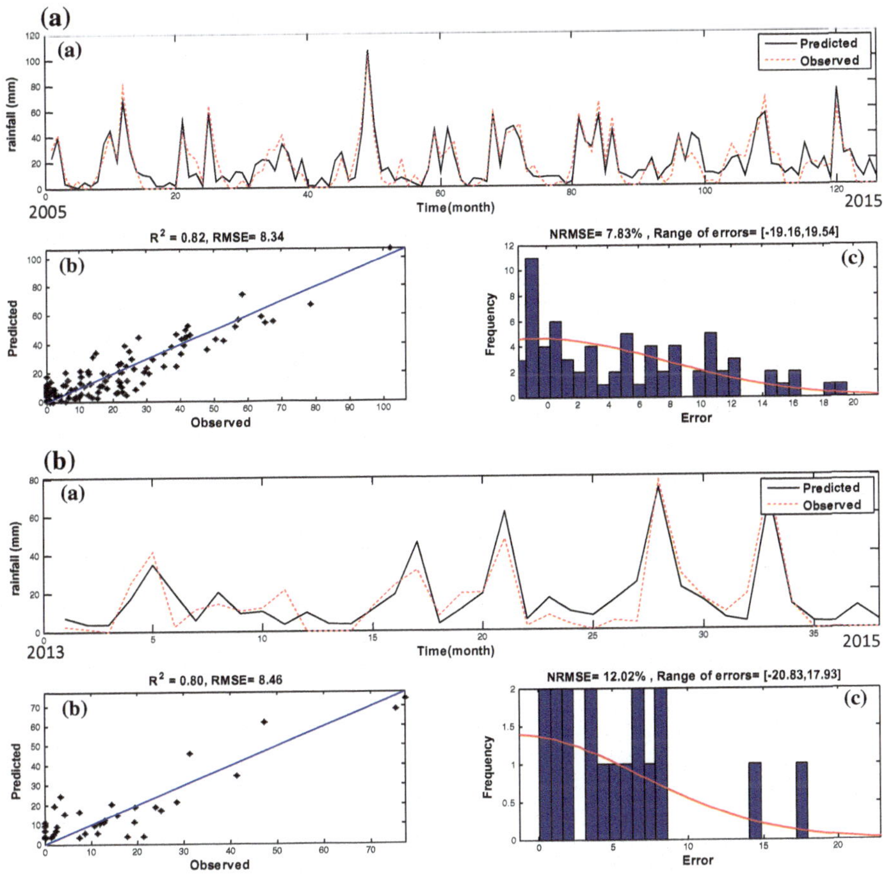

Fig. 17 Results of EEMD-GMDH modeling using rainfall data. **a** Mashahd, **b** Karaj, **c** Maragheh and **d** Ramhormoz

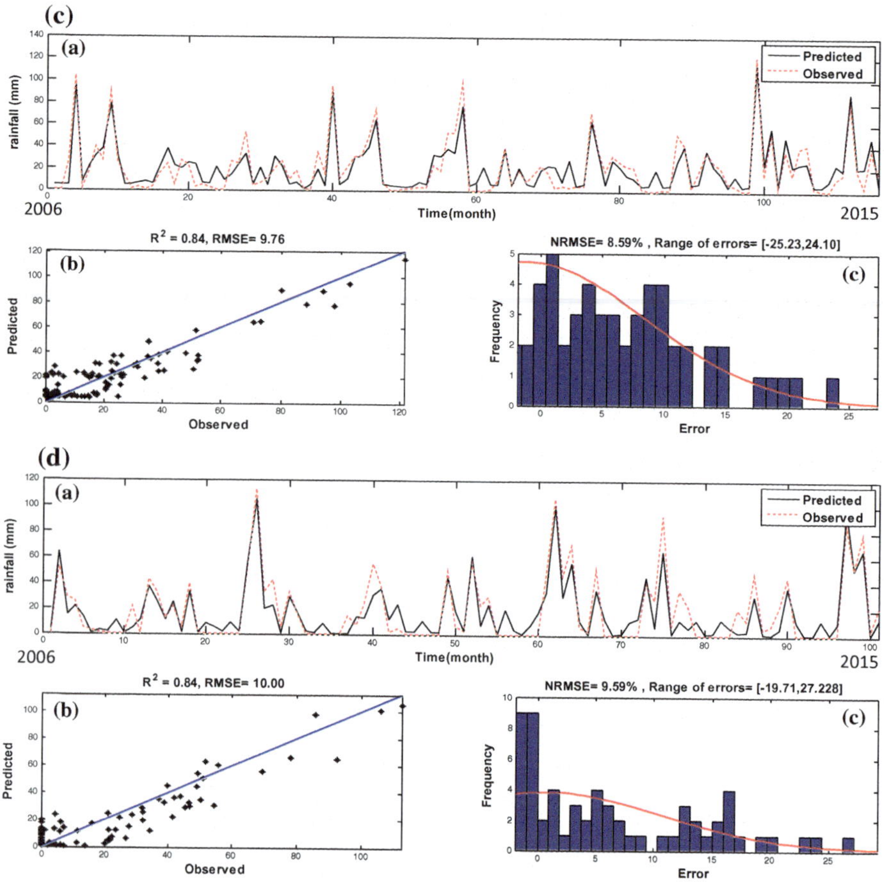

Fig. 17 (continued)

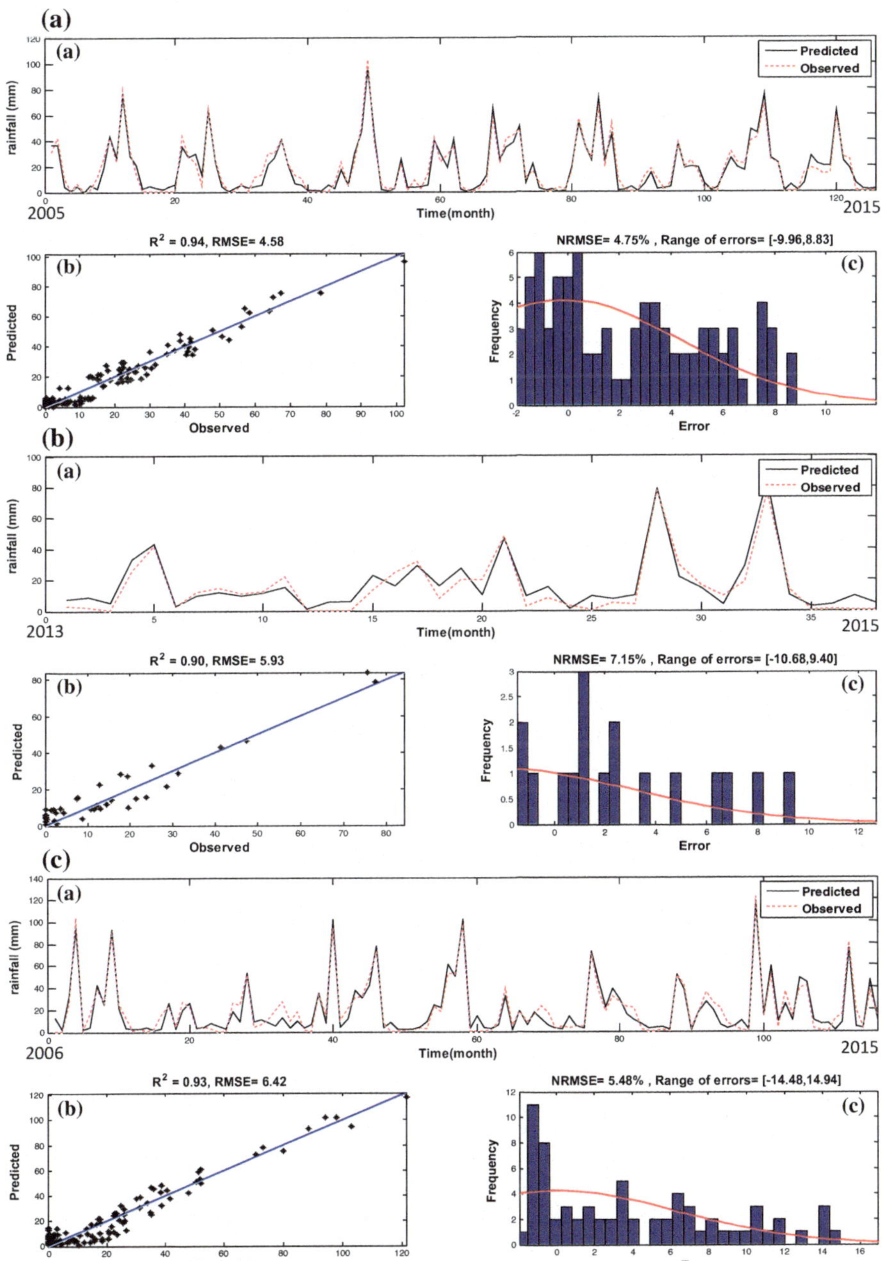

Fig. 18 Results of EEMD-GMDH modeling using rainfall data in addition to exogenous data. **a** Mashahd, **b** Karaj, **c** Maragheh and **d** Ramhormoz

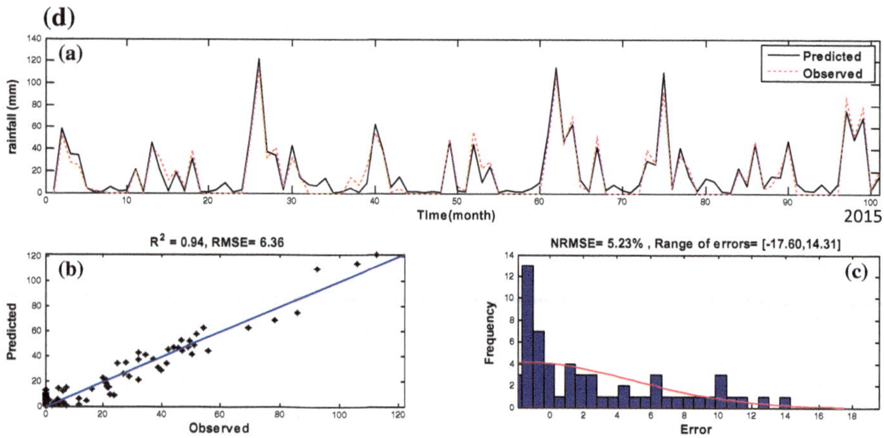

Fig. 18 (continued)

References

Adamowski J (2007) Development of a short-term river flood forecasting method based on wavelet analysis. Polish Academy of Sciences Publication, Warsaw, p 172

Adamowski J, Chan FH (2011) A wavelet neural network conjunction model for rainfall forecasting. J Hydrol 407:28–40

Adamowski J, Sun K (2010) Development of a coupled wavelet transform and neural network method for flow forecasting of non-perennial rivers in semi-arid watersheds. J Hydrol 390(1–2):85–91

Aghajani A, Kazemzadeh R, Ebrahimi A (2016) A novel hybrid approach for predicting wind farm power production based on wavelet transform, hybrid neural networks and imperialist competitive algorithm. Energ Convers Manageme 121:232–240

Alves MdC, Pozzab EA, Costac JCB, Carvalhod LG, Alvese LS (2011) Adaptive neuro-fuzzy inference systems for epidemiological analysis of soybean rust. Environ Modell Softw 26 (9):1089–1096

Alvisi S, Franchini M (2011) Fuzzy neural networks for water level and discharge forecasting with uncertainty. Environ Modell Softw 26(4):523–537. http://www.sciencedirect.com/science/journal/13648152/26/4

Amanifard N, Nariman-Zadeh N, Farahani MH, Khalkhali A (2008) Modelling of multiple short-length-scale stall cells in an axial compressor using evolved GMDH neural networks. J Energy Convers Manage 49(10):2588–2594

Breaker LC, Ruzmaikin A (2011) The 154-year record of sea level at San Francisco: extracting the long-term trend, recent changes, and other tidbits. Clim Dyn 36(3–4):545–559

Cannas B, Fanni A, Sias G, Tronei S, Zedda MK (2006) River flow forecasting using neural networks and wavelet analysis. In: Proceedings of the European Geosciences Union. 234–243

Christopoulou EB, Skodras AN, Georgakilas AA (2002) The "Trous" wavelet transform versus classical methods for the improvement of solar images. In: Proceedings of 14th international conference on digital signal processing, vol. 2, pp. 885–888

Daliakopoulos IN, Coulibaly P, Tsanis IK (2005) Rainfall forecasting using artificial neural networks. J of Hydrol 309:229–240

Daubechies L (1990) The wavelet transform, time-frequency localization and signal analysis. IEEE Trans Inform Theory 36:961–1005

Fallah-Mehdipour E, Bozorg Haddad O, Mariño MA (2013) Prediction and simulation of monthly rainfalls by genetic programming. J Hydro Environ Res 7(4):253–260
Farlow SJ (1984) Self -organizing method in modeling. Marcel Dekker Inc, GMDH Type Algorithm
Feng Z, Liang M, Zhang Y, Hou S (2012) Fault diagnosis for wind turbine planetary gearboxes via demodulation analysis based on ensemble empirical mode decomposition and energy separation. Renew Energy 47:112–126
Garcia C, Zikos G, Tziritas G (2000) Wavelet packet analysis for face recognition. Image Vision Comput 18:289–297
Guo W, Tse PW (2013) A novel signal compression method based on optimal ensemble empirical mode decomposition for bearing vibration signals. J Sound Vib 332:423–441
Han J, Zheng P, Wang H (2014) Structural modal parameter identification and damage diagnosis based on Hilbert-Huang transform. Earthquake Eng Vib 13:101–111
He Z, Wen X, Liu H, Du J (2014) A comparative study of artificial neural network, adaptive neuro fuzzy inference system and support vector machine for forecasting river flow in the semiarid mountain region. J Hydrol 509:379–386
Ivakhnenko AG (1966) Group method of data handling—a rival of the method of stochastic approximation. Sov Autom Cont 13:43–71
Ivakhnenko AG (1971) Polynomial theory of complex systems. IEEE Trans Syst Man Cybern Syst 1(4):364–378
Kalantary F, Ardalan H, Nariman-Zadeh N (2009) An investigation on the Su–NSPT correlation using GMDH type neural networks and genetic algorithms. Eng Geol 104:144–155
Kim Y, Shin HS, Plummer JD (2014) A wavelet-based autoregressive fuzzy model for forecasting algal blooms. Environ Modell Softw 62:1–10
Kisi O (2009) Neural networks and wavelet conjunction model for intermittent stream flow forecasting. J Hydrol Eng 14:773–782
Li X, Maier HR, Zecchin AC (2015) Improved PMI-based input variable selection approach for artificial neural network and other data driven environmental and water resource models. Environ Modell Softw 65:15–29
Liu H, Chen C, Hq Tian, Li Y (2012) A hybrid model for wind speed prediction using empirical mode decomposition and artificial neural networks. Renew Energ 48:545–556
Mackay JD, Jackson CR, Wang L (2014) A lumped conceptual model to simulate rainfall time-series. Environ Modell Softw 61:229–245
Manjula M, Sarma AVRS (2012) Comparison of Empirical Mode Decomposition and Wavelet Based Classification of Power Quality Events. Energy Procedia 14:1156–1162
Mariyappa N, Sengottuvel S, Parasakthi C (2014) Baseline drift removal and denoising of MCG data using EEMD: role of noise amplitude and the thresholding effect. Med Eng Phys 36 (10):1266–1276
Mehra RK (1977) Group method of data handling (GMDH): Review and experience. IEEE conf dec cont 16:29–34
Mehrara M, Moeini A, Ahrari M, Erfanifard A (2009) Investigating the efficiency in oil futures market based on GMDH approach. Expert Syst Appl 36(4):7479–7483
Millie DF, Weckman GR, Young WA, Ivey JE, Carrick HJ, Fahnenstiel GL (2012) Modeling microalgal abundance with artificial neural networks: Demonstration of a heuristic 'Grey-Box' to deconvolve and quantify environmental influences. Environ Modell Softw 27–39
Moosavi V, Niazi Y (2015) Development of Hybrid Wavelet Packet-Statistical Models (WP-SM) for landslide susceptibility mapping. Landslides 13(1):97–114. https://doi.org/10.1007/s10346-014-0547-0
Moosavi V, Talebi A, Mokhtari MH, Fallah Shamsi SR, Niazi Y (2015) A Wavelet-Artificial Intelligence Fusion Approach (WAIFA) for Blending Landsat and MODIS surface temperature. Remote Sens Environ 169:243–254
Moosavi V, Vafakhah M, Shirmohammadi B, Behnia N (2013) A Wavelet-ANFIS hybrid model for rainfall forecasting for different prediction periods. Water Resour Manag 27:1301–1321

Moosavi V, Vafakhah M, Shirmohammadi B, Ranjbar B (2014) Optimization of Wavelet-ANFIS and Wavelet-ANN Hybrid Models by Taguchi Method for Rainfall forecasting. Arab J Sci Eng 39:1785–1796

Najafzadeh M (2015) Neuro-fuzzy GMDH systems based evolutionary algorithms to predict scour pile groups in clear water conditions. Ocean Eng 99:85–94

Najafzadeh M, Barani GA, Hessami Kermani MR (2013) GMDH based back propagation algorithm to predict abutment scour in cohesive soils. Ocean Eng 59:100–106

Najafzadeh M, Lim SY (2015) Application of improved neuro-fuzzy GMDH to predict scour depth at sluice gates. Earth Sci Inform 8(1):187–196

Nourani V, Alami MT, Aminfar MH (2009) A combined neural-wavelet model for prediction of Ligvanchai watershed precipitation. Eng Appl Artif Intell 22:466–472

Nourani V, Kisi Z, Mehdi K (2011) Two hybrid artificial Intelligence approaches for modeling rainfall-runoff process. J Hydrol 402:41–59

Pal S, Devara PCS (2012) A wavelet-based spectral analysis of long-term time series of optical properties of aerosols obtained by lidar and radiometer measurements over an urban station in Western India. J Atmos Sol-Rerr Phy Solar-Terrestrial Phys 84–85:75–87

Quiroz R, Yarlequé Ch, Posadas A, Mares V, Immerzeel WW (2011) Improving daily rainfall estimation from NDVI using a wavelet transform. Environ Modell Softw 26(2):201–209

Ravisankar P, Ravi V (2010) Financial distress prediction in banks using Group Method of Data Handling neural network, counter propagation neural network and fuzzy ARTMAP. Knowl-Based Sys 23:823–831

Sànchez-Marrè M, Cortés U, Comas J (2004) Environmental sciences and artificial intelligence. Environ Modell Softw 19(9):761–762

Si J, Feng Q, Wen X, Xi H, Yu T, Li W, Zhao C (2015) Modeling soil water content in extreme arid area using an adaptive neuro-fuzzy inference system. J of Hydrol 527:679–687

Sreekanth P, Geethanjali DN, Sreedevi PD, Ahmed S, Kumar NR, Jayanthi PDK (2009) Forecasting rainfall using artificial neural networks. Current Sci 96(7):933–939

Talei A, Chua LHC, Wong TSW (2010) Evaluation of rainfall and discharge inputs used by Adaptive Network-based Fuzzy Inference Systems (ANFIS) in rainfall–runoff modeling. J of Hydrol 391(3–4):248–262

Volterra V (1959) Theory of functionals and of integrals and integro-differential equations. Madrid (Spanish), translated version reprinted New York: Dover Publications

Wieland R, Mirschel W, Zbell B, Krin Groth K, Pechenick A, Fukuda K (2010) A new library to combine artificial neural networks and support vector machines with statistics and a database engine for application in environmental modeling. Environ Modell Softw 25(4):412–420

Witczak M, Korbicz J, Mrugalski M, Patton R (2006) A GMDH neural networkbased approach to robust fault diagnosis: application to the DAMADICS benchmark problem. Control Eng Pract 14(6):671–683

Wu Zh, Huang Zh (2009) Ensemble empirical mode decomposition: a noise-assisted data analysis method. Adv Adapt Data Anal 1:1–41

Yoon H, Hyun Y, Ha K, Lee KK, Kim GB (2016) A method to improve the stability and accuracy of ANN- and SVM based time series models for long-term rainfall predictions. Comput Geosc 90:144–155

Yoon H, Jun SC, Hyun Y, Bae GO, Lee KK (2011) A comparative study of artificial neural networks and support vector machines for predicting rainfalls in a coastal aquifer. J of Hydrol 396:128–138

Young WA, Millie DF, Weckman GR, Anderson JS, Klarer DM, Fahnenstiel GL (2011) Modeling net ecosystem metabolism with an artificial neural network and Bayesian belief network. Environ Modell Softw 26(10):1199–1210
Yu Y, Li W, Sheng D, Chen J (2015) A novel sensor fault diagnosis method based on modified ensemble empirical mode decomposition and probabilistic neural network. Measurement 68:328–336
Zhan X (2005) Parallel Fortran-MPI software for numerical inversion of the Laplace transform and its application to oscillatory water levels in groundwater environments. Environ Modell Softw 20(3):279–284
Zhao JP, Huang DJ (2001) Mirror extending and circular spline function for empirical mode decomposition method. J Zhejiang Univ Sci 2(3):247–252

Landslide Inventory, Sampling and Effect of Sampling Strategies on Landslide Susceptibility/Hazard Modelling at a Glance

Isik Yilmaz and Murat Ercanoglu

Abstract Landslides have a significant portion of responsibility on the damages and losses caused by natural hazards such as earthquakes, floods, storms, and tsunamis all over the world. Thus, landslides and their consequences are of great importance among the scientists and authorities who want to minimize these effects for a long time. This procedure simply begins with the preparation of landslide database and inventory maps, which constitutes a fundamental basis for the further steps including landslide susceptibility, hazard, and risk assessments. In this aspect, this procedure can be considered as one of the most important stages for any landslide work to minimize the undesired consequences of landslides. This stage can be realized using some statistical techniques such as simple random, systematic, stratified and cluster sampling strategies in the literature. In this chapter, firstly, basic landslide definitions and concepts were discussed. Then, landslide inventory, susceptibility and hazard concepts were pointed out and linked to the sampling strategies with the recent literature. Although, every considered method has pros and cons, it could be concluded that the sampling carried out in the rupture zones of landslides as polygon features or seed cell approach representing the pre-failure conditions seem to be more realistic to obtain more accurate maps. The other important issue pointed out in this chapter is on the selection of data mining technique(s). Since landslides are complex processes and can be affected by many factors, this stage is very important to reflect the landslide conditions with huge amount of data. In many cases, the researchers generally encounter to struggle with huge amount of data related to the landslide initiation and/or mechanisms. Thus, the selection of data mining techniques deserve the necessary precaution and is elaborately discussed overall the chapter.

I. Yilmaz (✉)
Department of Geological Engineering Sivas, Cumhuriyet University, Sivas, Turkey
e-mail: iyilmaz@cumhuriyet.edu.tr

M. Ercanoglu
Geological Engineering Department, Hacettepe University, Ankara, Turkey
e-mail: murate@hacettepe.edu.tr

H. R. Pourghasemi and M. Rossi (eds.), *Natural Hazards GIS-based Spatial Modeling Using Data Mining Techniques*, Advances in Natural and Technological Hazards Research 48, https://doi.org/10.1007/978-3-319-73383-8_9

Keywords Data mining · GIS · Landslide inventory · Sampling strategy
Susceptibility/hazard mapping

1 Introduction

Outbreak consequences and undesired effects of landslides both on human beings and on environment are an ever increasing phenomenon throughout the world similar to the other natural hazards such as earthquakes, floods, storms, tsunamis etc. The main reasons for this situation can be sourced from the increase in population, unconsciously built settlements and the extraordinary climatologic changes. Unfortunately, people tend to live in hazardous areas due to economic reasons particularly in developing and/or under developed countries. As a matter of fact, these natural events have been occurring since the beginning of the Earth; but, if the human beings or the other living creatures are involved and suffered from these events, they are transformed into the natural hazards.

Landslides, one of the most hazardous natural events on the Earth, directly and/ or indirectly affect the human life and the inhabitants. Although it plays an important part on the evolution of the Earth, many countries have been suffering from landslides and their consequences. Generally speaking, governments, local or general authorities, institutions, and agencies have been struggling with these events all around the world. When compared with the 20 years' period before, it could be concluded that the people recently have been more informed and conscientious on combating with landslides.

One of the most important stages of landslide hazard mitigation efforts is the construction of the landslide database and preparation of landslide inventory maps. This could be done by different methods as will be explained in the following sections. But, the crucial point in this context is related to the selection of landslide and non-landslide data (e.g. point or polygon), which is mainly based on sampling strategy employed for the analyses. Actually, this stage is fundamentally linked to the inventory and susceptibility stages; but, landslide hazard and risk mapping stages are also influenced by this procedure. In other words, preparation of landslide inventory maps and database is the key point to any landslide mapping procedure. Particularly, during the database creation and analysis stage, statistical analyses have significant importance. In conjunction with the developments in GIS (Geographical Information Systems) and computer technology, it is now possible to assess huge data sets using some statistical techniques to create automatic, random or data-driven information related to landslide assessments. Thus, utilization of statistical techniques is indispensable during these stages. The statistical concepts related to this chapter can be found in Peck et al. 2008), Borzyszkowski and Sokolowski (1993), Pratt et al. (1995), Cochran (1977).

In this chapter, the readers will find some informative knowledge on landslides, landslide inventory mapping concepts related to landslide susceptibility/hazard and most recently used sampling strategies in landslide assessments. A detailed and recent literature survey was conducted on the subject and the outcomes were elaborately discussed in the following lines.

2 Landslides

Natural disasters which are resulted from the Earth's natural processes cause loss of life and damages to properties worldwide. The natural disasters include mainly earthquakes, landslides, hurricanes, tsunamis, floods, volcanic eruptions, and tornadoes. Landslides disturbing many parts of the world may be accepted to be one of the most common and the most crucial natural disasters. Therefore, the landslides are commonly answered for significant loss of money and life (Figs. 1 and 2).

Cruden and Varnes (1996) defines the landslide as "*it is the movement of a mass of rock, debris, or earth down a slope, under the influence of gravity*". Landslides can be classified depending on the type of movement (fall, topple, slide, spread, and flow) (Fig. 3) and type of material (rock, soil, or their combination) that failed as suggested by Varnes (1978) and Cruden and Varnes (1996) (Table 1). Herein this classification, rock is defined to be intact hard and/or firm bedrock before the slope movement. Soil is composed of poorly cemented, unconsolidated particles

Fig. 1 A landslide occurred in Koyulhisar (Sivas, Turkey) (photo taken by the author)

Fig. 2 Loss of life and damages to properties by the landslide occurred in Koyulhisar (Sivas, Turkey) (photos taken by the author)

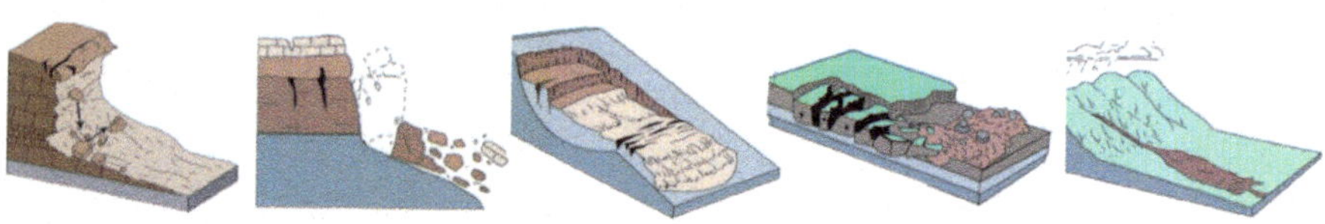

Fig. 3 Classification of slope movements (Varnes 1978)

and can be classified into two categories such as; coarse fragments (debris) and fine fragments (earth).

In fall (Fig. 4a), soil/rock masses displace in steep slopes and start to free fall, bounce/roll downslope. Topple (Fig. 4b) failure involve a forward rotation around an axis below the soil/rock mass and its movement. Lateral spreading (Fig. 4c) is a movement by horizontal extension, shear/tensile fractures which generally initiated

Table 1 Summary of Varnes' (1978) classification system (after Hungr et al. 2014)

Movement type	Rock	Debris	Earth
Fall	Rock fall	Debris fall	Earth fall
Topple	Rock topple	Debris topple	Earth topple
Rotational sliding	Rock slump	Debris slump	Earth slump
Translational sliding	Block slide	Debris slide	Earth slide
Lateral spreading	Rock spread	–	Earth spread
Flow	Rock creep	Talus flow	Dry sand flow
		Debris flow	Wet sand flow
		Debris avalanche	Quick clay flow
		Solifluction	Earth flow
		Soil creep	Rapid earth flow
			Loess flow
Complex	Rock slide-debris avalanche	Cambering, valley bulging	Earth slump-earth flow

by liquefaction of soil during earthquake and occur on gentle slopes. In slide (Fig. 4d), soil/rock masses displace along one or more discontinuity planes. In rotational slides, masses move along a curved and concave slide plane with a speed from extremely slow to extremely rapid. However, failure surface in translational sliding is more/less planar or wavy, and movement of the mass is generally parallel to the ground surface. Flow (Fig. 4e) is a rapid mass movement of mixture of water, soil, rock and moves in shear surfaces which are closely spaced and non-persistent. Sometimes, failures may occur to be combination of more than one type of movement (fall, topple, slide, spread, flow), and this is classified as complex movement (Fig. 4f).

Landslides can also be classified into mainly 2 categories of "*active*" and "*inactive*" (young inactive, mature inactive and old inactive) based on the activity according to the suggested classification system by Keaton and De Graff (1996). According to Wieczorek (1984), landslides can be classified to be active and dormant (young, mature and old) (Fig. 5).

Active landslide category includes landslides that currently moved or movement (s) have been recorded in the past and they have been still moving. The recent activities can be assessed by vegetation disruption, fresh cracks, etc. Water accumulation in depressions which is formed by mass movement and/or embankment of streams (Wieczorek 1984).

In *young dormant* landslides, relatively fresh landforms are observed, but historical movements were not recorded. Cracks cannot be observed because they are generally eroded. However, the scarps of landslides are observed to be rounded and sediments filled the depressions and/or landslide ponds.

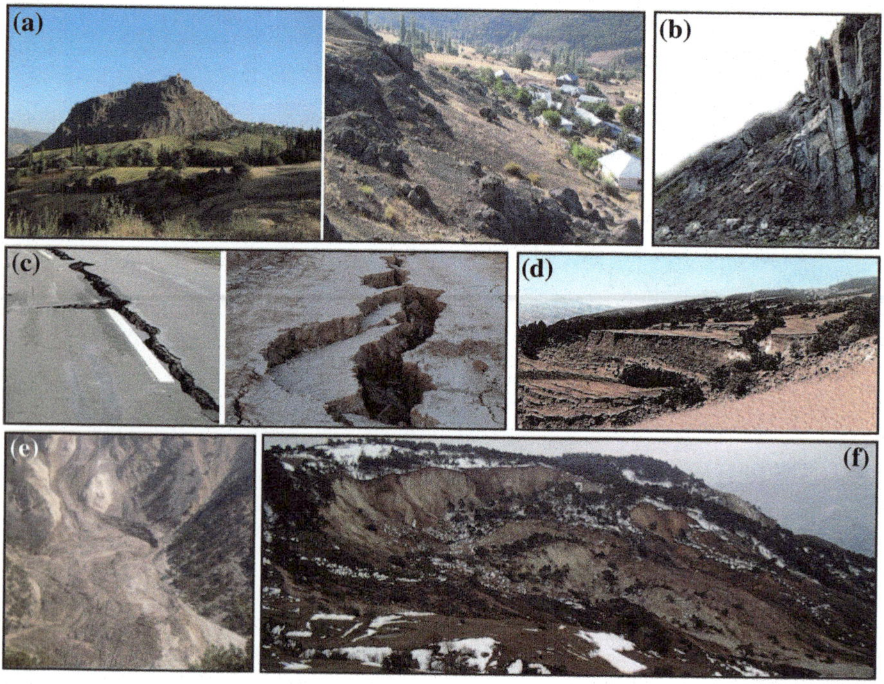

Fig. 4 Some example photos for; **a** fall, **b** topple, **c** lateral spreading, **d** slide, **e** flow, and **f** complex slide (photos taken by the author)

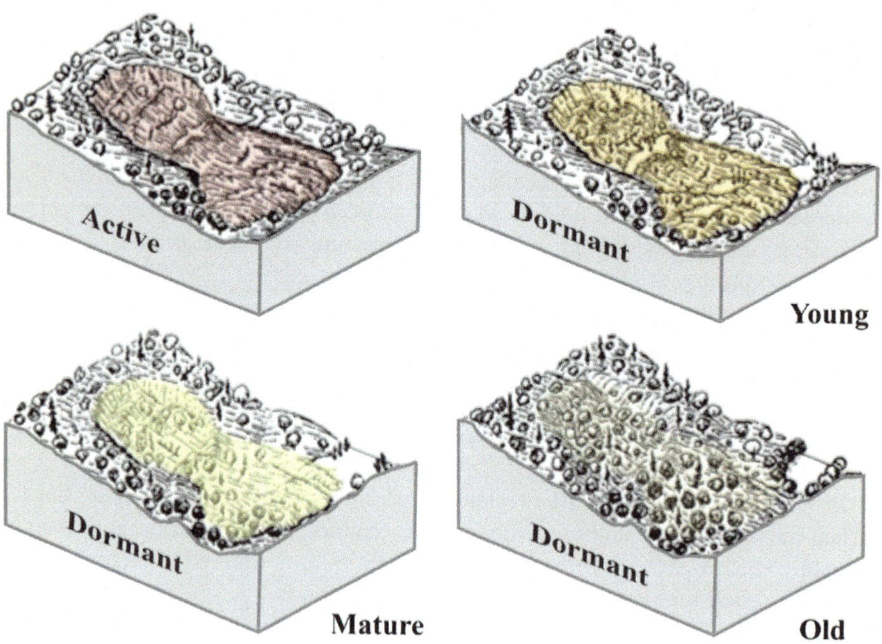

Fig. 5 Classification of the landslides according to the activity (after Wieczorek 1984)

In *mature dormant* landslides, landslide landforms are observed to be smoothed by erosion and/or revegetation. Rounded main scarp, eroded toe, new drainage occurred in landslide area is generally observed. Mounded topography and benches on slopes are covered by dense vegetation in a widespread manner (Wieczorek 1984).

Mostly rounded and subtle scarp, mounded topography and benches, greatly eroded landslide landforms with possible glens, canyons, closed and filled depressions, new and dense vegetation which is similar with the vegetation in outside of the landslide boundaries are observed in *old dormant* landslides (Wieczorek 1984).

3 Susceptibility/Hazard Assessment

The land use and urbanization strategies are also effected by the severity of landslide. It can be accepted that the knowledge is still incomplete according to the recent years' experiences related with recognition, understanding and treatment of landslide hazard. Combination or one of four main preventive measures such as drainage, modification of slope geometry, retaining structures and internal slope reinforcement are commonly used to control and prevent the landslides. In order to choose and design the favourable and cost effective remedial measures, conditions and processes caused the landslide occurrences must be clearly understood. It is hoped that forthcoming developed techniques which are new, cheaper and more efficient will minimize the impact of a landslide in the future.

However, there are few preventive measures for landslides, establishment of realistic reliable landslide prone zones on a map would be very useful tool in urban planning and effective land management in order to solve problems caused by landslides.

Assessment of landslide prone and susceptible areas is an essential prerequisite in hazard mitigation, disaster management and safe city and urban planning. Mainly three types of maps are qualitatively/quantitatively prepared such as landslide susceptibility, hazard and risk maps.

Unstable conditions sourced from presence or probable occurrence of slope failures in the future are described in landslide hazard maps. However, landslide susceptibility maps are prepared by real local and/or site properties to classify the relative probability of landsliding in the future. In the preparation of the landslide susceptibility maps, prior failures obtained from landslide inventory, factors of geological, geomorphological, hydrogeological, topographical, etc. should be considered. On the other hand, landslide potential with the expected losses of life and property by the occurrence of landslide are described on the landslide risk maps.

Landslide susceptibility maps which is the early stage of the assessment of landslides can serve to reduce the losses. However, there is no agreement on the standard assessment model type/procedure, many researchers used a number of various models in landslide susceptibility assessment and landslide susceptibility mapping. These models can be divided into the two main groups to be deterministic and non-deterministic which is more frequently used and known as probabilistic.

Probabilistic models had been frequently used and many methodologies based on inventory of landslides, geomorphological analysis, qualitative and statistical bivariate analysis, multivariate analysis in many published researches such as; Brabb et al. (1972), Degraff and Romersburg (1980), Carrara (1983), Carrara et al. (1991), Jade and Sarkar (1993), Baeza (1994), Chung et al. (1995), Irigaray (1995), Rengers et al. (1998), Chung and Fabbri (1999), Fernández et al. (2003), Yilmaz and Yildirim (2006), etc. And, many other researchers such as Ives and Messerli (1981), Ward et al. (1982), Rupke et al. (1988), Cascini et al. (1991), Van Westen (2000), Chacón et al. (1994, 1996), Gokceoglu and Aksoy (1996), Chung and Fabbri (1999), Barredo et al. (2000), Van Westen et al. (2000), Dai et al. (2001), Lee and Min (2001), Carrara et al. (2003), Yilmaz (2009a, b, 2010a), Yilmaz and Keskin (2009), Bednarik et al. (2012), Holec et al. (2013), etc. used various models such as heuristic, deterministic, statistical, fuzzy-logic, artificial neural networks, neuro-fuzzy, support vector machine, etc. in order to establish landslide susceptibility models.

In general, data mining can be defined as a tool describing and analysing a large amount of data set. It can be considered as one of the most useful techniques when analysing large amount of data and producing landslide susceptibility, hazard and risk maps. When the recent landslide literature has been examined, of the above mentioned data mining techniques, it is clear that ANN (Artificial Neural Network), DT (Decision Tree), LR (Logistic Regression) and SVM (Support Vector Machine) techniques have been more commonly used among the researchers. In the authors' opinion, the main reasons behind this situation are sourced from the characteristics of these methods such as their high capability of reflecting the nonlinear features of landslide occurrences and the complex relations of the considered parameters. Of course, all these techniques have advantages and disadvantages. For example, ANN has a black box nature and may include overfitting or underfitting problems during the analyses. However, when used correctly, it gives by far the most powerful results. It is very clear that it is not easy, to some extent, impossible, to solve these complicated natural processes by the linear models. In addition, landslide researchers generally select the pixels or points for evaluating the landslide process for the analyses instead of using polygons or sub-basins/catchments as the slope units.

GIS is a very important tool when analysing the landslide events and producing landslide maps. However, many GIS programs have no capability of performing such analyses by themselves. Therefore, the researchers have to use external statistical packages and have to transform the outputs into a GIS platform. Perhaps, the most lacking part of the GIS programs is the absence of these data mining techniques although its utilization is indispensable in these analyses.

4 Landslide Inventory Maps

A landslide inventory is a data set presenting a single event, a regional event, or multiple events (Yilmaz 2010b) which are very crucial clues for future landsliding. Landslide inventory map which is the most basic landslide map shows the locations and outlines of landslides. Some examples for landslide inventory maps are shown in Figs. 6 and 7.

As it was reported by Guzzetti et al. (2012); landslide inventory maps can be prepared for various aims (Brabb 1991). One of these aims is documentation of extend of landslides in areas small to large watersheds and from regions to states or national scales (Cardinali et al. 2001; Brabb and Pampeyan 1972; Antonini et al. 1993; Duman et al. 2005; Delaunay 1981; Radbruch-Hall et al. 1982; Brabb et al. 1989; Cardinali et al. 1990; Trigila et al. 2010). Another aim can be defined to be preliminary step in landslide susceptibility, hazard, risk assessment (Cardinali et al. 2002, 2006; Guzzetti et al. 2005, 2006a, b; Van Westen et al. 2006, 2008; Balteanu et al. 2010). By investigating the types, distribution and patterns of landslides, inventory maps are prepared for explanation of geological and/or morphological characteristics (Guzzetti et al. 1996). The landslide inventory maps also serve to the determination of the landscape evolution dominated by mass wasting processes (Hovius et al. 1997, 2000; Malamud et al. 2004, 2004; Guzzetti et al. 2008, 2009; Parker et al. 2011).

The first and crucial step in landslide susceptibility/hazard mapping is definition of areas and compilation of a landslide map (Galli et al. 2008; Van Westen et al. 2008; Fell et al. 2008; De Graff et al. 2012). The landslide susceptibility/hazard mapping is impossible without a complete landslide inventory map, and the quality of resulting map is strongly influenced by quality, reliability, and completeness of the landslide inventory.

The compulsory component of the assessment of landslide susceptibility, hazard and/or risk is the preparation of consistent and precise landslide inventory map. However, landslide inventory maps have been prepared in all over the world for many years, the appropriate method, reliability, eligibility, entirety and resolution of the landslide inventory maps are rarely agreed. The deficiencies on the appropriate information related with the value of inventory maps and the consistency of the methods for completing inventories may make the hazard or risk assessment highly questionable.

However, the standard mapping method is realized in the field, a landslide inventory map can be prepared by interpretation of aerial photographs and/or satellite imageries (Soeters and Van Westen 1996; Van Westen et al. 2006) with the detail information about type of movement, dimension, activity, morphology, etc. for each landslides (Holec et al. 2013; McCalpin 1984). Because mapping by field works may sometimes be handicapped by few difficulties such as; old and very large size landslides, the landslides covered by vegetation, forest, agricultural lands and disrupted by other landslides and/or erosion, etc. Nevertheless, the prepared map must then be checked by field works in order to ensure the completeness and reliability of the inventory for landslide susceptibility/hazard assessment.

During the mapping of landslides, classification of the landslides according to the activity must also be taken under consideration and must be indicated on the map. Activity categories allow to estimate how recently slopes moved and the type of hazard represented. As it was known that the movement/failure probability in active landslides is more than older landslides which may move/fail less likely as single and small slide masses.

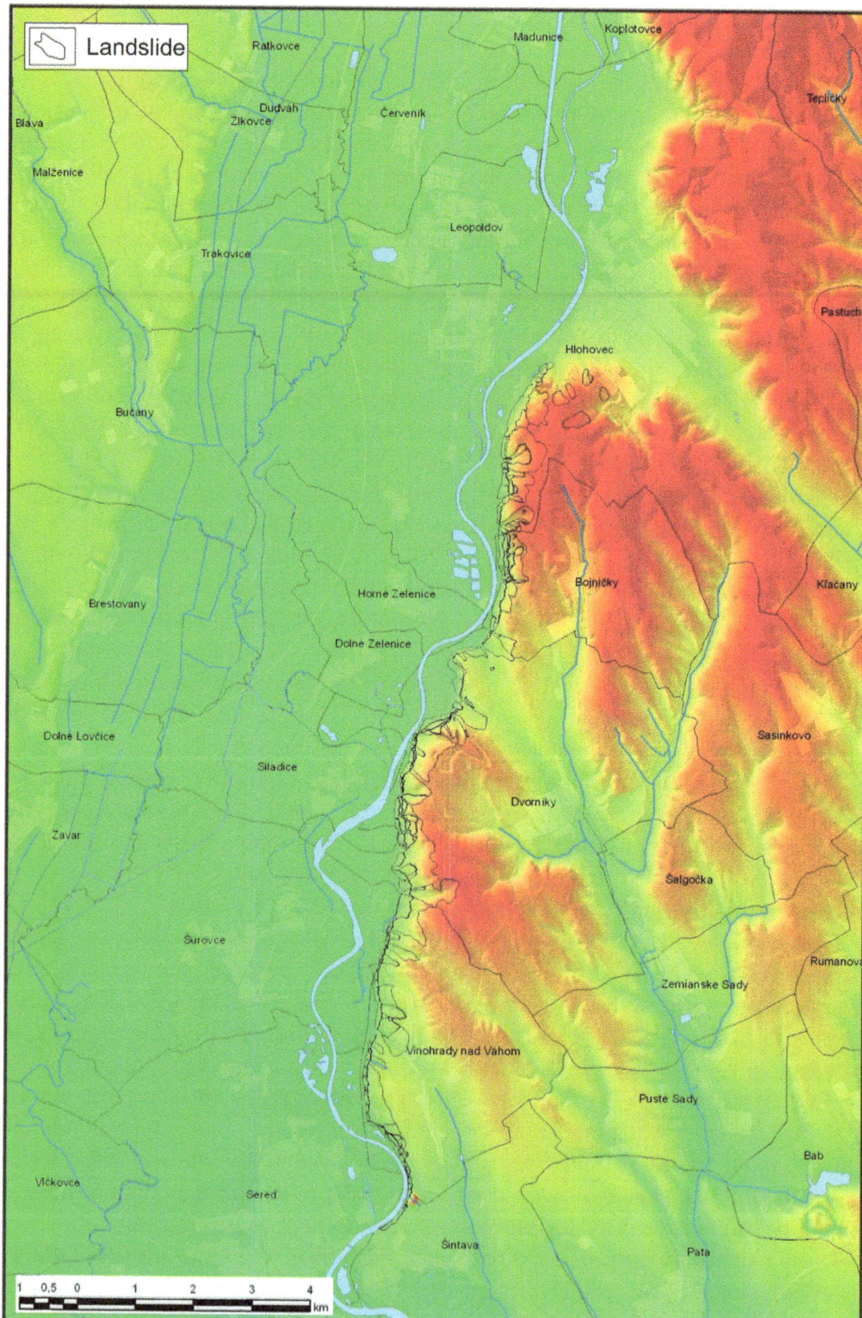

Fig. 6 Landslide inventory map which was used to be an input parameter in landslide hazard mapping in Hlohovec–Sered' landslide area in south-west Slovakia (Bednarik et al. 2012)

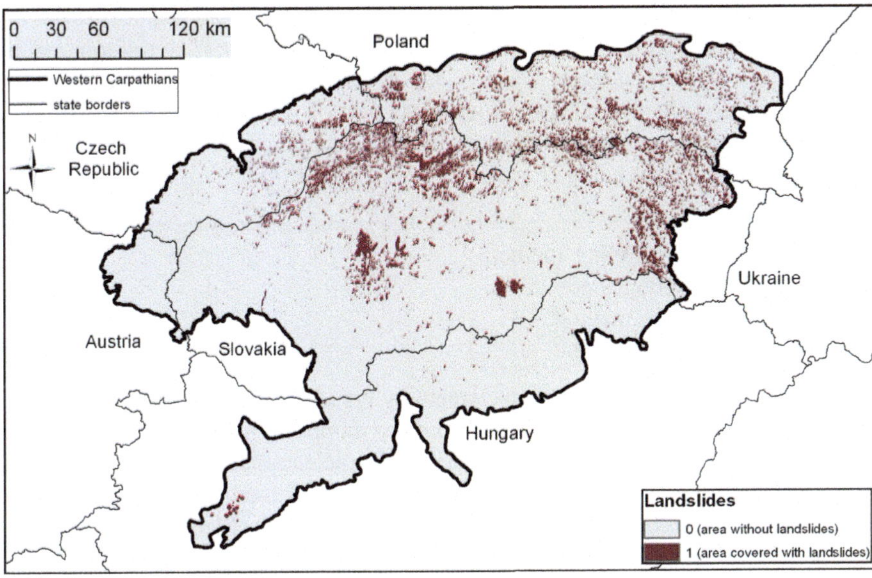

Fig. 7 Landslide inventory map which was used to be an input parameter in small-scale landslide susceptibility assessment for the territory of Western Carpathians (Holec et al. 2013)

5 Sample and Sampling in Statistics

The key is to have representative samples of location specific condition and of course landslides. And, they must be selected randomly. The two main questions are which part of the landslides and how they will be randomly sampled.

In statistics, samples are parts of a population and represent a subset of manageable size. Sample points, sampling units and observations are defined to be elements of sample (Peck et al. 2008). Because of the size of the population which is very large, sample census and/or listing all values of the population is mostly impossible and not practical (Borzyszkowski and Sokolowski 1993). Therefore, selection and/or collection of a set of data from statistical population is required. As it was defined by Pratt et al. (1995); a complete sample can be defined to be set of objects selected from the main population which contain objects that meet well defined criteria of the selection.

The most important thing in sampling is to avoid selecting an unrepresentative and/or biased sample(s). In order to have a representative and unbiased sample, it must be selected from a population by a method which is independent with the features of objects. The main difference between random and non-random sampling techniques is related with their probability sampling and non-probability sampling characteristics. Non-random sampling techniques use non-probabilistic sampling, and sampling is effected by judgement of researcher. Therefore, non-random sampling is generally very biased and non-representative. However, random sampling techniques are called as probability sampling and reduces bias and samples are representative. As it was defined by Cochran (1977); random sample can be defined to be sample where every discrete member of population has a known, non-zero chance of being selected as part of sample.

Random sampling method, random number generation and sampling frame is required in random sampling. Figure 8 shows random sampling methods and can be classified mainly:

a. Simple random sampling method,
b. Systematic sampling method,
c. Stratified sampling method,
d. Cluster sampling method.

A simple random sampling is an elementary type of sampling methods and it can be accepted to be an element of further complex sampling methods. The probability to be chosen of all objects are the same in simple random sampling. Systematic sampling method is a kind of simple random sampling method where samples are selected in an ordered systematic way. Stratified Sampling Method can be applied by dividing the population into homogeneous subgroups and sample is selected from each group using random simple or systematic sampling method. As in stratified sampling method, population is similarly divided into clusters in Cluster Sampling Method and a sample is then randomly selected from clusters. If the individuals in the population are not known, but groups in population are known, cluster sampling method is used.

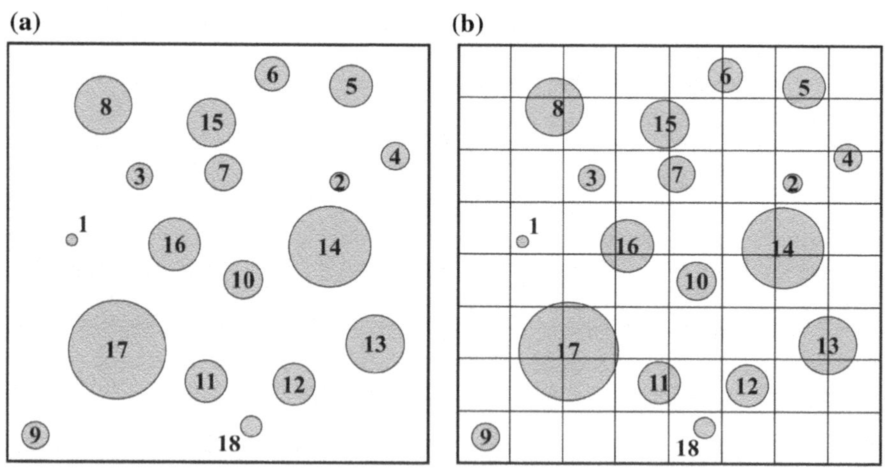

Fig. 8 Sampling from the circles numbered 1–18; **a** simple random sampling: 7 circles are selected at random, **b** cluster sampling method: circles lying within the squares are selected until reaching 7 circles

6 Sampling Strategy

Different inventory maps for sampling where the landslides are generally shown/ drawn to be point, scarp and seed-cell (Fig. 9) are in use by researchers, however, there is no agreement. In the landslide literature, there are mainly four sampling strategies and these are represented in Fig. 9 as for the different landslide locations. Of these, the seed cell approach was introduced by Suzen and Doyuran (2004) by applying a buffer zone ("4 pixel*resolution" meter sized) of a landslide body (Fig. 9, Landslide A). This approach considers approximately 2/3 of the zone of depletion starting from just out of the scarp, drawn by a buffer zone outside the landslide location, expressing the pre-failure conditions. The second commonly used approach is that a polygon is drawn of the main scarp which is distinguished from the accumulation/depletion zone approach (Fig. 9, Landslide B). The other most commonly used approach is to consider all point samples from the whole depletion zone of the landslide bodies (Fig. 9, Landslide C). This approach is generally used when there are deep-seated large landslides. Finally, the fourth approach is that the sampling could be done by as points at the upper part of the scar (Fig. 9, Landslide D).

Almost every sampling strategy has pros and cons, but the main principle is that the sampling strategy taken into account should be representative for the area studied. As it was reported by Nefeslioglu et al. (2008) that the conceptive distinction related with the various sampling strategies applied is frequently ignored and not stated anymore. Only a few study emphasized on this difference in literature. In the analyses by Dai and Lee (2003), the source area was used after the separation of source area and run-out zone. Fernandez et al. (2003) had explained the analysis of rupture zone in mapping of landslide susceptibility. Remondo et al. (2003) proposed the hypothesis of landslide rupture, and the rupture zone was considered by Santacana et al. (2003) in susceptibility assessment of landslide. Suzen and Doyuran

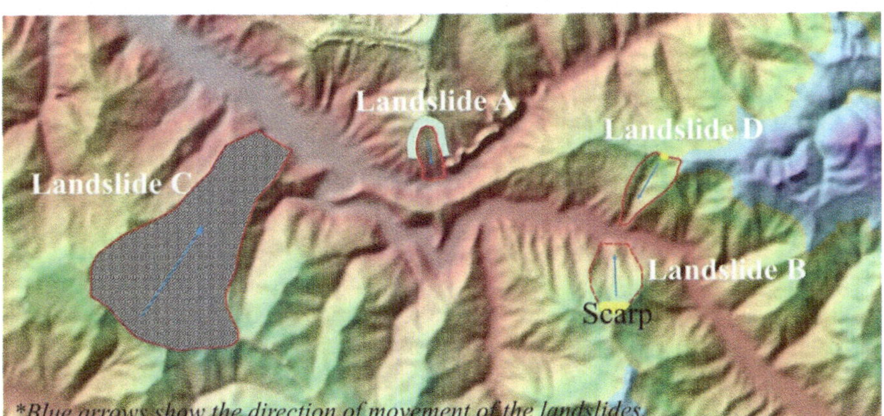

Fig. 9 A schematic representation of different sampling strategies

(2004) suggested the extraction of undisturbed morphological conditions from the close vicinity of the landslide polygon itself. Selected factor cells defined on the upper edge of the main scarps of landslides were included in the landslide susceptibility assessment by Clerici et al. (2006) and Nefeslioglu et al. (2008).

Yilmaz (2010b) had compared the results from different sampling strategies such as; point, scarp and seed cells in production of the landslide susceptibility maps for a case location in Turkey. The comparison obtained from the study allowed to the quantitative estimation of differences between inventories of point, scarp and seed cells. In the study of Yilmaz (2010b), 3 different inventory maps were firstly prepared by different strategies for sampling. The first one was prepared by drawing polygons of main scarp which is visible from accumulation/depletion or rupture zone. Second map was constructed by considering the suggested method of seed-cell (Suzen and Doyuran 2004). The last map was produced by plotting of locations to be a point selected on upper part of scar. As the main result of the analysis, Yilmaz (2010b) reported that the unreliable result was obtained from "point sampling" while the most realistic result was obtained from "scarp". The accuracies obtained from "scarp" was relatively similar with the result obtained from "seed-cell" because of the sampled areas of both of them are very close one another. Similarly, Nefeslioglu et al. (2008) had also pointed out that the sampling procedure applied in the presence data gathered from which the samples represent pre-failure conditions (the seed cell concept in that study) of landslides were produced more realistic landslide susceptibility maps. As discussed previously, the most important issue in sampling strategy procedure is the representation capacity of the selected elements on the overall data, whether they are point or polygon. However, as stated by Yilmaz (2010b), the areal extent or coverage of the landslides is also a significant parameter when selecting the type of sampling element (i.e. point or polygon). In other words, since a point can be defined by a single X, Y coordinate, it has no capability of representing the whole landslide body itself. On the other hand, particularly for small-sized landslides, it could be more logical to use point data when the areal extent is too small to draw by considering the scale of the map.

In the sampling strategy concept, there is one more important issue, which should be considered before applying any method to produce landslide susceptibility, hazard or risk map, called data partitioning. When partitioning the data in landslide analyses, there is no rule of thumb either for landslides or for nonlandslide data. Indeed, this partition procedure is needed in many applications to represent how well the applied methodology works. In other words, the researchers generally divide the overall data for training and validation stages to evaluate the efficiency of the so-produced maps. A recent work is performed on this subject by Ercanoglu et al. (2016) what the ratio should be on partitioning of data for training and validation stages. In general, the researchers subjectively select this ratio ranging from 50 to 90% for training, while the left parts (ranging from 50 to 10%, summing up to 100%) of the data were considered for the validation stage. Ercanoglu et al. (2016) evaluated that the partitioning the overall data as 75% for training and 25% for validation stages produced more reliable and powerful results.

7 GIS and Data Mining in Landslide Assessments

In the beginning of the 1980s, the computer technology and the GIS applications and software have witnessed a real "boom" with respect to the development and usefulness. When considered with the old technologies, the speed and capability of the GIS components have been emerged extensively. Today, it could be concluded that the utilization of GIS in any landslide assessment (i.e. susceptibility, hazard or risk) is indispensable. Main reasons for this situation comes from the fact that it provides many advantages in data mining, organization and analysis as well as the representation of the results. There are too many parameters and data sources and layers, sometimes reaching up to hundreds of millions pixels according to the scale and the resolution, in landslide assessments and mapping. Thus, there is a necessity of managing and organizing the data to produce an output (i.e. a map of susceptibility, hazard or risk) in such a work. In addition, there are many uncertainties related to the landslide assessments sourced from the nature of the landslides. Therefore, to model or to assess any landslide mapping work, some traditional modelling approaches such as geomorphological assessments, basic overlying techniques etc. were disappeared in the recent landslide literature. New techniques, commonly data driven methods, such as SVM (Support Vector Machines), ANN (Artificial Neural Networks), MLM (Machine Learning Methods), come to the fore based on the landslide inventory and the considered parameters expressed by the GIS layers in the recent landslide literature. Perhaps, the data mining concept is much more important in such a big data environment to model the landslides and their future prediction since the susceptibility, hazard and risk of the landslides are related to the future conditions. In other words, it could be concluded that this necessity was sourced from the number of data and the uncertainty since the landslides were seldom linked to a single cause. Linear models have no capability of modelling such this case, but, nonlinear ones have the opportunity to reflect the actual conditions and lacking of the data related to the uncertainties.

Since the landslide related studies are very difficult tasks due to the huge spatial and temporal variables and the considered parameters, data mining techniques and GIS may be helpful in assessing the landslide phenomenon since they explain the collected or gathered data in many ways. For example, these methods show us how to find the useful knowledge or information in such a big database related to the landslides including the millions of pixels. In addition, it helps us to remove inconsistent and noisy data to represent more important and beneficial way to solve the complex landslide phenomenon.

8 Conclusions

To minimize the effects of landslides on lives, properties and the environment, the first crucial point is to prepare landslide inventory map and to construct a reliable landslide database. This stage also influences the results of the further assessments such as landslide susceptibility, hazard or risk. Thus, in every landslide study, the required attention should be paid for this stage. Of course, the produced maps should be considered by the decision makers and should be used in engineering applications before they were built.

Generally, the other important aspect of this chapter was revealed that there were two major approaches for sampling procedures to build a landslide database. The first one is based on the sampling strategy carried out in the zone of depletion or rupture zone as landslide polygons or point samplings. The other one is based on the pre-failure conditions such as seed cell approach. In general, the researchers use simple random, systematic, stratified and cluster samplings in landslide assessments. Non-random sampling procedure is rarely preferred since it is a completely subjective procedure. Whether it shows pre-failure or after failure conditions, based on the landslide literature, the researchers generally obtain reliable results reflecting the landslide locations. There is no a generally accepted approach or methodology for this subject. However, it could be concluded that both types of sampling could be used, but the selection depends upon the performances of the final maps. In other words, sampling strategy representing the pre-failure conditions is very logical way to represent the landslide conditions will be occurred in the future (i.e. for landslide susceptibility mapping) because of the fact that they were selected from very close vicinity of the landslide bodies before they moved. Contrary, based on the very well-known principle of uniformitarianism ("today and past are key to the future") in geology science, the samples taken from inside the landslided bodies represent the failure conditions, and they will reflect the same conditions will be occurred in the future for a landslide. The landslide type is also a significant issue herein this subject. If an earth flow type landslide is considered for a landslide analysis, utilization of any point sampling strategy taken from the scarp would be appropriate to reflect the landslide initiation conditions. However, for a deep-seated large sized earth slide, it would not be wise to represent this landslide with a single point. Consequently, it should be noted that the selected samples should represent the population (i.e. landslided and nonlandslided areas) they belong in any sampling strategy. Since the landslide researchers struggle with huge number of data, this stage should be performed accurately. Utilization of GIS and data mining techniques may be helpful in this context to solve huge number of data and complexity problems. By doing so, more reliable and powerful landslide maps could be produced.

References

Antonini G, Cardinali M, Guzzetti F, Reichenbach P, Sorrentino A (1993) Carta Inventario dei Fenomeni Franosi della Regione Marche ed aree limitrofe. CNR, Gruppo Nazionale per la Difesa dalle Catastrofi Idrogeologiche, Publication n. 580, 2 sheets, scale 1:100,000, (in Italian)

Baeza C (1994) Evaluación de las condiciones de rotura y la movilidad de los deslizamientos superficiales mediante el uso de técnicas de análisis multivariante, Tesis Univ. Pol. Catalunya

Balteanu D, Chendeş V, Sima M, Enciu P (2010) A country-wide spatial assessment of landslide susceptibility in Romania. Geomorphology 124:102–112

Barredo JJ, Benavides A, Hervas J, Van Westen CJ (2000) Comparing heuristic landslide hazard assessment techniques using GIS in the Trijana basin, Gran Canaria Island, Spain. JAG 2(1):9–23
Bednarik M, Yilmaz I, Marschalko M (2012) Landslide hazard and risk assessment: a case study from the Hlohovec-Sered landslide area in south-west Slovakia. Nat Hazards 64(1):547–575
Borzyszkowski AM, Sokolowski S (eds) (1993) Mathematical foundations of computer science 1993. in 18th international symposium, MFCS'93 Gdansk, Poland, August 30–September 3, 1993 Proceedings, Lecture Notes in Computer Science, vol 711, pp 281–290
Brabb EE (1991) The world landslide problem. Episodes 14(1):52–61
Brabb EE, Pampeyan EH (1972) Preliminary map of landslide deposits in San Mateo County, California. U.S. Geological Survey Miscellaneous Field Studies Map, MF-344
Brabb EE, Pampeyan EH, Bonilla M (1972) Landslide susceptibility in the San Mateo County, California, scale 1: 62.500, U.S. Geol. Survey Misc. Field Studies Map MF344
Brabb EE, Wieczorek GF, Harp EL (1989) Map showing 1983 landslides in Utah. U.S. Geological Survey Miscellaneous Field Studies Map MF-1867
Cardinali M, Guzzetti F, Brabb EE (1990) Preliminary map showing landslide deposits and related features in New Mexico. U.S. Geological Survey Open File Report 90/293, 4 sheets, scale 1:500,000
Cardinali M, Antonini G, Reichenbach P, Guzzetti F (2001) Photo geological and landslide inventory map for the Upper Tiber River basin. CNR, Gruppo Nazionale per la Difesa dalle Catastrofi Idrogeologiche, Publication n. 2116, scale 1:100,000
Cardinali M, Carrara A, Guzzetti F, Reichenbach P (2002) Landslide hazard map for the Upper Tiber River basin. CNR, Gruppo Nazionale per la Difesa dalle Catastrofi Idrogeologiche, Publication n. 2634, scale 1:100,000
Cardinali M, Galli M, Guzzetti F, Ardizzone F, Reichenbach P, Bartoccini P (2006) Rainfall induced landslides in December 2004 in south-western Umbria, central Italy: types, extent, damage and risk assessment. Nat Hazards Earth Syst Sci 6:237–260
Carrara A (1983) Multivariate models for landslide hazard evaluation. Math Geol 15(3):403–426
Carrara A, Cardinalli M, Detti R, Guzzetti F, Pasqui V, Reichenbach P (1991) GIS techniques and statitistical models in evaluating landslide hazards. Earth Surf Proc Land 16:427–445
Carrara A, Crosta G, Frattini P (2003) Geomorphological and historical data in assessing landslide hazard. Eart Surf Processes Land 28:1125–1142
Cascini L, Critelli S, Gulla G, Di Nocera S (1991) A methodological approach to landslide hazard assessment: a case history. In: Proceedings of 16th international landslide conference. Balkema, Rotterdam, pp 899–904
Chacón J, Irigaray C, Fernández T (1994) Large to middle scale landslide inventory, analysis and mapping with modelling and assessment of derived susceptibility, hazards and risks in a GIS. In: Proceedings of 7th IAEG congress, Balkema, Rotterdam, Holland, pp 4669–4678
Chacón J, Irigaray C, Fernández T (1996) From the inventory to the risk analysis: improvements to a large scale GIS method. In: Chacón J, Irigaray C, Fernández T (eds), Proceedings of 8th international conference and field workshop on landslides, Balkema, Rotterdam, Holland, pp 335–342
Chung CF, Fabbri AG (1999) Probabilistic prediction models for landslide hazard mapping. Photogram Eng Remote Sens 65(12):1389–1399
Chung CF, Fabbri AG, Van Westen CJ (1995) Multivariate regression analysis for landslide hazard zonalition. In: Carrara A, Guzetti F (eds) Geographical informations systems in assessing natural hazards. Kluwer Publishers, Dordrecht
Clerici A, Perego S, Tellini C, Vescovi P (2006) A GIS-based automated procedure for landslide susceptibility mapping by the conditional Analysis method: the Baganza valley case study (Italian Northern Apennines). Environ Geol 50(7):941–961
Cochran WG (1977) Sampling techniques, 3rd edn. Wiley. ISBN 0-471-16240-X
Cruden DM, Varnes DJ (1996) Landslide types and processes. In: Turner AK, Schuster RL (eds) Landslides investigation and mitigation. Transportation research board, US National Research Council. Special Report 247, Washington, DC, Chapter 3, pp 36–75

Dai FC, Lee CF (2003) A spatiotemporal probabilistic modelling of storm induced shallow landsliding using aerial photographs and logistic regression. Earth Surf Proc Land 28:527–545
Dai FC, Lee CF, Zhang XH (2001) GIS-based geo-environmental evaluation for urban land-use planning: a case study. Eng Geol 61:257–271
De Graff JV, Romesburg HC, Ahmad R, McCalpin JP (2012) Producing landslide-susceptibility maps for regional planning in data-scarce regions. Nat Hazards 64:729–749
DeGraff J, Romesburg H (1980) Regional landslide-susceptibility assessment for wildland management: a matrix approach. In: Coates D, Vitek J (eds) Thresholds in geomorphology. George Allen and Unwin, London, pp 401–414
Delaunay J (1981) Carte de France des zones vulnèrables a des glissements, écroulements, affaissements et effrondrements de terrain. Bureau de Recherches Géologiques et Minières, 81. SGN 567 GEG, 23 p., (in French)
Duman TY, Çan T, Emre Ö, Keçer M, Doğan A, Şerafettin A, Serap D (2005) Landslide inventory of northwestern Anatolia, Turkey. Eng Geol 77(1–2):99–114
Ercanoglu M, Dagdelenler G, Özsayin E, Alkevli T, Sönmez H, Özyurt NN, Kahraman B, Uçar İ, Çetinkaya S (2016) Application of Chebyshev theorem to data preparation in landslide susceptibility mapping studies: an example from Yenice (Karabük, Turkey) region. J Mt Sci 13 (11):1923–1940
Fell R, Corominas J, Bonnard C, Cascini L, Leroi E, Savage WZ (2008) Guidelines for landslide susceptibility, hazard and risk zoning for land use planning. Eng Geol 102:85–98
Fernández T, Irigaray C, Hamdouni RE, Chacón J (2003) Methodology for landslide susceptibility mapping by means of a GIS. Application to the Contraviesa area (Granada, Spain). Nat Hazards 30:297–308
Galli M, Ardizzone F, Cardinali M, Guzzetti F, Reichenbach P (2008) Comparing landslide inventory maps. Geomorphology 94:268–289
Gokceoglu C, Aksoy H (1996) Landslide susceptibility mapping of the slopes in the residual soils of the Mengen region (Turkey) by deterministic stability analyses and image processing technique. Eng Geol 44:147–161
Guzzetti F, Cardinali M, Reichenbach P (1996) The influence of structural setting and lithology on landslide type and pattern. Environ Eng Geosci 2(4):531–555
Guzzetti F, Reichenbach P, Cardinali M, Galli M, Ardizzone F (2005) Probabilistic landslide hazard assessment at the basin scale. Geomorphology 72:272–299
Guzzetti F, Galli M, Reichenbach P, Ardizzone F, Cardinali M (2006a) Landslide hazard assessment in the Collazzone area, Umbria, central Italy. Nat Hazards Earth Syst Sci 6:115–131
Guzzetti F, Reichenbach P, Ardizzone F, Cardinali M, Galli M (2006b) Estimating the quality of landslide susceptibility models. Geomorphology 81:166–184
Guzzetti F, Ardizzone F, Cardinali M, Galli M, Reichenbach P (2008) Distribution of landslides in the Upper Tiber River basin, central Italy. Geomorphology 96:105–122
Guzzetti F, Ardizzone F, Cardinali M, Galli M, Rossi M, Valigi D (2009) Landslide volumes and landslide mobilization rates in Umbria, central Italy. Earth Planet Sci Lett 279:222–229
Guzzetti F, Mondini AC, Cardinali M, Fiorucci F, Santangelo M, Chang KT (2012) Landslide inventory maps: new tools for an old problem. Earth Sci Rev 112:42–66
Holec J, Bednarik M, Sabo M, Minar J, Yilmaz I, Marschalko M (2013) A small-scale landslide susceptibility assessment for the territory of Western Carpathians. Nat Hazards 69(1):1081–1107
Hovius N, Stark CP, Allen PA (1997) Sediment flux from a mountain belt derived by landslide mapping. Geology 25:231–234
Hovius N, Stark CP, Hao-Tsu C, Jinn-Chuan L (2000) Supply and removal of sediment in a landslide-dominated mountain belt: Central Range, Taiwan. J Geol 108:73–89
Hungr O, Leroueil S, Picarelli L (2014) The Varnes classification of landslide types, an update. Landslides 11:167–194
Irigaray C (1995) Movimientos de ladera: inventoria, analisis y cartografaa de susceptibilidad mediante un Sistema de Informacion Geografica. Aplicacion a las zonas de Colmenar (Ma), Rute (Co) y Montefrio (Gr). Thesis Doctoral, University Granada

Ives JD, Messerli B (1981) Mountain hazard mapping in Nepal: introduction to an applied mountain research project. Mt Res Dev 1(3–4):223–230
Jade S, Sarkar S (1993) Statistical models for slope instability classification. Eng Geol 36:91–98
Keaton JR, DeGraff JV (1996) Surface observation and geologic mapping. In: Turner AK, Schuster RL (eds) Landslides investigation and mitigation: National Research Council Transportation Research Board Special Report, vol 247, pp 178–230
Lee S, Min K (2001) Statistical analyses of landslide susceptibility at Yongin, Korea. Environ Geol 40:1095–1113
Malamud BD, Turcotte DL, Guzzetti F, Reichenbach P (2004a) Landslides, earthquakes and erosion. Earth Planet Sci Lett 229:45–59
Malamud BD, Turcotte DL, Guzzetti F, Reichenbach P (2004b) Landslide inventories and their statistical properties. Earth Surf Proc Land 29(6):687–711
McCalpin J (1984) Preliminary age classification of landslides for inventory mapping. In: Proceedings of the 21st engineering geology and soils engineering symposium, University, Moscow, ID, pp 99–120
Nefeslioglu HA, Gokceoglu C, Sonmez H (2008) An assessment on the use of logistic regression and artificial neural networks with different sampling strategies for the preparation of landslide susceptibility maps. Eng Geol 97:171–191
Parker RN, Densmore AL, Rosser NJ, de Michele M, Li Y, Huang R, Whadcoat S, Petley DN (2011) Mass wasting triggered by the 2008 Wenchuan earthquake is greater than orogenic growth. Nat Geosci 4(7):449–452
Peck R, Olsen C, Devore JL (2008) Introduction to statistics and data analysis, 3rd edn. Cengage Learning. ISBN 0-495-55783-8
Pratt JW, Raiffa H, Schaifer R (1995) Introduction to statistical decision theory. MIT Press, Cambridge, MA. MR1326829
Radbruch-Hall DH, Colton RB, Davies WE, Lucchitta I, Skipp BA, Varnes DJ (1982) Landslide overview map of the conterminous United States. U.S. Geological Survey Professional Paper, 1183. WWW page http://pubs.usgs.gov/pp/p1183/pp1183.html. 25 p
Remondo J, Gonzalez-Diez A, Teran JRD, Cendrero A (2003) Landslide susceptibility models utilising spatial data analysis techniques. A case study from the lower Deba Valley, Guipúzcoa (Spain). Nat Hazards 30:267–279
Rengers N, Van Westen CJ, Chacón J, Irigaray C (1998) Draft for the chapter on the application of digital techniques for natural hazard zonation, Report on Mapping of Natural Hazards, International Association of Engineering Geology. Commission No. 1 on Engineering Geological Mapping
Rupke J, Cammeraat E, Seijmonsbergen AC, Van Westen CJ (1988) Engineering geomorphology of the widentobel catchment, Switzerland: a geomorphological inventory system applied to geotechnical appraisal of the slope stability. Eng Geol 26:33–68
Santacana N, Baeza B, Corominas J, Paz A, Marturia J (2003) A GIS based multivariate statistical analysis for shallow landslide susceptibility mapping in la Pobla de Lillet area (Eastern Pyrenees, Spain). Nat Hazards 30:281–295
Soeters R, Van Westen CJ (1996) Slope instability, recognition, analysis, and zonation. In: Turner AK, Schuster RL (eds) Landslides—investigation and mitigation, transportation research board special report 247. National Academy Press, Washington, DC, pp 129–177
Suzen ML, Doyuran V (2004) Data driven bivariate landslide susceptibility assessment using geographical information systems: a method and application to Asarsuyu catchment, Turkey. Eng Geol 71:303–321
Trigila A, Iadanza C, Spizzichino D (2010) Quality assessment of the Italian landslide inventory using GIS processing. Landslides 7:455–470
Van Westen CJ, Soeters R, Sijmons K (2000) Digital geomorphological landslide hazard mapping of the Alpago area, Italy. Int J Appl Earth Obs Geoinf 2(1):51–59
Van Westen CJ, van Asch TWJ, Soeters R (2006) Landslide hazard and risk zonation—why is it still so difficult? Bull Eng Geol Environ 65:167–184

Van Westen CJ, Castellanos Abella EA, Sekhar LK (2008) Spatial data for landslide susceptibility, hazards and vulnerability assessment: an overview. Eng Geol 102:112–131

Varnes DJ (1978) Slope movement types and processes. In: Schuster RL, Krizek RJ (eds) Landslides, analysis and control, special report 176: transportation research board. National Academy of Sciences, Washington, DC, pp 11–33

Ward T, Ruh-Ming L, Simons D (1982) Mapping landslide hazard in forest watershed. J Geotech Eng Div 108(GT2):319–324

Wieczorek GF (1984) Preparing a detailed landslide-inventory map for hazard evaluation and reduction. Assoc Eng Geol Bull 21(3):337–342

Yilmaz I (2009a) a. Landslide susceptibility mapping using frequency ratio, logistic regression, artificial neural networks and their comparison: a case study from Kat landslides (Tokat–Turkey). Comput Geosci 35(6):1125–1138

Yilmaz I (2009b) b. A case study from Koyulhisar (Sivas–Turkey) for landslide susceptibility mapping by artificial neural networks. Bull Eng Geol Environ 68(3):297–306

Yilmaz I (2010a) Comparison of landslide susceptibility mapping methodologies for Koyulhisar, Turkey: conditional probability, logistic regression, artificial neural networks, and support vector machine. Environ Earth Sci 61(4):821–836

Yilmaz I (2010b) The effect of the sampling strategies on the landslide susceptibility mapping by conditional probability (CP) and artificial neural networks (ANN). Environ Earth Sci 60(3): 505–519

Yilmaz I, Keskin I (2009) GIS based statistical and physical approaches to landslide susceptibility mapping (Sebinkarahisar, Turkey). Bull Eng Geol Env 68(4):459–471

Yilmaz I, Yildirim M (2006) Structural and geomorphological aspects of the Kat landslides (Tokat —Turkey), and susceptibility mapping by means of GIS. Environ Geol 50(4):461–472

GIS-Based Landslide Susceptibility Evaluation Using Certainty Factor and Index of Entropy Ensembled with Alternating Decision Tree Models

Wei Chen, Hamid Reza Pourghasemi, Aiding Kornejady and Xiaoshen Xie

Abstract Up to now, numerous models have been developed and put to use by modelers to portray susceptibility of an area to landsliding. What keep them going might be the slightest differences in performance. These differences, however small, still would surprisingly make huge progress in identifying well suited areas for strategic planning. This kept in mind, we aimed to map landslide susceptibility over a critical landslide prone area, the Longhai Region, Baoji City, in China, using two models namely certainty factor (CF) and index of entropy (IOE) ensemble with alternating decision tree (ADTree). As inputs, 93 landslides together with 14 predisposing factors were mapped. Both CF and IOE models pointed at three main factors as the most important ones including residential land use, areas nearby roads, and normalized difference vegetation index (NDVI). Although obtained ADTrees for both models were similar, slightly different results were obtained. IOE-ADTree was more practical, since it better predicts highly susceptible areas. The receiver operating characteristic (ROC) curve cleared further the differences so that IOE-ADTree with 84% fitting ability and 85.3% generalization capacity outperformed CF-ADTree with the respective values of 83.9 and 83.8%. Therefore, the IOE-ADTree exhibits as a promising ensemble model for the study area.

Keywords Ensemble modeling · Generalization · Land use planning
Longhai region

W. Chen · X. Xie
College of Geology and Environment, Xi'an University of Science and Technology, Xi'an 710054, China

H. R. Pourghasemi (✉)
Department of Natural Resources and Environmental Engineering, College of Agriculture, Shiraz University, Shiraz, Iran
e-mail: hr.pourghasemi@shirazu.ac.ir

A. Kornejady
Department of Watershed Management Engineering, Gorgan University of Agricultural Sciences and Natural Resources, Gorgan, Iran

H. R. Pourghasemi and M. Rossi (eds.), *Natural Hazards GIS-based Spatial Modeling Using Data Mining Techniques*, Advances in Natural and Technological Hazards Research 48, https://doi.org/10.1007/978-3-319-73383-8_10

1 Introduction

Landslides, as geological hazards, have caused notable losses to industrial and agricultural productions as well as people's lives and property worldwide (Khavaninzadeh et al. 2010). Also, it is likely to cause great economic losses even slow down the pace of society progress (Singh 2010). In China, around 9710 landslides took place in 2016, which caused a sizable damage to 614 people, either dead or injured, and $0.45 billion worth of economic loss (http://www.cigem.gov.cn). Therefore, landslide spatial prediction is not only very meaningful and helpful to the research field, but also is convenient to the decision makers from all over the world.

In the past few decades, various approaches have been widely used for landslide susceptibility mapping including qualitative and quantitative methods, respectively (Kayastha et al. 2013). The qualitative method is subjective and expert knowledge-based (Kayastha et al. 2013), such as analytical hierarchy process (Kumar and Anbalagan 2016; Pourghasemi et al. 2012c). Quantitative method is objective which produces the probabilities of landslides occurrence in an area (Guzzetti et al. 1999; Kayastha et al. 2013). Nowadays, rapid development of GIS has motivated many researchers to map landslide susceptibility using various statistical models such as certainty factor (Devkota et al. 2013; Kanungo et al. 2011; Pourghasemi et al. 2013d; Prefac et al. 2016; Sujatha et al. 2012), frequency ratio (Kayastha 2015; Kumar and Anbalagan 2015; Ramesh and Anbazhagan 2015; Shahabi et al. 2015), index of entropy (Constantin et al. 2011; Devkota et al. 2013; Jaafari et al. 2014; Wang et al. 2015), evidential belief function (Althuwaynee et al. 2012; Zhang et al. 2016), and weights of evidence (Chen et al. 2016a; Kayastha et al. 2012).

In addition, various data mining techniques have been used to map landslide susceptibility, such as artificial neural network (Chen et al. 2017e; Lee et al. 2004; Pham et al. 2017; Zare et al. 2013), support vector machines (Chen et al. 2016b, d; Kavzoglu et al. 2014), decision trees (Lombardo et al. 2015; Saito et al. 2009), maximum entropy (Chen et al. 2017b; Davis and Blesius 2015; Felicísimo et al. 2013), alternating decision tree (Chen et al. 2017f; Hong et al. 2015; Pham et al. 2016), random forests (Chen et al. 2017d; Pourghasemi and Kerle 2016; Trigila et al. 2015), and classification and regression trees (Chen et al. 2017g; Youssef et al. 2015b).

The present study aimed at producing an ensemble of certainty factor and index of entropy with a data mining technique namely alternating decision tree to predict the landslide susceptibility for the Longhai area (China). The novelty of this study is that hybrid integration approach of alternating decision tree and bivariate certainty factor and index of entropy models is a relatively new contribution that has been rarely used for landslide spatial prediction.

2 Study Area

The study area is situated at the Longhai Region of the Baoji City, China (Fig. 1). It accounts for an area of 1186 km^2, within latitudes of 34° 16′N to 34° 40′N and longitudes of 106° 18′E to 106°56′E. The Longhai Railway traffic line and the roads extend 330 km across the study area. The altitude of the study area ranges from 626 to 2410 m asl. On average, the area experiences rainy days from July to September. The average annual rainfall is more than 600 mm. The study area has many steep slopes with angles up to 68° in mountainous regions.

3 Materials and Methods

3.1 Making a List of Landslides

Correct understanding the interconnection between the condition predictors and a failure event is very important to predict landslide susceptibility (Devkota et al. 2013; Pourghasemi et al., 2013a). In general, the information involves the landslide location, type, and the time of occurrence. This activities are crucial when preparing a reliable and accurate landslide inventory map (Chen et al. 2016d). In this study, firstly, according to historical records, interpretation of aerial photographs, and several field inspections, 93 landslides were recorded and mapped (Fig. 1). Secondly, 70% (65) and 30% (28) of the landslides were randomly split and used respectively for training and validation of the models using ArcGIS 10 software (Hussin et al. 2016).

3.2 Landslide Predictors

Reviewing different studies, 14 landslide predictors were selected for assessing landslide spatial prediction, including slope degree, altitude, slope angle, STI (Sediment Transport Index), SPI (Stream Power Index), TWI (Topographic Wetness Index), plan curvature, profile curvature, NDVI, land use, lithological units, distance to roads, distance to rivers, and distance to faults.

Slope Aspect
Slope aspect, as the direction that slopes face, is an essential factor to analyze landslides susceptibility (Çevik and Topal 2003). This factor often controls the lineament, rainfall, wind effect, and exposition to sunlight (Dahal et al. 2008; Ercanoglu et al. 2004). We divided slope aspect into nine different directional classes: flat, north, north-east, north-west, south-east, south, south-east, west, and east (Fig. 2a).

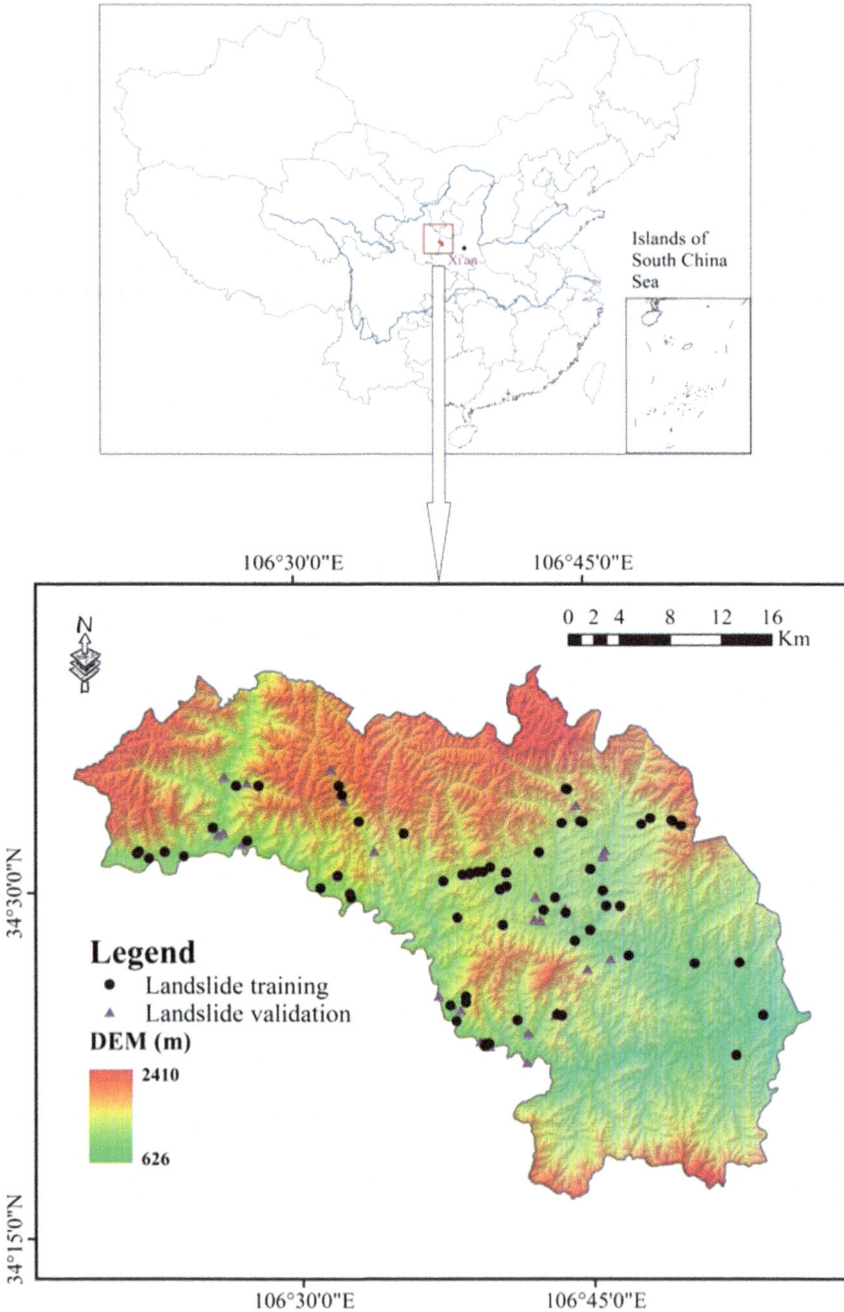

Fig. 1 The map of the study area

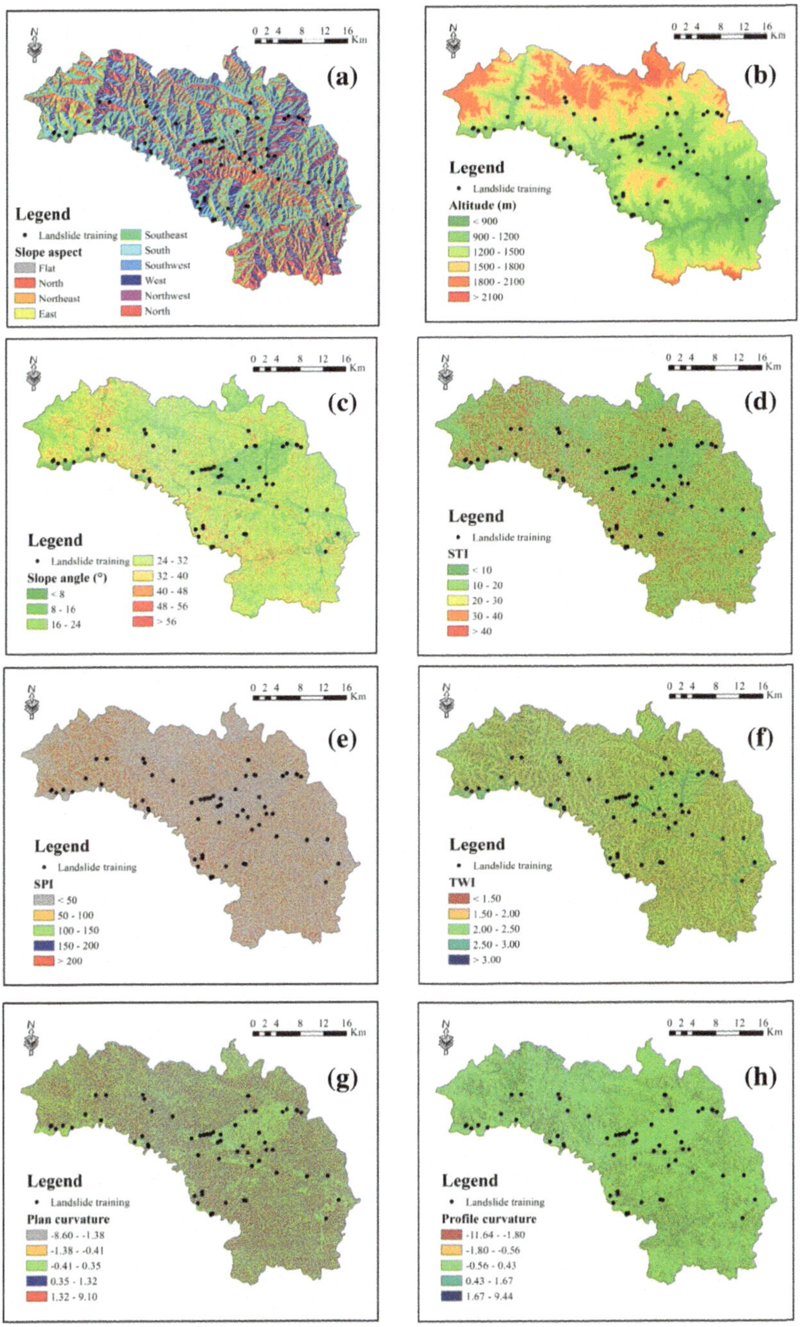

Fig. 2 The map of landslide predictors: **a** slope aspect, **b** altitude, **c** slope angle, **d** STI, **e** SPI, **f** TWI, **g** plan curvature, **h** profile curvature, **i** NDVI, **j** land use, **k** lithological units, **l** distance to roads, **m** distance to rivers, and **n** distance to faults

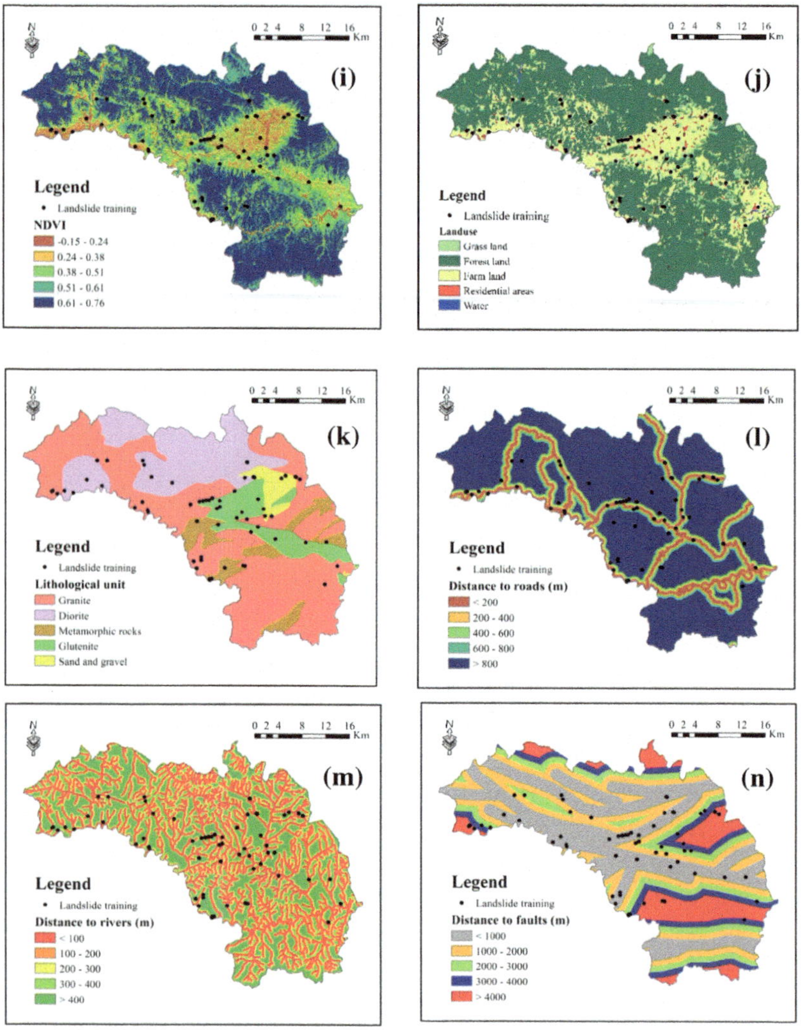

Fig. 2 (continued)

Altitude

Altitude is in charge of the phreatic water level, vegetation, and overall geography. Commonly, the potential of landslides occurrence is high at intermediate elevations (Oh and Pradhan 2011; Wang et al. 2015). The ASTER GDEM data (http://www.gscloud.cn) with 30 m spatial resolution was used to produce the altitude map. Then, it was categorized into 6 classes with 300 m intervals as follows: <900, 900–1200, 1200–1500, 1500–1800, 1800–2100 and >2100 m (Fig. 2b).

Slope Angle
Slope angle, as a description of the degree of inclination, has been always contributed to analyze landslide susceptibility. Generally, the grater the slope angle, the higher of the failure potential of slope (Dahal et al. 2008; Ercanoglu and Gokceoglu 2002). The slope degree was obtained from the 30 m × 30 m grid size DEM in ArcGIS 10.0 and values of slope angle were reclassified into eight groups with an interval of 8°, such as <8°, 8°–16°, 16°–24°, 24°–32°, 32°–40°, 40°–48°, 48°–56° and >56° (Fig. 2c).

STI
The STI shows the effect of erosion and deposition. Its values were adopted to analyze the landslides susceptibility by many researchers (Conforti et al. 2011; Yilmaz 2009). Generally, he higher the values have the higher the possibility of landslides occurrence. For this study, the STI was prepared from the DEM and was categorized as <10, 10–20, 20–30, 30–40 and >40 with an interval of 10, respectively (Fig. 2d).

SPI
As another important factor influencing landslide occurrence, the SPI represents the erosion power of stream (Moore and Grayson 1991). The erosion will break the structure and decrease the strength of the rocks in the toe, and then landslides will occur easily (Conforti et al. 2011). We obtained the SPI map from DEM, and regrouped it into five categories with an interval of 50, namely <50, 50–100, 100–150, 150–200 and >200 (Fig. 2e).

TWI
The values of TWI are influenced by the condition of soil, geography, and volume of runoff (Wang et al. 2016). Thus, it was considered as another factor influencing landslide susceptibility. We categorized TWI values into five groups with 0.5 intervals, viz. <1.5, 1.5–2.0, 2.0–2.5, 2.5–3.0 and >3.0 (Fig. 2f).

Plan Curvature
The plan curvature signifies the concavity and convexity of a slope in parallel with the valley (Erener and Düzgün 2010; Lee and Min 2001). We obtained the plan curvature map from the DEM layer, and categorized it into five ranges using natural break (NB) method as follows: −8.60 to −1.38, −1.38 to −0.41, −0.41 to 0.35, 0.35–1.32 and 1.32–9.10 (Fig. 2g).

Profile Curvature
The profile curvature indicates the curvature of a slope perpendicular to the valley (Chen et al. 2016a; Yesilnacar and Topal 2005). We derived profile curvature map from DEM layer and categorized it into five classes using NB scheme: −11.64 to −1.80, −1.80 to −0.56, −0.56 to 0.43, 0.43–1.67 and 1.67–9.44 (Fig. 2h).

NDVI
NDVI is related to the transpiration of plants, the sunlight, and photosynthesis (Pourghasemi et al. 2013b, c). Therefore, NDVI also provides a crude estimate of the vegetation. We produced the NDVI map using the LANDSAT-8 images (http://

www.gscloud.cn) using Eq. (1). Five different groups were prepared using natural break method, such as −0.15 to 0.24, 0.24–0.38, 0.38–0.51, 0.51–0.61 and 0.61–0.76 (Fig. 2i).

$$\mathrm{NDVI} = \frac{B5 - B4}{B5 + B4} \tag{1}$$

where, *B5* and *B4* are the infrared and red bands imbedded in LANDSAT-8 satellite (Chen et al. 2017a; Pourghasemi et al. 2014).

Land Use
Land use is constantly used to predict landslide spatial pattern (Leventhal and Kotze 2008). To some extent, the landslide stability is related to the types of vegetation and the covering dimensions (Fell et al. 2008). Commonly, a lower landslide potential is always associated with the more types and the larger covering area (Leventhal and Kotze 2008). In the current study, the following five classes were considered: grass land, forest land, farm land, residential areas, and water bodies (Fig. 2j).

Lithology Units
Lithology is pivotal to analyze landslide spatial pattern. Different rocks have different mechanical properties and can manifest different stability states (Constantin et al. 2011; Das et al. 2012). The lithological units were grouped into five groups namely granite, diorite, metamorphic rocks, glutenite, and sand and gravel (Fig. 2k).

Distance to Roads
According to recent studies, the road construction, as a human-made agent, can sometimes trigger slope failures (Jaafari et al. 2014; Zhao et al. 2015). Therefore, involving such influential factor is substantial to this study. The map of distance to roads was divided into five buffers with an interval of 200 m, including <200, 200–400, 400–600, 600–800 and >800 m (Fig. 2l).

Distance to Rivers
The river erosion has significant impacts on the evolving process of landslides (Chen et al. 2016c; Tien Bui et al. 2012a). Hence, we prepared the map of proximity to rivers and then categorized the values into five ranges with 100 m intervals: <100, 100–200, 200–300, 300–400 and >400 m (Fig. 2m).

Distance to Faults
Fault is the external manifestation of stress being broken (Tien Bui et al. 2012b). This process will lessen the strength of soil and rocks. We obtained the distribution of faults from geological maps and then categorized the proximity values to faults into five ranges with 1000 m intervals: <1000, 1000–2000, 2000–3000, 3000–4000 and >4000 m (Fig. 2n).

3.3 Modeling of Landslide Susceptibility

3.3.1 Certainty Factor

Certainty factor, as a GIS-based bivariate statistical model, has been attracted many researchers mainly due to both simplicity and robust factors integration algorithm (Devkota et al. 2013; Kanungo et al. 2011). First, the model rates the thematic layers through an equation based on the conditional probability and the prior probability assumptions (Eq. 2). The calculated CF values range between +1 and −1, representing the model being certain of recognizing the landslide and non-landslide areas (Heckerman 1985). So that, higher CF values reflect that the model is more certainty about recognizing landslide locations correctly and vice versa. The medium ranges indicate the uncertainty of the model to judge about the susceptibility of a specific area (Devkota et al. 2013).

$$CF = \begin{cases} \frac{ppa-pps}{ppa\times(1-pps)} & \text{if ppa}\rangle\text{pps} \\ \frac{ppa-pps}{pps\times(1-ppa)} & \text{if ppa}\langle\text{pps} \end{cases} \tag{2}$$

where, ppa and pps are the respective landslide conditional and prior probability which can be expressed as follows:

$$ppa = \frac{S_L}{S_C}, \quad pps = \frac{S_{Lt}}{S_{Bt}} \tag{3}$$

where, S_L is the landslide areas in a particular class, S_C is the class area within a particular factor, S_{LT} is the whole area of the landslides in the basin, and S_{BT} is the basin whole area.

The predictors' classes were rated following the equations above (Chen et al. 2016c). To combine the map of predictors, we used the following expression by renaming factors as X and Y (Pourghasemi et al. 2013d; Chen et al. 2016c; Devkota et al. 2013; Hong et al. 2017):

$$\mathrm{Z} = \begin{cases} X+Y-XY & X,Y \geq 0 \\ \frac{X+Y}{1-\min(|X|,|Y|)} & X*Y<0 \\ X+Y+XY & X,Y<0 \end{cases} \tag{4}$$

3.4 Index of Entropy

Many authors have discussed and studied entropy, each of which pointed out a specific aspect of it. Shannon, by relying on Boltzmann law, first raised this issue to explain the pattern of an unknown phenomenon (Shannon 1948). Shannon's entropy, also known as the index of entropy, as an inherent property stored inside the environmental data lies in information theory. He pointed out that the more entropy exists in our data, there equivalently would be more information to explore. Nowadays, this feature is increasingly being used in different scientific fields, mostly environmental sciences and natural processes (Pourghasemi et al. 2012b) with different metaphorical interpretations of the entropy such as instability, imbalance, and uncertainty in a system (Shi and Jin 2009). Apart from the fundamentals, the main strength of the model is the ability to compute the classes' rate and factors' weights itself as an ideal package which makes researchers free and needles of other expertise and expert-knowledge-based methods. Instead, the calculation process draws heavily on the actual recorded data. The mathematics follow:

$$FR_{ij} = \frac{b_{ij}}{a_{ij}} \tag{5}$$

$$P_{ij} = \frac{FR_{ij}}{\sum_{j-1}^{N_j} FR_{ij}} \tag{6}$$

$$H_j = -\sum_{i=1}^{N_j} P_{ij} \log_2 P_{ij}, \quad j = 1, \ldots, n \tag{7}$$

$$H_{j\max} = \log_2 N_j \tag{8}$$

$$I_j = \frac{H_{j\max} - H_j}{H_{j\max}}, \quad I = (0, 1), \quad j = 1, \ldots, n \tag{9}$$

$$W_j = I_j \times FR_{ij} \tag{10}$$

where, b is landslide area in each class divided by the landslides whole area and a is the class area within a particular factor divided by the basin whole area. More details on the mathematical process are given in Pourghasemi et al. (2012a), Youssef et al. (2015a), Chen et al. (2017c) and Hong et al. (2017). The final susceptibility value was calculated by Eq. 11 (Devkota et al. 2013).

$$IOE = \sum_{i=1}^{n} \frac{Z}{m_i} \times C \times W_j \tag{11}$$

where, i is the total number of conditioning factors, Z is the number of classes of the factor that own the most classes, m_i is the number of classes within each specific factor, C is the calculated rate of each class, and W_j is the final weight of each factor.

3.5 Alternating Decision Tree

The concept of machine learning methods argue the things to be learned including classification learning, association learning, clustering, and numeric prediction (Witten et al. 2011). Machine learning methods not only are being used for prediction purposes, but also offer a bright insight into the data and the problem space. Decision trees, as profound machine learning methods, aim at producing an intelligible yet operational result of the learning process. Inputs are sets of instances that are to be classified, associated, and clustered to predict a target phenomenon. Each instance is exclusive in essence, in that, it is formed with unique attributes and features. Decision trees try to recast the relationship between nodes of a tree as independent instances to summarize combination of these instances and the attributes therein, and to classify them as to landslide or non-landslide localities. However, finding a flat-file (summarized laws) in the tree may cause some oversimplification especially when encountering large amount of data and attributes. Hence, modelers offered recursive pattern learning. Test set are the data to analyze how well the concept has been learned. In general, decision trees, as an output representation style, follows a technique called "divide-and-conquer" (Witten et al. 2011). It is simply a visualized set of rules, like what goes on in the mind of a chess player about moves and endgame probabilities, but more summarized, more organized, more structured, and more analyzed due to the recursive property of it.

Alternating decision trees combine boosting, as an algorithm for reducing bias and converting weak learning process to strong one, and decision trees to produce decision rules (Hong et al. 2015). ADTs' graphical rule sets form leafs. Each branch ends in an outcome and goes for another rule until it reaches the root (Holmes et al. 2002; Pham et al. 2016). Although the simpler ADT, a heavily pruned one, will end in low performance, but an unpruned tree submissively follows the training set ending in overfitting problems and a weak generalized tree. But, ADT claims that it can make a simpler tree, yet with less errors and more interpretable results (Freund and Mason 1999; Pfahringer et al. 2001). Each node begins performing a test on one or more attributes based on a function or predefined constant value such as the average of the training set values. Then, nodes split successively throughout the leaf node based on a weight number proportional to the number of training instances. Attributes are tested over and over in a path with different constants (rules) (Rokach 2010). Once a set of instances reach the leaf node, a classification will be established over them. The final prediction probability forms from summation of all the weights contributed to the root (Freund and Mason 1999).

3.6 Descriptions of Ensemble Modeling

Ensemble models have reportedly outperformed the traditional statistical methods (Tien Bui et al. 2012a, 2013). They give more insights on the nature of the

Table 1 The rates of factors' classes derived from CF and IOE models

Conditioning factors	Classes	No. of pixels in domain	No. of landslide	FR(b/a)	Certainty factor	H_j	H_{jmax}	I_j	W_j	IOE
Slope aspect	Flat	65	0	0.000	−1.000	2.909	3.170	0.082	0.072	0.000
	North	134254	5	0.755	−0.245					0.054
	Northeast	159058	5	0.638	−0.362					0.046
	East	182848	7	0.777	−0.224					0.056
	Southeast	177187	10	1.145	0.126					0.082
	South	159382	11	1.400	0.286					0.101
	Southwest	172493	13	1.529	0.346					0.110
	West	173380	10	1.170	0.145					0.084
	Northwest	159744	4	0.508	−0.492					0.037
Altitude (m)	<900	90136	8	1.800	0.445	1.779	2.585	0.312	0.255	0.459
	900–1200	379991	33	1.761	0.432					0.449
	1200–1500	380603	20	1.066	0.062					0.272
	1500–1800	282134	4	0.288	−0.712					0.073
	1800–2100	175809	0	0.000	−1.000					0.000
	>2100	9738	0	0.000	−1.000					0.000
Slope angle (°)	<8	81015	8	2.003	0.501	2.594	3.000	0.135	0.128	0.256
	8–16	209091	13	1.261	0.207					0.161
	16–24	322617	23	1.446	0.308					0.185
	24–32	351902	10	0.576	−0.424					0.074
	32–40	244729	9	0.746	−0.254					0.095
	40–48	92338	1	0.220	−0.780					0.028
	48–56	15662	1	1.295	0.228					0.166
	>56	1057	0	0.000	−1.000					0.000

(continued)

Table 1 (continued)

Conditioning factors	Classes	No. of pixels in domain	No. of landslide	FR(b/a)	Certainty factor	H_j	H_{jmax}	I_j	W_j	IOE
STI	<10	533812	34	1.292	0.226	2.175	2.322	0.063	0.056	0.072
	10–20	386943	19	0.996	−0.004					0.056
	20–30	181624	2	0.223	−0.777					0.013
	30–40	84851	4	0.956	−0.044					0.054
	>40	131181	6	0.928	−0.072					0.052
SPI	<50	921658	45	0.990	−0.010	2.271	2.322	0.022	0.024	0.024
	50–100	180897	9	1.009	0.009					0.024
	100–150	69144	3	0.880	−0.120					0.021
	150–200	35574	3	1.711	0.415					0.041
	>200	111138	5	0.913	−0.087					0.022
TWI	<1.50	106069	1	0.191	−0.809	1.981	2.322	0.147	0.150	0.029
	1.50–2.00	529421	18	0.690	−0.310					0.103
	2.00–2.50	382756	23	1.219	0.180					0.183
	2.50–3.00	160267	18	2.278	0.561					0.342
	>3.00	139898	5	0.725	−0.275					0.109
Plan curvature	−8.60 to −1.38	84871	3	0.717	−0.283	2.282	2.322	0.017	0.016	0.011
	−1.38 to −0.41	305987	17	1.127	0.113					0.018
	−0.41 to 0.35	484941	29	1.213	0.176					0.019
	0.35–1.32	337646	11	0.661	−0.339					0.011
	1.32–9.10	104966	5	0.966	−0.034					0.015
Profile curvature	−11.64 to −1.80	79006	1	0.257	−0.743	2.143	2.322	0.077	0.064	0.016
	−1.80 to −0.56	279100	11	0.799	−0.201					0.051
	−0.56 to 0.43	519823	25	0.975	−0.025					0.062

(continued)

phenomenon and corresponding factors as they solve the problem of time-consuming process of running several methods (Nefeslioglu et al. 2010; Jebur et al. 2014; Umar et al. 2014). The secondary maps obtained from CF and IoE models were used as inputs to ADTree model in ArcGIS and Weka software.

4 Results and Discussions

4.1 Inferences of CF-ADTree Model

After preparing thematic layers in ArcGIS environment, the CF values were calculated according to Eqs. 2–4 which is presented in Table 1. As shown, the range between 0 to 200 meters from roads, residential areas in land use map, and the range between 0.24 to 0.38 in NDVI map with the respective CF values of 0.78, 0.777, and 0.738 had the highest landslide susceptibility, likewise the highest importance in the modeling process. These results support the fact that areas nearby roads can be more susceptible to landslide more mainly due to the unsupervised constructions without a proper foundation and compactness. Similarly, residential areas are responsible for the redistribution of the old landslides. The CF values correspond to NDVI are one of the unique examples of vegetation-soil mutual relationship in which besides the positive functionality of roots in reinforcement of soils, they can adversely exert a downward force on soil as an extra load when facing high-velocity winds. The factor importance results (CF values) were followed by lithological units (sand and gravel; 0.645), TWI (2.5–3; 0.561), slope angle (<8; 0.501), altitude (<900; 0.445), distance to rivers (<100, 0.435), SPI (150–200; 0.415), slope aspect (southwest; 0.346), profile curvature (0.43–1.67; 0.321), distance to faults (<1000; 0.242), STI (<10; 0.226), and plan curvature (−0.97 to −0.76; 0.176). Unexpected results, such as slope angle, have roots in the fact that the stronger factors (those with more tangible and understandable landslide distribution pattern) outweigh the weaker, less important, and less influential ones.

After calculation of CF values and factors' secondary maps, the Weka software was used to implement the ADTree algorithm running for 10 iterations on the landslide dataset (Fig. 3). The positive and negative values in Fig. 3 correspond to landslide and non-landslide area, respectively. The alternating tree consists of 10 decision nodes. Parallel decision nodes, those in the same level specifically in the first level, represent the little or no interaction. That is, NDVI values less than −0.714 increases the landslide susceptibility irrespective to DEM values less than −0.469, but with a different contribution (weights) to the final output (the root). But the lower nodes should be interpreted correspond with their ancestral decision nodes (Freund and Mason 1999). The reached path to root is shown in Fig. 3 as NDVI (≥ -0.714)—slope aspect (<0.215)—STI (≥ -0.425)—TWI. The final landslide susceptibility indices were categorized into four classes based on equal area classification method (Tien Bui et al. 2016b), including very high (VH) (10%), high (H) (20%), moderate (M) (20%), and low (L) (50%) (Fig. 4).

Table 1 (continued)

Conditioning factors	Classes	No. of pixels in domain	No. of landslide	FR(b/a)	Certainty factor	H_j	H_{jmax}	I_j	W_j	IOE
	0.43–1.67	344097	25	1.474	0.321					0.094
	1.67–9.44	96385	3	0.631	−0.369					0.040
NDVI	−0.15 to 0.24	61001	11	3.658	0.727	1.753	2.322	0.245	0.474	1.734
	0.24–0.38	154301	29	3.812	0.738					1.807
	0.38–0.51	199686	16	1.625	0.385					0.770
	0.51–0.61	300813	8	0.539	−0.461					0.256
	0.61–0.76	602610	1	0.034	−0.966					0.016
Land use	Grass land	39987	1	0.507	−0.493	1.468	2.322	0.368	0.582	0.295
	Forest land	926590	19	0.416	−0.584					0.242
	Farm land	330522	41	2.516	0.603					1.464
	Residential areas	18110	4	4.480	0.777					2.607
	Water	3202	0	0.000	−1.000					0.000
Lithological units	Granite	720275	30	0.845	−0.155	2.125	2.322	0.085	0.126	0.106
	Diorite	320273	11	0.697	−0.303					0.088
	Metamorphic rocks	109578	6	1.111	0.100					0.140
	Glutenite	125102	12	1.946	0.486					0.245
	Sand and gravel	43183	6	2.818	0.645					0.355
Distance to roads (m)	<200	125052	28	4.542	0.780	1.785	2.322	0.231	0.414	1.880
	200–400	97930	8	1.657	0.397					0.686
	400–600	87798	1	0.231	−0.769					0.096

(continued)

Table 1 (continued)

Conditioning factors	Classes	No. of pixels in domain	No. of landslide	FR(b/a)	Certainty factor	H_j	H_{jmax}	I_j	W_j	IOE
	600–800	77969	8	2.081	0.520					0.862
	>800	929662	20	0.436	−0.564					0.181
Distance to rivers (m)	<100	332387	29	1.770	0.435	2.201	2.322	0.052	0.049	0.087
	100–200	266118	9	0.686	−0.314					0.034
	200–300	253417	11	0.880	−0.120					0.043
	300–400	170620	6	0.713	−0.287					0.035
	>400	295869	10	0.686	−0.314					0.034
Distance to faults (m)	<1000	507526	33	1.319	0.242	2.241	2.322	0.035	0.034	0.045
	1000–2000	307478	7	0.462	−0.538					0.016
	2000–3000	182199	7	0.779	−0.221					0.026
	3000–4000	135981	8	1.193	0.162					0.041
	>4000	185227	10	1.095	0.087					0.037

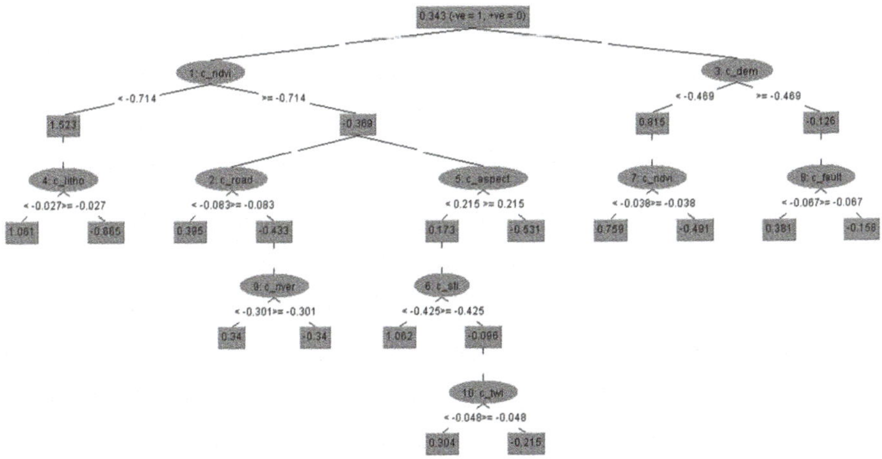

Fig. 3 Decision tree for classifying landslide susceptibility using CF-ADTree model

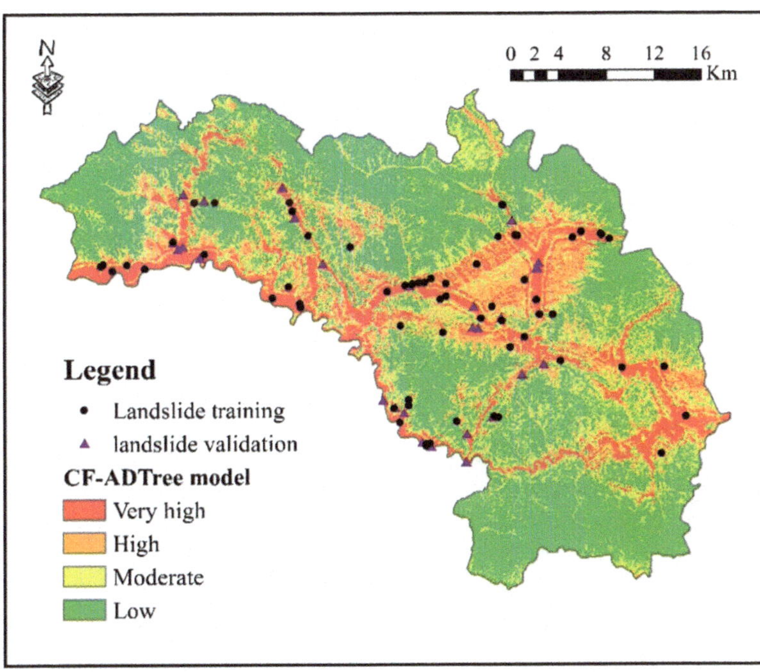

Fig. 4 Landslide spatial prediction map generated from CF-ADTree model

4.2 Inferences of IOE-ADTree Model

Equations 5–10 were applied to the dataset to calculate classes' rate and factors' weight which are summarized in Table 1. As a result, residential areas in land use map had the highest importance order with the IOE values of (2.607), followed by distance to roads (0–200; 1.88), NDVI (0.24–0.38; 1.807), altitude (<900; 0.459), lithological unit (sand and gravel; 0.355), TWI (2.5–3; 0.342), slope angle (<8; 0.256), slope aspect (southwest; 0.11), profile curvature (0.43–1.67; 0.094), distance to rivers (<100; 0.087), STI (<10; 0.072), distance to faults (<1000; 0.045), SPI (150–200; 0.041), and plan curvature (−0.97 to −0.76; 0.019). This is exactly in line with the W_j values where land use factor had highest weight as the highest importance (0.582), followed by NDVI (0.474) and distance to roads (0.414). In this regard, plan curvature had the lowest importance in susceptibility modeling. Apparently, both models (CF and IOE) are speaking in the same way in which the orders of factors are almost alike. Although some slight differences in orders rest on the fact that different mathematical algorithm may infer differently, it could not stop the influential classes from gleaming, so that, the same exact classes were identified as highly important ones in both models. This, as an interesting result, can also lead to the same ADTrees and correspondingly to nearly identical susceptibility maps which makes it hard to find the premier model.

Expectedly, the IOE-ADTree was the same as CF-ADTree; same nodes, same instances, same path, and even same weights (Fig. 5). The final landslide susceptibility indices were also categorized into four classes based on equal area classification method (Tien Bui et al. 2016b), including very high (10%), high (20%), moderate (20%), and low (50%) (Fig. 6).

4.3 Model Performance and Comparison

The ROC curve was adopted here to validate the models' results. It is a graphical plot that illustrates the performance of the model. The area under the curve (AUC) is the measure of differences between models. When using training set, the curve reflects the goodness-of-fit of a model (known as AUSRC), while test set validates the models in terms of prediction power and generalization of a model (known as AUPRC) (Jr and Schneider 2001; Pearce and Ferrier 2000). ROC curve was made plotting "sensitivity" (correctly detected landslide locations) as vertical axis versus the "100-specificity" (correctly detected non-landslide locations) as horizontal axis (Fig. 7).

As shown in Fig. 7a, CF-ADTree, and IOE-ADTree with the respective AUSRC values of 0.839 and 0.840 are well qualified in terms of fitting well on the training set. So far, IOE-ADTree is comparatively performing better than CF-ADTree regarding practicality and goodness-of-fit. According to Fig. 7b, IOE-ADTree interestingly has even higher predictive power than CF-ADTree. This property, a

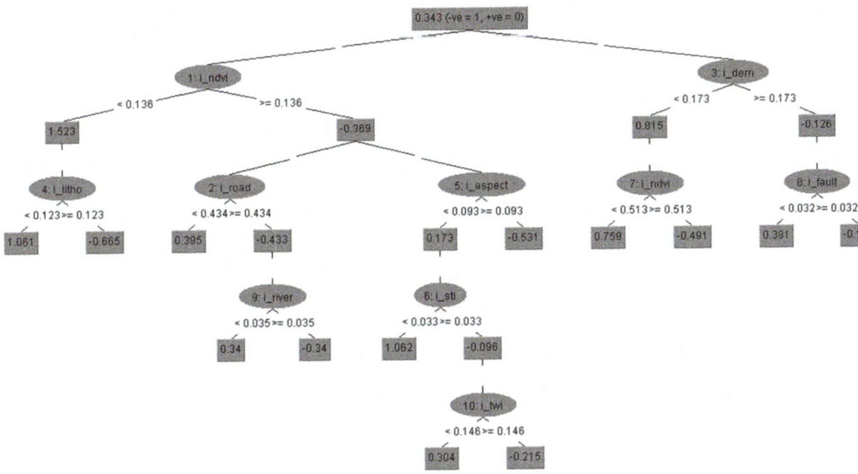

Fig. 5 Decision tree for classifying landslide susceptibility using IOE-ADTree model

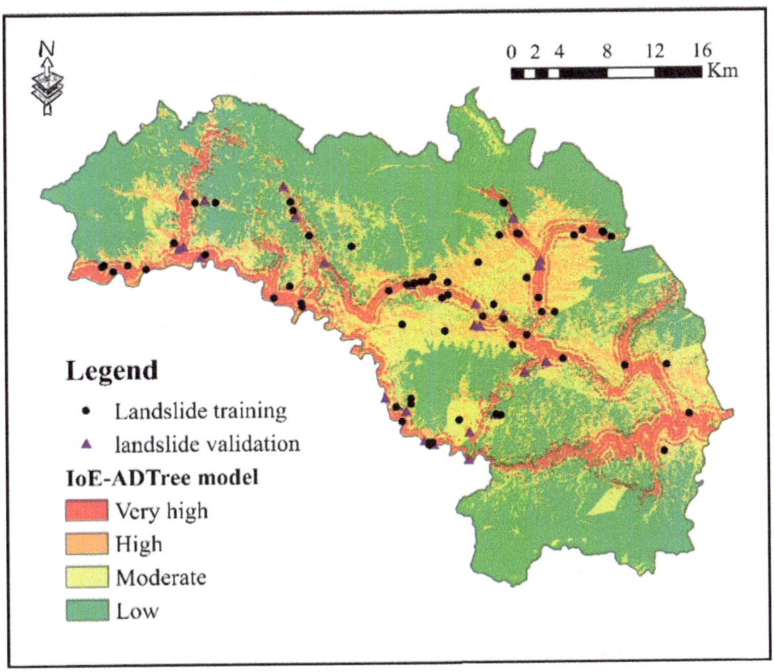

Fig. 6 Landslide spatial prediction map derived from IOE-ADTree model

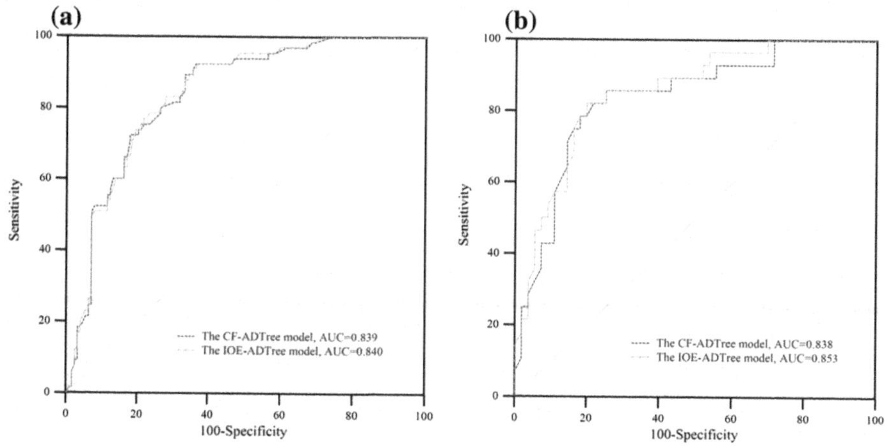

Fig. 7 ROC curves of CF-ADTree and IOE-ADTree models: **a** training accuracy; **b** predictive accuracy

Table 2 Comparison of the two models using Wilcoxon signed-rank test (two-tailed)

Parameters	CF-ADTree versus IOE-ADTree
Z value	−2.537
p value	0.011
Significance	Yes

model with a high fitting ability and even higher generalization, is very special among modelers. So, now it is possible to introduce the IOE-ADTree as the premier ensemble model as to assessing landslide susceptibility in the study area which supports its beneficial property of computing weights and rates congruently.

In addition to the AUSRC and AUPRC in the validation process, the Wilcoxon signed-rank (Wilcoxon 1945) tests was also used to test the significant differences between the two models. According to the results (Table 2), the obtained *P*-value is less than 0.05 and the z values exceeded the critical values (−1.96 to +1.96), which indicates that the two landslide models is clearly significantly different.

In order to demonstrate the superiority of hybrid models, the susceptibility maps obtained from two single bivariate models were reclassified into four classes using the same method (Figs. 8 and 9). The relative frequency ratio (FR) values were calculated for each susceptibility class. According to the study reviews, these values should follow a decreasing pattern from VH to L class (Pradhan and Lee 2010; Tien Bui et al. 2016a). The results summarized in Table 3 attested to such decreasing pattern. Results also show that the ensemble models could concentrate more landslides in the very high class, and the IOE-ADTree model could improve the performance of IOE model more significantly than the CF-ADTree ensemble model. Therefore, it can be concluded that the result of the present study is reasonable.

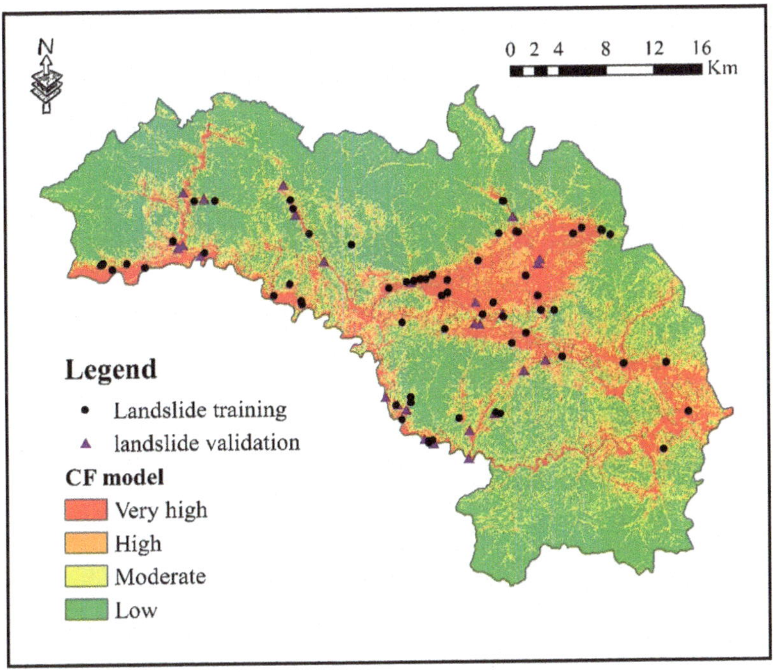

Fig. 8 Landslide spatial prediction map generated from CF model

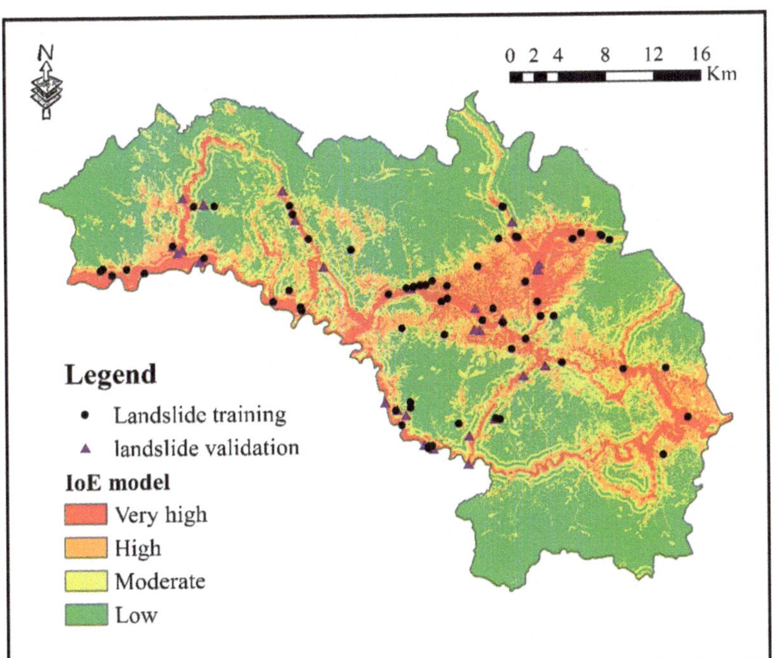

Fig. 9 Landslide spatial prediction map derived from IOE model

Table 3 Frequency ratio analysis for landslide susceptibility maps

Class	CF		CF-ADTree		IOE		IOE-ADTree	
	Landslide (%)	FR	Landslide (%)	FR	Landslide (%)	FR	Landslide (%)	FR
Very high (VH)	51.61	5.16	56.99	5.70	55.91	5.59	63.44	6.34
High (H)	36.56	1.83	32.26	1.61	32.26	1.61	25.81	1.29
Moderate (M)	10.75	0.54	9.68	0.48	10.75	0.54	8.60	0.43
Low (L)	1.08	0.02	1.08	0.02	1.08	0.02	2.15	0.04

5 Conclusions

Since producing landslide susceptibility maps is the primary prescription to address most hazardous areas, it is of prime importance to choose a more realistic method and also to compile more relative landslide-conditioning factors. Nowadays, ensemble modeling remains an active area of interest in different scientific fields, in a sense that modeling community is shifting towards hybrid models. In the current study, two ensemble models with a robust computational algorithm (CF-ADTree and IOE-ADTree) together with 14 landslide-conditioning factors were employed. As the first analysis, CF and IOE models both suggested three main factors namely land use (residential areas), distance to roads (0–200 m), and NDVI (0.24–0.38) as highly important ones in modeling process and susceptibility of the area to landsliding. Therefore, both human-made and natural agents are simultaneously contributing to landslide occurrence. Alternating decision trees helped abovementioned bivariate statistical models fuse to a summarized yet powerful machine learning technique to recursively learn the spatial pattern, so that only 10 nodes were established and tested. Relying on AUSRC and AUPRC tests, IOE-ADTree ensemble model with the respective values of 0.840 and 0.853 was found to be more fitted and more generalized in the study area, whose 39% was identified as highly susceptible to landslide occurrence. Since both models had the same figuration of decision tress, so the successful results of the IOE-ADTress may lies in the more powerful and complicated algorithm of IOE model. Finally, more pragmatic actions are required to implement in the area of concern.

Acknowledgements This research was supported by China Postdoctoral Science Foundation funded project (Grant No. 2017M613168), Scientific Research Program Funded by Shaanxi Provincial Education Department (Program No. 17JK0511), and College of Agriculture, Shiraz University (Grant No. 96GRD1M271143).

References

Althuwaynee OF, Pradhan B, Lee S (2012) Application of an evidential belief function model in landslide susceptibility mapping. Comput Geosci 44:120–135

Çevik E, Topal T (2003) GIS-based landslide susceptibility mapping for a problematic segment of the natural gas pipeline, Hendek (Turkey). Environ Geol 44:949–962

Chen W, Chai H, Sun X, Wang Q, Ding X, Hong H (2016a) A GIS-based comparative study of frequency ratio, statistical index and weights-of-evidence models in landslide susceptibility mapping. Arab J Geosci 9:1–16

Chen W, Chai H, Zhao Z, Wang Q, Hong H (2016b) Landslide susceptibility mapping based on GIS and support vector machine models for the Qianyang County, China. Environ Earth Sci 75:1–13

Chen W, Li W, Chai H, Hou E, Li X, Ding X (2016c) GIS-based landslide susceptibility mapping using analytical hierarchy process (AHP) and certainty factor (CF) models for the Baozhong region of Baoji City, China. Environ Earth Sci 75:1–14

Chen W, Panahi M, Pourghasemi HR (2017a) Performance evaluation of GIS-based new ensemble data mining techniques of adaptive neuro-fuzzy inference system (ANFIS) with genetic algorithm (GA), differential evolution (DE), and particle swarm optimization (PSO) for landslide spatial modelling. Catena 157:310–324

Chen W, Pourghasemi HR, Kornejady A, Zhang N (2017b) Landslide spatial modeling: Introducing new ensembles of ANN, MaxEnt, and SVM machine learning techniques. Geoderma 305:314–327

Chen W, Pourghasemi HR, Naghibi SA (2017c) A comparative study of landslide susceptibility maps produced using support vector machine with different kernel functions and entropy data mining models in China. Bull Eng Geol Environ 1–18. https://doi.org/10.1007/s10064-017-1010-y

Chen W, Pourghasemi HR, Naghibi SA (2017d) Prioritization of landslide conditioning factors and its spatial modeling in Shangnan County, China using GIS-based data mining algorithms. Bull Eng Geol Environ 1–19. https://doi.org/10.1007/s10064-017-1004-9

Chen W, Pourghasemi HR, Zhao Z (2017c) A GIS-based comparative study of Dempster-Shafer, logistic regression and artificial neural network models for landslide susceptibility mapping. Geocarto Int 32:367–385

Chen W, Wang J, Xie X, Hong H, Trung NV, Tien Bui D, Wang G, Li X (2016d) Spatial prediction of landslide susceptibility using integrated frequency ratio with entropy and support vector machines by different kernel functions. Environ Sci 75

Chen W, Xie X, Peng J, Wang J, Duan Z, Hong H (2017f) GIS-based landslide susceptibility modelling: a comparative assessment of kernel logistic regression, Naïve-Bayes tree, and alternating decision tree models. Geomatics, Nat Hazards Risk 1–24. https://doi.org/10.1080/19475705.2017.1289250

Chen W, Xie X, Wang J, Pradhan B, Hong H, Tien Bui D, Duan Z, Ma J (2017f) A comparative study of logistic model tree, random forest, and classification and regression tree models for spatial prediction of landslide susceptibility. Catena 151:147–160

Conforti M, Aucelli PPC, Robustelli G, Scarciglia F (2011) Geomorphology and GIS analysis for mapping gully erosion susceptibility in the Turbolo stream catchment (Northern Calabria, Italy). Nat Hazards 56:881–898

Constantin M, Bednarik M, Jurchescu MC, Vlaicu M (2011) Landslide susceptibility assessment using the bivariate statistical analysis and the index of entropy in the Sibiciu Basin (Romania). Environ Earth Sci 63:397–406

Dahal RK, Hasegawa S, Nonomura A, Yamanaka M, Masuda T, Nishino K (2008) GIS-based weights-of-evidence modelling of rainfall-induced landslides in small catchments for landslide susceptibility mapping. Environ Geol 54:311–324

Das I, Stein A, Kerle N, Dadhwal VK (2012) Landslide susceptibility mapping along road corridors in the Indian Himalayas using Bayesian logistic regression models. Geomorphology 179:116–125

Davis J, Blesius L (2015) A hybrid physical and maximum-entropy landslide susceptibility model. Entropy 17:4271–4292

Devkota KC, Regmi AD, Pourghasemi HR, Yoshida K, Pradhan B, Ryu IC, Dhital MR, Althuwaynee OF (2013) Landslide susceptibility mapping using certainty factor, index of entropy and logistic regression models in GIS and their comparison at Mugling-Narayanghat road section in Nepal Himalaya. Nat Hazards 65:135–165

Ercanoglu M, Gokceoglu C (2002) Assessment of landslide susceptibility for a landslide-prone area (north of Yenice, NW Turkey) by fuzzy approach. Environ Geol 41:720–730

Ercanoglu M, Gokceoglu C, Asch TWJV (2004) Landslide susceptibility zoning north of Yenice (NW Turkey) by multivariate statistical techniques. Nat Hazards 32:1–23

Erener A, Düzgün HSB (2010) Improvement of statistical landslide susceptibility mapping by using spatial and global regression methods in the case of More and Romsdal (Norway). Landslides 7:55–68

Felicísimo ÁM, Cuartero A, Remondo J, Quirós E (2013) Mapping landslide susceptibility with logistic regression, multiple adaptive regression splines, classification and regression trees, and maximum entropy methods: a comparative study. Landslides 10:175–189

Fell R, Corominas J, Bonnard C, Cascini L, Leroi E, Savage WZ (2008) Guidelines for landslide susceptibility, hazard and risk zoning for land use planning. Eng Geol 102:85–98

Freund Y, Mason L (1999) The alternating decision tree learning algorithm. icml. pp 124–133

Guzzetti F, Carrara A, Cardinali M, Reichenbach P (1999) Landslide hazard evaluation: a review of current techniques and their application in a multi-scale study, Central Italy. Geomorphology 31:181–216

Heckerman D (1985) Probabilistic interpretation for MYCIN's certainty factors. UAI '85. In: Proceedings of the first conference on uncertainty in artificial intelligence, Los Angeles, CA, USA, pp 167–196

Holmes G, Pfahringer B, Kirkby R, Frank E, Hall M (2002) Multiclass alternating decision trees. Proceedings of machine learning: ECML 2002, European conference on machine learning, Helsinki, Finland, 19–23 Aug 2002. pp 161–172

Hong H, Chen W, Xu C, Youssef AM, Pradhan B, Tien Bui D (2017) Rainfall-induced landslide susceptibility assessment at the Chongren area (China) using frequency ratio, certainty factor, and index of entropy. Geocarto Int 32:139–154. https://doi.org/10.1080/10106049.2015.1130086

Hong H, Pradhan B, Xu C, Tien Bui D (2015) Spatial prediction of landslide hazard at the Yihuang area (China) using two-class kernel logistic regression, alternating decision tree and support vector machines. Catena 133:266–281

Hussin HY, Zumpano V, Reichenbach P, Sterlacchini S, Micu M, van Westen C, Bălteanu D (2016) Different landslide sampling strategies in a grid-based bi-variate statistical susceptibility model. Geomorphology 253:508–523

Jaafari A, Najafi A, Pourghasemi HR, Rezaeian J, Sattarian A (2014) GIS-based frequency ratio and index of entropy models for landslide susceptibility assessment in the Caspian forest, northern Iran. Int J Environ Sci Technol 11:1–18

Jebur MN, Pradhan B, Tehrany MS (2014) Optimization of landslide conditioning factors using very high-resolution airborne laser scanning (LiDAR) data at catchment scale. Remote Sens Environ 152:150–165

Jr RGP, Schneider LC (2001) Land-cover change model validation by an ROC method for the Ipswich watershed, Massachusetts, USA. Agric Ecosyst Environ 85:239–248

Kanungo DP, Sarkar S, Sharma S (2011) Combining neural network with fuzzy, certainty factor and likelihood ratio concepts for spatial prediction of landslides. Nat Hazards 59:1491–1512

Kavzoglu T, Sahin EK, Colkesen I (2014) Landslide susceptibility mapping using GIS-based multi-criteria decision analysis, support vector machines, and logistic regression. Landslides 11:425–439

Kayastha P (2015) Landslide susceptibility mapping and factor effect analysis using frequency ratio in a catchment scale: a case study from Garuwa sub-basin, East Nepal. Arabian Journal of Geosciences:1–13

Kayastha P, Dhital MR, De Smedt F (2013) Application of the analytical hierarchy process (AHP) for landslide susceptibility mapping: a case study from the Tinau watershed, west Nepal. Comput Geosci 52:398–408. https://doi.org/10.1016/j.cageo.2012.11.003

Kayastha P, Dhital MR, Smedt FD (2012) Landslide susceptibility mapping using the weight of evidence method in the Tinau watershed, Nepal. Nat Hazards 63:479–498

Khavaninzadeh N, Motagh M, Sharifi M, Alipour S (2010) C-band and L-band InSAR for recognition and monitoring of landslides in Taleghan, Central Iran. J Hellenic Stud 26:304–308

Kumar R, Anbalagan R (2015) Landslide susceptibility zonation in part of Tehri reservoir region using frequency ratio, fuzzy logic and GIS. J Earth Syst Sci 124:431–448. https://doi.org/10.1007/s12040-015-0536-2

Kumar R, Anbalagan R (2016) Landslide susceptibility mapping using analytical hierarchy process (AHP) in Tehri reservoir rim region, Uttarakhand. J Geol Soc India 87:271–286

Lee S, Min K (2001) Statistical analysis of landslide susceptibility at Yongin, Korea. Environ Geol 40:1095–1113

Lee S, Ryu JH, Won JS, Park HJ (2004) Determination and application of the weights for landslide susceptibility mapping using an artificial neural network. Eng Geol 71:289–302

Leventhal AR, Kotze GP (2008) Landslide susceptibility and hazard mapping in Australia for land-use planning—with reference to challenges in metropolitan suburbia. Eng Geol 102:238–250

Lombardo L, Cama M, Conoscenti C, Märker M, Rotigliano E (2015) Binary logistic regression versus stochastic gradient boosted decision trees in assessing landslide susceptibility for multiple-occurring landslide events: application to the 2009 storm event in Messina (Sicily, southern Italy). Nat Hazards 79:1621–1648

Moore ID, Grayson RB (1991) Terrain-based catchment partitioning and runoff prediction using vector elevation data. Water Resour Res 27:1177–1191

Nefeslioglu HA, Sezer E, Gokceoglu C, Bozkir AS, Duman TY (2010) Assessment of landslide susceptibility by decision trees in the metropolitan area of Istanbul, Turkey. Math Probl Eng 242–256

Oh HJ, Pradhan B (2011) Application of a neuro-fuzzy model to landslide-susceptibility mapping for shallow landslides in a tropical hilly area. Comput Geosci 37:1264–1276

Pearce J, Ferrier S (2000) Evaluating the predictive performance of habitat models developed using logistic regression. Ecol Model 133:225–245

Pfahringer B, Holmes G, Kirkby R (2001) Optimizing the induction of alternating decision trees. Springer, Berlin Heidelberg

Pham BT, Bui DT, Prakash I, Dholakia MB (2017) Hybrid integration of Multilayer Perceptron Neural Networks and machine learning ensembles for landslide susceptibility assessment at Himalayan area (India) using GIS. CATENA 149:52–63

Pham BT, Tien Bui D, Dholakia M, Prakash I, Pham HV (2016) A comparative study of least square support vector machines and multiclass alternating decision trees for spatial prediction of rainfall-induced landslides in a tropical cyclones area. Geotech Geol Eng 34:1807–1824

Pourghasemi HR, Jirandeh AG, Pradhan B, Chong XU, Gokceoglu C (2013a) Landslide susceptibility mapping using support vector machine and GIS at the Golestan Province, Iran. J Earth Syst Sci 122:349–369

Pourghasemi HR, Kerle N (2016) Random forests and evidential belief function-based landslide susceptibility assessment in Western Mazandaran Province, Iran. Environ Sci 75:1–17

Pourghasemi HR, Mohammady M, Pradhan B (2012a) Landslide susceptibility mapping using index of entropy and conditional probability models in GIS: Safarood Basin, Iran. Catena 97:71–84

Pourghasemi HR, Mohammady M, Pradhan B (2012b) Landslide susceptibility mapping using index of entropy and conditional probability models in GIS: Safarood Basin, Iran. Catena 97:71–84

Pourghasemi HR, Moradi HR, Aghda SF, Gokceoglu C, Pradhan B (2014) GIS-based landslide susceptibility mapping with probabilistic likelihood ratio and spatial multi-criteria evaluation models (North of Tehran, Iran). Arab J Geosci 7:1857–1878

Pourghasemi HR, Moradi HR, Aghda SMF (2013b) Landslide susceptibility mapping by binary logistic regression, analytical hierarchy process, and statistical index models and assessment of their performances. Nat Hazards 69:749–779

Pourghasemi HR, Moradi HR, Aghda SMF, Gokceoglu C, Pradhan B (2013c) GIS-based landslide susceptibility mapping with probabilistic likelihood ratio and spatial multi-criteria evaluation models (North of Tehran, Iran). Arab J Geosci 7:1857–1878

Pourghasemi HR, Pradhan B, Gokceoglu C (2012c) Application of fuzzy logic and analytical hierarchy process (AHP) to landslide susceptibility mapping at Haraz watershed, Iran. Nat Hazards 63:1–32

Pourghasemi HR, Pradhan B, Gokceoglu C, Mohammadi M, Moradi HR (2013d) Application of weights-of-evidence and certainty factor models and their comparison in landslide susceptibility mapping at Haraz watershed, Iran. Arab J Geosci 6:2351–2365

Pradhan B, Lee S (2010) Landslide susceptibility assessment and factor effect analysis: backpropagation artificial neural networks and their comparison with frequency ratio and bivariate logistic regression modelling. Environ Model Softw 25:747–759

Prefac Z, Dumitru S, Chendeş V, Sîrodoev I, Cracu G (2016) Assessment of landslide susceptibility using the certainty factor model: Răşcuţa catchment (Curvature Subcarpathians) case study 11:617–626

Ramesh V, Anbazhagan S (2015) Landslide susceptibility mapping along Kolli hills Ghat road section (India) using frequency ratio, relative effect and fuzzy logic models. Environ Earth Sci 73:8009–8021

Rokach L (2010) Ensemble-based classifiers. Artif Intell Rev 33:1–39

Saito H, Nakayama D, Matsuyama H (2009) Comparison of landslide susceptibility based on a decision-tree model and actual landslide occurrence: the Akaishi Mountains, Japan. Geomorphology 109:108–121

Shahabi H, Hashim M, Ahmad BB (2015) Remote sensing and GIS-based landslide susceptibility mapping using frequency ratio, logistic regression, and fuzzy logic methods at the central Zab basin, Iran. Environ Earth Sci 73:8647–8668

Shannon CE (1948) A mathematical theory of communication. Bell Syst Tech J 5:3–55

Shi Y, Jin F (2009) Landslide stability analysis based on generalized information entropy. In: International conference on environmental science and information application technology, pp 83–85

Singh AK (2010) Bioengineering techniques of slope stabilization and landslide mitigation. Disaster Prev Manag 19:384–397

Sujatha ER, Rajamanickam GV, Kumaravel P (2012) Landslide susceptibility analysis using probabilistic certainty factor approach: a case study on Tevankarai stream watershed, India. J Earth Syst Sci 121:1337–1350

Tien Bui D, Pham BT, Nguyen QP, Hoang N-D (2016a) Spatial prediction of rainfall-induced shallow landslides using hybrid integration approach of least-squares support vector machines and differential evolution optimization: a case study in Central Vietnam. Int J Digital Earth 9:1077–1097. https://doi.org/10.1080/17538947.2016.1169561

Tien Bui D, Pradhan B, Lofman O, Revhaug I (2012a) Landslide susceptibility assessment in Vietnam using support vector machines, decision tree, and Naive Bayes Models. Math Problems Eng

Tien Bui D, Pradhan B, Lofman O, Revhaug I, Dick OB (2012b) Landslide susceptibility assessment in the Hoa Binh province of Vietnam: a comparison of the Levenberg–Marquardt and Bayesian regularized neural networks. Geomorphology s 171–172:12–29

Tien Bui D, Pradhan B, Lofman O, Revhaug I, Dick ØB (2013) Regional prediction of landslide hazard using probability analysis of intense rainfall in the Hoa Binh province, Vietnam. Nat Hazards 66:707–730. https://doi.org/10.1007/s11069-012-0510-0

Tien Bui D, Tuan TA, Klempe H, Pradhan B, Revhaug I (2016b) Spatial prediction models for shallow landslide hazards: a comparative assessment of the efficacy of support vector machines, artificial neural networks, kernel logistic regression, and logistic model tree. Landslides 13:361–378. https://doi.org/10.1007/s10346-015-0557-6

Trigila A, Iadanza C, Esposito C, Scarascia-Mugnozza G (2015) Comparison of logistic regression and random forests techniques for shallow landslide susceptibility assessment in Giampilieri (NE Sicily, Italy). Geomorphology 249:119–136

Umar Z, Pradhan B, Ahmad A, Jebur MN, Tehrany MS (2014) Earthquake induced landslide susceptibility mapping using an integrated ensemble frequency ratio and logistic regression models in West Sumatera Province, Indonesia. Catena 118:124–135

Wang LJ, Guo M, Sawada K, Lin J, Zhang J (2016) A comparative study of landslide susceptibility maps using logistic regression, frequency ratio, decision tree, weights of evidence and artificial neural network. Geosci J 1–20

Wang Q, Li W, Chen W, Bai H (2015) GIS-based assessment of landslide susceptibility using certainty factor and index of entropy models for the Qianyang County of Baoji city, China. J Earth Syst Sci 124:1399–1415

Wilcoxon F (1945) Individual comparisons by ranking methods. Biometrics bulletin 1:80–83

Witten IH, Frank E, Mark AH (2011) Data mining: practical machine learning tools and techniques, 3rd edn. Morgan Kaufmann, Burlington, USA

Yesilnacar E, Topal T (2005) Landslide susceptibility mapping: a comparison of logistic regression and neural networks methods in a medium scale study, Hendek region (Turkey). Eng Geol 79:251–266

Yilmaz I (2009) Landslide susceptibility mapping using frequency ratio, logistic regression, artificial neural networks and their comparison: a case study from Kat landslides (Tokat—Turkey). Comput Geosci 35:1125–1138

Youssef AM, Al-Kathery M, Pradhan B (2015a) Landslide susceptibility mapping at Al-Hasher Area, Jizan (Saudi Arabia) using GIS-based frequency ratio and index of entropy models. Geosci J 19:113–134

Youssef AM, Pourghasemi HR, Pourtaghi ZS, Al-Katheeri MM (2015b) Landslide susceptibility mapping using random forest, boosted regression tree, classification and regression tree, and general linear models and comparison of their performance at Wadi Tayyah Basin, Asir Region, Saudi Arabia. Landslides 13:1–18

Zare M, Pourghasemi HR, Vafakhah M, Pradhan B (2013) Landslide susceptibility mapping at Vaz Watershed (Iran) using an artificial neural network model: a comparison between multilayer perceptron (MLP) and radial basic function (RBF) algorithms. Arab J Geosci 6:2873–2888

Zhang Z, Yang F, Chen H, Wu Y, Li T, Li W, Wang Q, Liu P (2016) GIS-based landslide susceptibility analysis using frequency ratio and evidential belief function models. Environ Earth Sci 75:1–12

Zhao C, Chen W, Wang Q, Wu Y, Yang B (2015) A comparative study of statistical index and certainty factor models in landslide susceptibility mapping: a case study for the Shangzhou District, Shaanxi Province, China. Arab J Geosci 8:1–10

Evaluation of Sentinel-2 MSI and Pleiades 1B Imagery in Forest Fire Susceptibility Assessment in Temperate Regions of Central and Eastern Europe. A Case Study of Romania

Bogdan-Andrei Mihai, Ionuţ Săvulescu, Marina Vîrghileanu and Bogdan Olariu

Abstract Romania is a Carpathian country that experiences an increasing number of wildfire events. The production of a reliable model for the zonation of the monthly forest fire susceptibility degrees with a National scale coverage was the target of the SIAFIM project. Our approach is oriented towards the integration of complementary satellite imagery in the evaluation of forest fire susceptibility with the help of data mining techniques. A complex of ground reflectance calibrated spectral data and vegetation radiometric-biophysical indices is produced at two different scales and spectral resolutions from Sentinel-2 MSI multispectral imagery and Pleiades 1B ortho imagery from the month of August, in the region of Domogled-Valea Cernei, south western Romania. The main objective is the production and the evaluation of the representative indices from the available satellite imagery for the mapping of the forested surfaces sensitive to wildfire hazards. The analysis confirmed the reliability of some indices for the assessment of forest fire susceptibility in temperate regions of Central and Eastern Europe: LAI, SAVI, RedNDVI, Cab. Leaf Area Index (LAI) offer interesting information for the selected forest stands, between 0.06 and 0.2: pine stands on limestone steep slopes, Banat black pine stands and beech on shallow soil.

B.-A. Mihai (✉) · I. Săvulescu · M. Vîrghileanu
Faculty of Geography, University of Bucharest, 1, Nicolae Bălcescu Blvd., 010041 Bucharest, Romania
e-mail: bogdanandrei0771@gmail.com

I. Săvulescu
e-mail: savulescu@geo.unibuc.ro

M. Vîrghileanu
e-mail: rujoiumarina@yahoo.com

B. Olariu
Faculty of Geography, Simion Mehedinţi Doctoral School, University of Bucharest, 1, Nicolae Bălcescu Blvd., 010041 Bucharest, Romania
e-mail: bogdanolariu28@yahoo.com

H. R. Pourghasemi and M. Rossi (eds.), *Natural Hazards GIS-based Spatial Modeling Using Data Mining Techniques*, Advances in Natural and Technological Hazards Research 48, https://doi.org/10.1007/978-3-319-73383-8_11

Keywords Forest fires · Forest stands · Sentinel-2 MSI · Pleiades 1B Biophysical indices · Radiometric indices

1 Introduction

Wildfire in temperate regions of Central and Eastern Europe is a natural hazard and a risk factor, on the background of climatic changes (Flannigan et al. 2000; Dale et al. 2001; Bowman et al. 2009). Romania is a country with an increasing rate of the extreme natural phenomena, including wildfires in forested regions and neighbouring rural/urban areas (Joint Research Center 2014). Particularly, after 2000, the National database from the IGSU (National Emergency Inspectorate) shows an increasing rate of forest fire events. Between 2000 and 2004, the average surface affected by wildfires in Romania was 7.77 ha, while the period 2010–2014 is featured by 20.19 ha. The economic losses for 2001–2014 are evaluated at 815 thousand euros.

For the first time in Romania, the SIAFIM project—Satellite Image Analysis for Fire Monitoring (http://www.intergraph.ro/siafim/index.htm), financed by European Space Agency/Romanian Space Agency (2012–2015), produced a set of 12 monthly maps of forest fire susceptibility levels at National scale, at 100 m resolution, as a base layer for a specific product to be developed and used by forest and environmental managers.

Forest fire hazard modelling is a difficult task because it integrates a big amount of processed data, with an increased variety and complexity (Chuvieco et al. 2010, 2014). Most of the geospatial information cannot be produced directly from satellite imagery (Chuvieco et al. 2004; Chuvieco 2000). Data mining techniques can help finding adequate solutions to map these features from remote sensing data (Datcu et al. 2003). The large data volumes available in digital format, from simple grids to multi-channel satellite imagery opened new directions in forest fires analysis and monitoring (Mithal et al. 2011), employing a complex set of algorithms, from simple spatial statistics and predictive model development (Han et al. 2003) to machine learning like Support Vector Machine/SVM, random forest/RF (Cortez and Morais 2007; Arpaci et al. 2014) and neural networks/ANN (Cheng and Wang 2008) etc. An interesting approach is the integration of satellite imagery in forest fire related data mining approaches (Hsu et al. 2002), at different scales and resolutions. This started mainly with the development of data clustering and then with the spectral indices calculation on multispectral data, using NOAA/AVHRR multi-temporal imagery and the NDVI (Fernandez et al. 1997), and continued with the development of object base image analysis strategies (Datcu and Seidel 2000). Different contributions focus on the application of data mining on satellite imagery in: forest fire zones pattern recognition on NOAA/AVHRR data validated with SPOT XS imagery (Tay et al. 2003), forest fire prediction with weather data and MODIS data in Slovenia (Stojanova et al. 2006) and with fuzzy sets on MODIS imagery (Angayarkkani and Radhakrishnan 2009), fire scar object-oriented approach on multi-temporal geostationary satellite imagery on METEOSAT-SEVIRI data in Portugal, validated with AQUA and MODIS data (Umamaheshwaran et al. 2007), modelling of fire effects like smoke plumes using data mining on MISR-MODIS multi-temporal data (Mazzoni et al. 2007), fire scar mapping from imagery in Mediterranean region with MODIS data (Quintano et al. 2011). A general remark is that the approaches focus on more complex data layer structure with different weights in forest fire susceptibility (Pourtaghi et al. 2016; Pourghasemi 2016), while the use of satellite data is more or less adapted to larger regions and lower spatial resolution, as the wildfire phenomena need imagery of higher spectral and temporal resolution like MODIS data. This is the reason of our approach, covering a smaller mountain region in temperate forest zone (about 5600 ha), which need the integration of higher spatial resolution imagery in forest fire prediction.

The recent advances in Earth Observation domain (European Forest Fire Information System EFFIS ESA COPERNICUS programmehttp://effis.jrc.ec.europa.eu/) and the launch of a large family of complementary optical sensors with improved resolutions during the last five years, creates new opportunities for the production of imagery and derived data for natural hazard evaluation, mapping and monitoring.

The 13-band multispectral imagery from Sentinel-2A MSI instrument active from 2015 (Drusch et al. 2012; Frampton et al. 2013; Agapiou et al. 2014; Main-Knorn et al. 2015) and continued with S-2B, in 2017, is already explored in the field of wildfire hazards modelling (Huang et al. 2016) and radiometric-biophysical indices production at Global and regional scales (Delegido et al. 2011; Richter et al. 2012; Majasalmi and Rautiainen 2016). All this open access data offers good opportunities for large fire monitoring, but there are a lot of events, mainly in temperate regions, affecting smaller areas of few hectares which can be mapped and monitored only with very high resolution imagery.

Pleiades 1B ortho imagery is an example of VHR (Very High Resolution) images (2 m resolution multispectral data, 0.5 m resolution panchromatic data) with a superior pixel depth at 12 bits, especially designed for security and defence purposes, including natural hazard monitoring during the extreme events (Maxant et al. 2013). This commercial imagery is produced on pre-order unlike Sentinel-2 MSI (5-day temporal resolution from 2017), but it can provide the key features of the event effects, if obtained immediately afterwards (ex. burned area detailed mapping and inventorying). Usually, the Pleiades 1B products are integrated into multi-sensor approach related to wildfire hazards (see http://www.bigdataearth.com/major-events/2016-fort-mcmurray-wildfire-satellite-imagery/).

The objective of this paper is the evaluation of some representative radiometric and biophysical indices to be used for the identification and mapping of the highly susceptible forest stands to forest fires. These data layers derived from complementary satellite imagery are analysed together with the recorded burned area polygons at local scale, in order to extract the significant thresholds for these indices that could be assigned to other stands on larger areas.

1.1 Study Area

The assessment of the forest fire susceptibility for the entire territory of Romania is a difficult task, particularly when the susceptibility model training is performed selecting random areas with point samples on different imagery coverage. This is almost impossible for many of the regions where wildfire events in the official records are those that affect more than one hectare.

This was the reason we select a test area from southwestern part of the Romanian Carpathians, where forest fire affects every summer some stands like the protected Banat black pine (*Pinus nigra* ssp. *Banatica*) and the beech stands (*Fagus silvatica moesiaca*). This region (Fig. 1) is a group of mountains and plateaus featured by an intensive tectonical fragmentation along the Cerna River Valley, Northern from the Danube River Defile. It is centred on the Mehedinți Mountains ridge (1466 m a.s.l. in Vârful lui Stan Peak), a limestone ridge on granitic bedrock, with a lot of protected landscape elements (around Domogled Peak—1105 m, there are sharp ridges with Banat black pine stands). This area from the Danube River to the north and the northeast is affected by longer dry periods between July and September, under the Submediterranean climatic influence (Pătroescu et al. 2007; Clima României 2008). This creates favourable conditions for wildfire ignition and spreading, on the forest stands covering shallow soils with little clay horizon and especially on limestone rocks (Török-Oance and Török-Oance 2002). It is the effect of high temperatures combined with the anthropogenic activity like grazing, traffic and tourism, as in Băile Herculane resort area, famous for thermal and mineral springs since Roman times. The local forestry district official statistics (2000–2013) records a lot of forest fires in Domogled Mountain protected area (297 ha). After the year 2000, the magnitude and the recurrence of forest fires increased to 4–5 bigger events per summer.

2 Satellite Images Processing

The analysis is focused on the integration of two complementary satellite images (Table 1), centred on the same study area, in order to produce a reliable set of indices and to map the vegetation types, especially the exposed forest stands and other features usually affected by wildfires.

Both datasets were calibrated for the production of the upper level products containing the ground surface reflectance. Pleiades ortho image features a limited spectral resolution, offering only the reflectance in four spectral channels, from visible to near infrared. After the atmospheric correction using FLAASH processor, the topographic correction was not performed, because the available elevation dataset, produced from topographic map is limited to a resolution of only 10 m. In this context we used visual interpretation combined with field observation for the mapping of the fire affected stands.

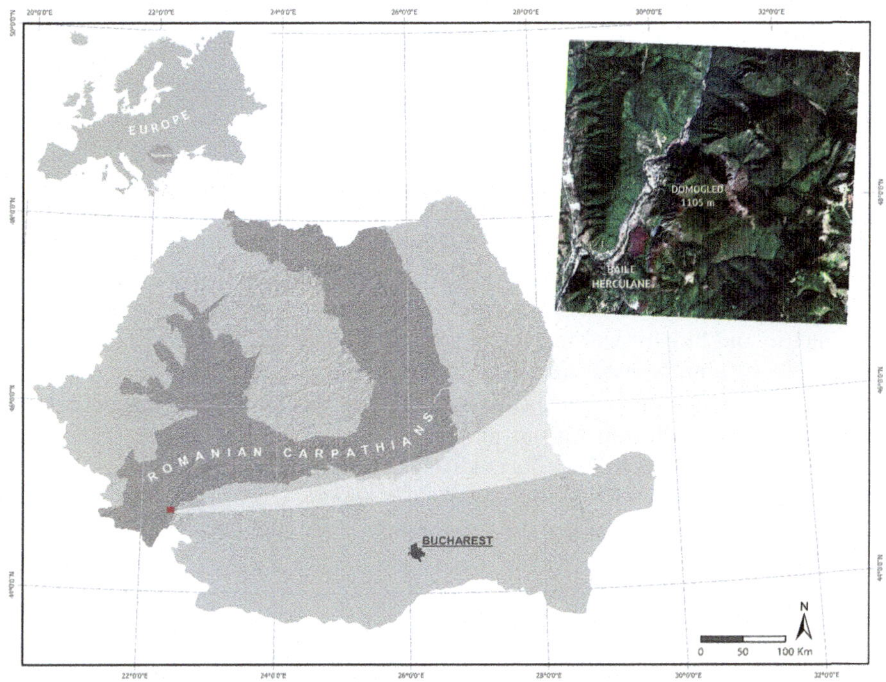

Fig. 1 The study area, centred on Mehedinți Mountains—Domogled Ridge, south-western Romanian Carpathians. Sentinel-2A MSI from August 31, 2015, natural colours (ESA-Copernicus Scientific Hub)

Table 1 Basic features of satellite data sources

Satellite image product	Source	Date	Spatial resolution	Spectral resolution	Radiometric resolution (pixel depth)	Processing level
Pleiades 1B Ortho	Astrium/ Airbus Defence and Space TradsymSrl Bucharest	2013/ 08/20	2.0 m MS 0.5 m Pan	4 bands (visible, NIR)	12 bits	L1B Orthorectified UTM/ WGS-84
Sentinel-2A MSI L1C	ESA Copernicus Scientific hub	2015/ 08/31	10, 20, 60 m MS —native	13 bands (visible, NIR, SWIR)	12 bits	L1C-TOA calibrated tile Orthorectified UTM/ WGS-84

The Sentinel-2 MSI, L1C product from ESA Copernicus archive, is limited to 10 m spatial resolution but it features a higher spectral resolution (13 channels), with an adequate coverage of the peak reflectance intervals for vegetation canopies (Drusch et al. 2012). After the atmospheric correction of the product using the

Sen2Cor processor (ESA SNAP 5.0), the dataset is resampled to the essential ground reflective bands, covering visible, near infrared and short wave infrared intervals. Topographically normalised L2A data is obtained after the illumination correction process using a 10 m spatial resolution DEM (digital elevation model), generated from contour lines (Romania Topographic Map, scale 1:25,000, produced by Romanian Military Topographic Directorate). Because of the higher spectral resolution in infrared (7 bands), the L2A product is used as a key data source in vegetation radiometric and biophysical indices production.

The processing workflow follows two directions adapted to the integration of data extracted from the complementary satellite imagery. First, a complex of radiometric and biophysical indices is generated from the ground reflectance calibrated data of both type of imageries. The 13-band Sentinel-2 imagery allows the calculation of a high number of radiometric, as well as biophysical indices, but only some of them are relevant for our purpose. Although it offers a fine spatial resolution, the indices derived from Pleiades 1B ortho image are limited because of the four spectral channels with a high correlation degree.

Secondly is the evaluation of the separability of the land cover classes and forest types, in order to select the relevant bands and indices for the identification of the areas potential susceptible to wildfires.

For data processing purpose, a separate layer is derived from the vector file representing the forest fire scars inventory 2000–2013, collected from the archives of the local forestry district of Băile Herculane. The polygons are spatially and statistically overlaid with each radiometric and biophysical index. Zonal statistics tool made possible the production of two parameters like mean and standard deviation for the areas where fires have occurred. These values are essential in building a Gaussian distribution of the indices values for the thresholds definition in forest stand fire susceptibility modelling for the entire area of study. The first standard deviation values were used to define the specific thresholds for each index according to the forest fire susceptibility level.

Finally, a thematic classification of land cover correlated to forest stands typology is obtained from Sentinel-2 MSI calibrated data in order to validate the characteristic thresholds of some indices.

3 Derived Products and Evaluation

Decorrelated datasets help to discriminate between forest stand classes. Indeed, each index should bring different information about the components of the canopies, from greenness to moisture content, chlorophyll content and quality etc. The available datasets represented by the spectral bands of the images were stacked together with the derived data like different radiometric and biophysical indices. The primary reflectance data is usually limited in information and correlated, like in the case of visible bands of Pleiades 1B image and for a smaller part of the Sentinel-2 MSI image (bands 1, 2, 3 and 4).

Figure 2 shows the separability diagrams between thematic classes (forest stands and land cover types) in different spectral bands, radiometric and biophysical indices. It is easy to note that forest stands, including Banat black pine and the sparse beech stands of shallow soil or bedrock (limestone, sandstone, schyst and granite) have a good separability on Sentinel-2 red-edge bands and some normalised indices. The best separability is reflected by the standard deviation shifting featuring the spectral reflectance for the selected classes.

From the complex of radiometric and biophysical indices created on calibrated Sentinel-2 and Pleiades 1B data, only those having the highest separability and being significant for the determination of the areas susceptible to wildfires are selected (Table 2). Biophysical (LAI, Cab) and radiometric (RedNDVI, SAVI) indices have been proved to be the most useful according to the temperate climatic conditions of Romania. Each of those datasets is the subject of a zonal statistics approach, that return the mean and standard deviation in and outside of the burned area polygons, collected from the records of the local forestry district at Băile Herculane, 2000–2013. Each of the fire affected forest stand is almost homogenous as regards the species and the environmental conditions. For example, forest fire areas correspond very well with Banat black pine and beech canopies on shallow soil or bedrock. The polygons can be assigned to data samples with a Gaussian distribution of the environmental parameters. This is a step in drawing the characteristic thresholds for the segmentation of each index layer following the Gaussian distribution of the extracted data, at one and two standard deviations from the mean value (Table 2).

A set of 3723 random points (with a density of one point per hectare) is used for the extraction of the corresponding values of each raster-index. From those, 292 points overlap the burned areas polygons recorded between 2000 and 2013. The mean and standard deviation parameters of the indices values within the burned areas are used for the thresholds establish. The level which shows the most exposed vegetation areas to fire is situated in the first standard deviation interval, as observed from the overlay between the largest burned area polygons and the layers featuring the selected indices. The segmentation process of the indices derived for the entire study area image, based on the thresholds, is structured in five classes corresponding to the interval between the mean values and the first of the three standard deviations (Fig. 3).

For most of the selected indices—LAI, SAVI, Cab, RedNDVI—the first standard deviation interval indicates the vegetation areas with high burning potential. High exposure is revealed for areas covered with black pine (in all vegetation cases related with the slope—Fig. 4), pastures or even sparse beech forest developed on soils having high porosity, high level of stoniness and low water retention potential, according to the documentation from the Mehedinţi county Office for Pedological Studies and Agrochemistry. From these vegetation categories the pine remarks itself as being the best fuel, containing a high level of resin and growing on high level of stoniness soils (Pătroescu et al. 2007). This aspect is very well captured by SAVI, an index showing the canopy typology and state, taking into account the soil spectral properties. The beech forest developed on deep soils which contain a high

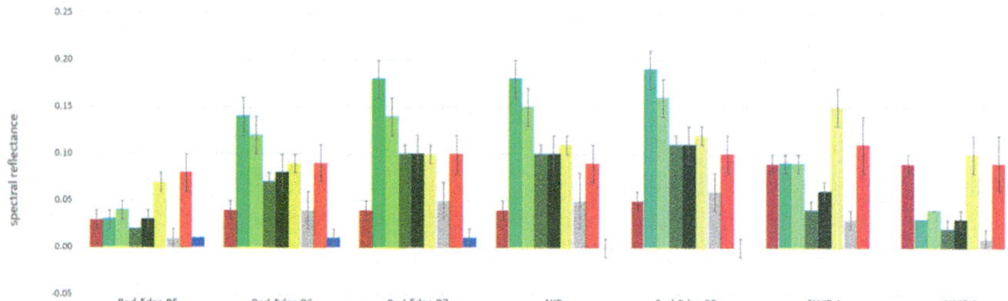

	Red-Edge B5	Red-Edge B6	Red-Edge B7	NIR	Red-Edge B8a	SWIR 1	SWIR 2
burnt area	0.03	0.04	0.04	0.04	0.05	0.09	0.09
beech forest	0.03	0.14	0.18	0.18	0.19	0.09	0.03
sparse beech forest	0.04	0.12	0.14	0.15	0.16	0.09	0.04
spruce fir	0.02	0.07	0.1	0.1	0.11	0.04	0.02
black pine	0.03	0.08	0.1	0.1	0.11	0.06	0.03
herbaceous	0.07	0.09	0.1	0.11	0.12	0.15	0.1
bare land	0.01	0.04	0.05	0.05	0.06	0.03	0.01
built-up area	0.08	0.09	0.1	0.09	0.1	0.11	0.09
water	0.01	0.01	0.01	0	0	0	0

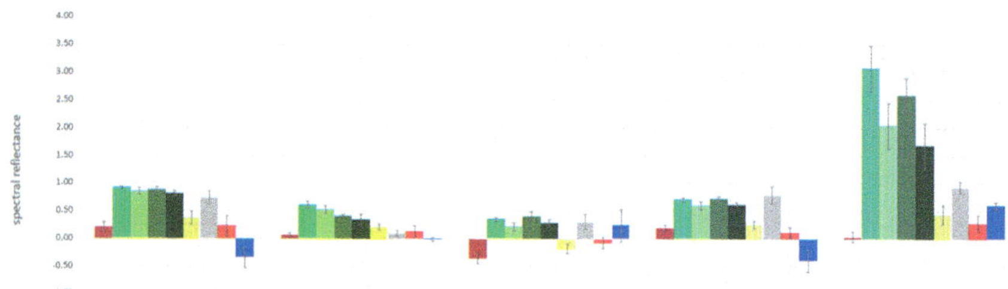

	NDVI	SAVI	NDMI	RED-NDVI 8a/5	LAI
burnt area	0.22	0.07	-0.36	0.20	0.04
beech forest	0.91	0.61	0.35	0.71	3.08
sparse beech forest	0.85	0.53	0.22	0.6	2.05
spruce fir	0.87	0.41	0.4	0.72	2.6
black pine	0.81	0.35	0.28	0.62	1.69
herbaceous	0.37	0.22	-0.18	0.26	0.44
bare land	0.74	0.1	0.31	0.79	0.93
built-up area	0.23	0.13	-0.07	0.12	0.29
water	-0.35	-0.02	0.24	-0.4	0.6

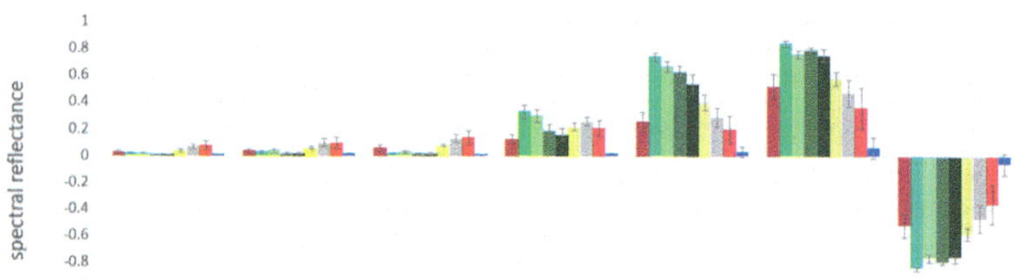

	Blue	Green	Red	NIR	NDVI	GNDI	NDWI
burnt area	0.03	0.04	0.06	0.13	0.26	0.52	-0.52
dense deciduous	0.02	0.03	0.02	0.33	0.74	0.84	-0.84
dispersed deciduous	0.02	0.04	0.03	0.3	0.67	0.76	-0.76
spruce fir	0.01	0.02	0.02	0.19	0.63	0.79	-0.79
black pine	0.01	0.02	0.02	0.16	0.54	0.75	-0.75
herbaceous	0.04	0.06	0.08	0.22	0.4	0.58	-0.58
bare land	0.07	0.1	0.13	0.26	0.29	0.47	-0.47
built-up area	0.08	0.1	0.14	0.21	0.2	0.36	-0.36
water	0.01	0.02	0.01	0.02	0.03	0.06	-0.06

Fig. 2 The separability of thematic classes evaluated for spectral data and derived indices for Sentinel-2 (**a** and **b**) and Pleiades 1B (**c**), using the mean spectral reflectance and standard deviation values

Table 2 Representative radiometric and biophysical indices derived from satellite images and evaluated for the definition of threshold values of forest fire susceptibility in the study area

Satellite images	Index	Vegetation properties	Expression	Reference	Statistical parameters—forest fire susceptible areas (mean/standard deviation)
Sentinel-2A MSI	NDVI	Canopy condition	$(\rho_8 - \rho_4)/(\rho_8 + \rho_4)$	(Rouse et al. 1973)	0.62/0.19
	RedNDVI	Canopy health condition	$(\rho_{8a} - \rho_6)/(\rho_{8a} + \rho_6)$	(Gitelson and Merzlyak 1994; Huete 1988)	0.47/0.14
	SAVI	Canopy condition without soil brightness effect	$1.5(\rho_8 - \rho_4)/(\rho_8 + \rho_4 + 0.16)$	(Huete 1988)	0.23/0.10
	LAI	Green leaf canopy ground coverage degree	$8.452((\rho_5 - \rho_4)/(\rho_5 + \rho_4))$	(Whittaker and Niering 1975; Wang et al. 2010; Delegido et al. 2011)	0.13/0.07
	Cab	Chlorophyll content in the leaf (correspond to the three variables: chlorophyll a, b and carotenoids)	The expression is integrated in *SNAP > Biophysical processor* and it is based on PROSPECT model	(Jacquemoud and Baret 1990; Baret and Fourty 1997)	0.12/0.06
Pleiades 1B ortho	NDVI	Canopy condition	$(\rho_4 - \rho_3)/(\rho_4 + \rho_3)$	(Rouse et al. 1973)	0.44/0.13
	GNDVI	Canopy greenness and photosynthetic activity	$(\rho_4 - \rho_3)/(\rho_4 + \rho_3)$	(Gitelson and Merzlyak 1998)	0.65/0.10
	NDWI	Difference between canopy and water bodies	$(\rho_2 - \rho_4)/(\rho_2 + \rho_4)$	(Gao 1996)	0.65/0.10

Where, ρ is the ground reflectance at the base of atmosphere or BOA in the mentioned spectral band

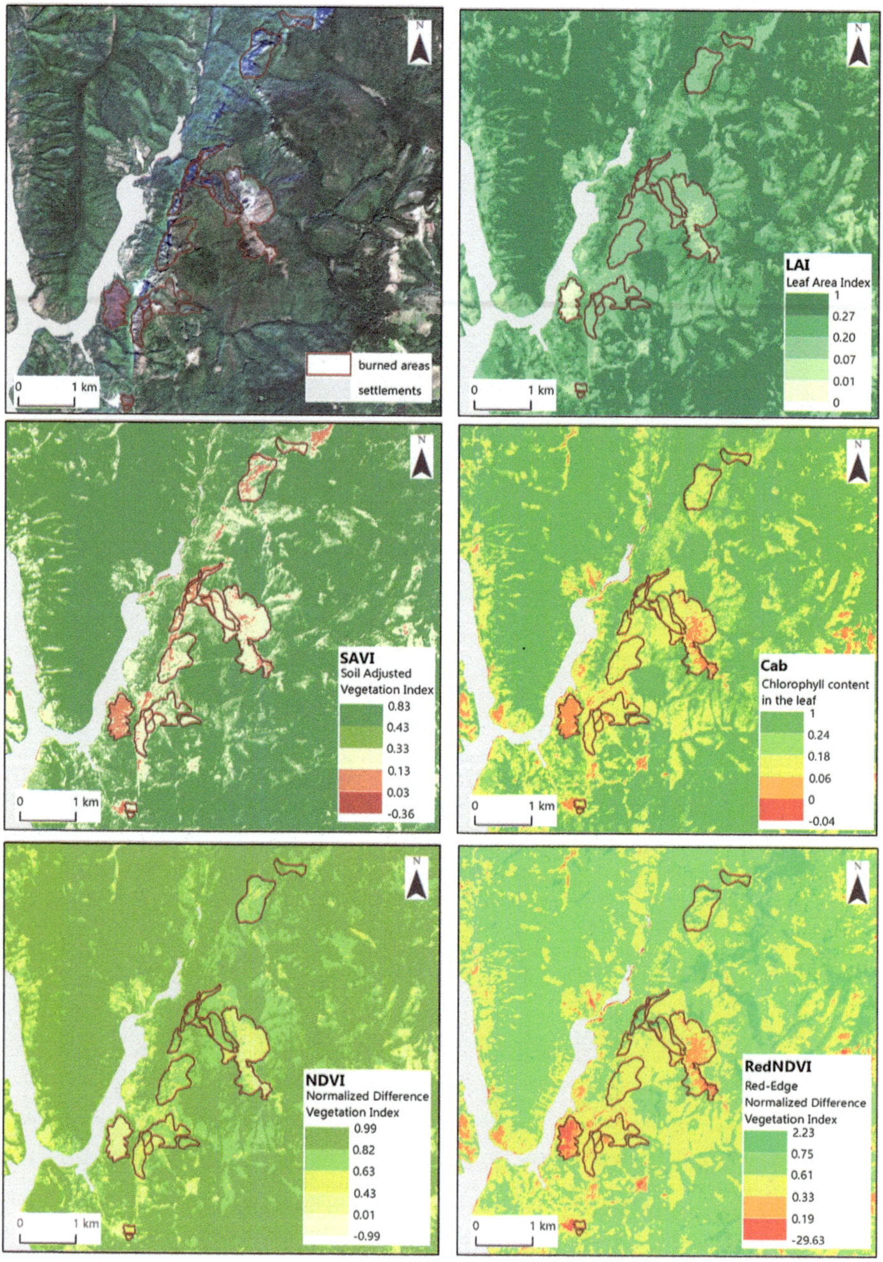

Fig. 3 Radiometric and biophysical indices produced from Sentinel-2A MSI imagery. The segmentation is based on the standard deviation defined threshold

Fig. 4 Black pine forest stands, highly susceptible to forest fires on the steep slopes of Domogled Mountain, near Băile Herculane resort (*Photo* I. Săvulescu, April 2014)

level of clay and water, are very resistant to fire ignition and spreading (Mihai and Săvulescu 2014). These forest stands falls between the second and the third standard deviation values of all the indices. Low burning potential is also specific to areas where the fuel has been exhausted by fire events in the last year (2015 for Sentinel-2 and 2013 for Pleiades) and where the vegetation didn't have enough time to recover.

Because of the limited spectral resolution of the Pleiades imagery, the number of derived indices is low: NDVI, GNDI and NDWI (Fig. 5). The same as Sentinel-2, mean and standard deviation values are used for segmentation threshold definition. According to different authors (Gitelson et al. 1996), in comparison with the NDVI, where the first two standard deviations are used for the definition of the burning potential, GNDVI is five time more sensitive to the chlorophyll content, being useful for differencing the vegetation stress and senescence.

The thematic classification generated from Sentinel-2 image (Fig. 6) can help the identification of forest canopies with a higher susceptibility to forest fires, using the spatial overlay between the most representative indices and the forest stand-land cover. The thematic classification used the SVM (Support Vector Machine) algorithm, based on an adequate volume of ROI polygons: five degree polynomial kernel with gamma factor of 0.05 for the production of the hyperplanes between classes.

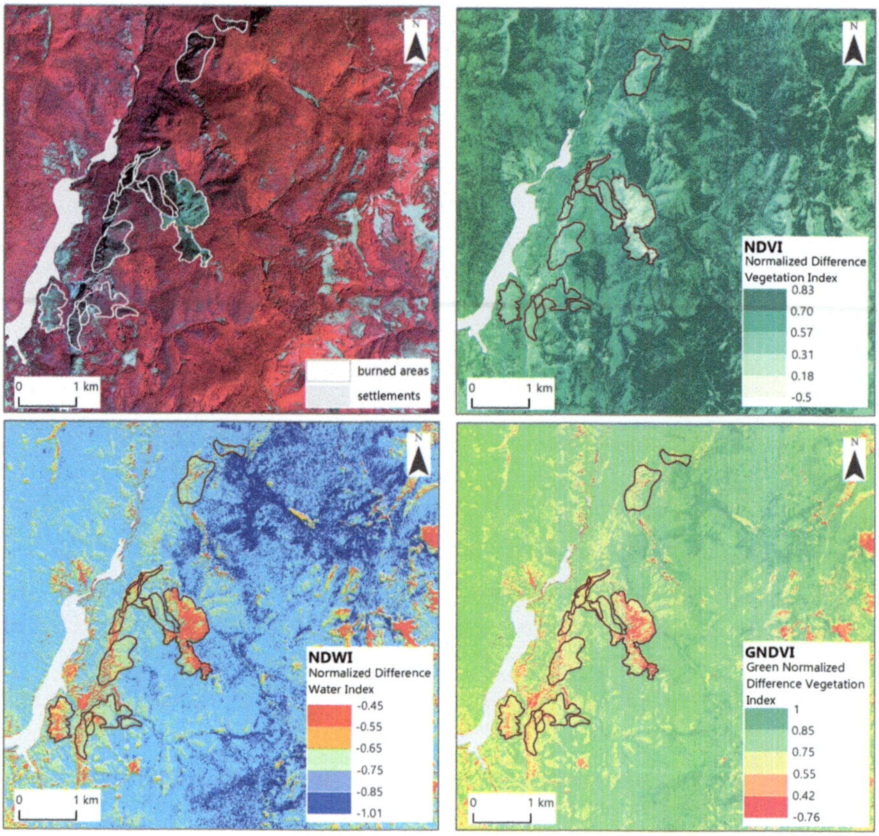

Fig. 5 Pleiades 1B ortho imagery, for the study area of Domogled-Băile Herculane and the three derived spectral indices

4 Discussion and Conclusion

The statistical approach shows the distribution of forest stand data and land cover features according to the values of the indices. Figure 7 is an example of statistical correlation between mean and standard deviation of the vegetation classes and LAI values, in this case. This example shows that the mean and the standard deviation of the Banat black pine stands class have a good correspondence with the interval defined by the first standard deviation of LAI. This demonstrates the reliability and the validation of the characteristic segmentation thresholds.

The current approach covers a limited region with protected forest stands and integrates recent satellite data offering reliable information on burned areas, from the months of August, featured by the most frequent events. In this context, we produced data at higher spatial resolution (10 m and, even 2 m), in comparison with other approaches focused on larger areas like tropical forests from Amazonia (Anderson

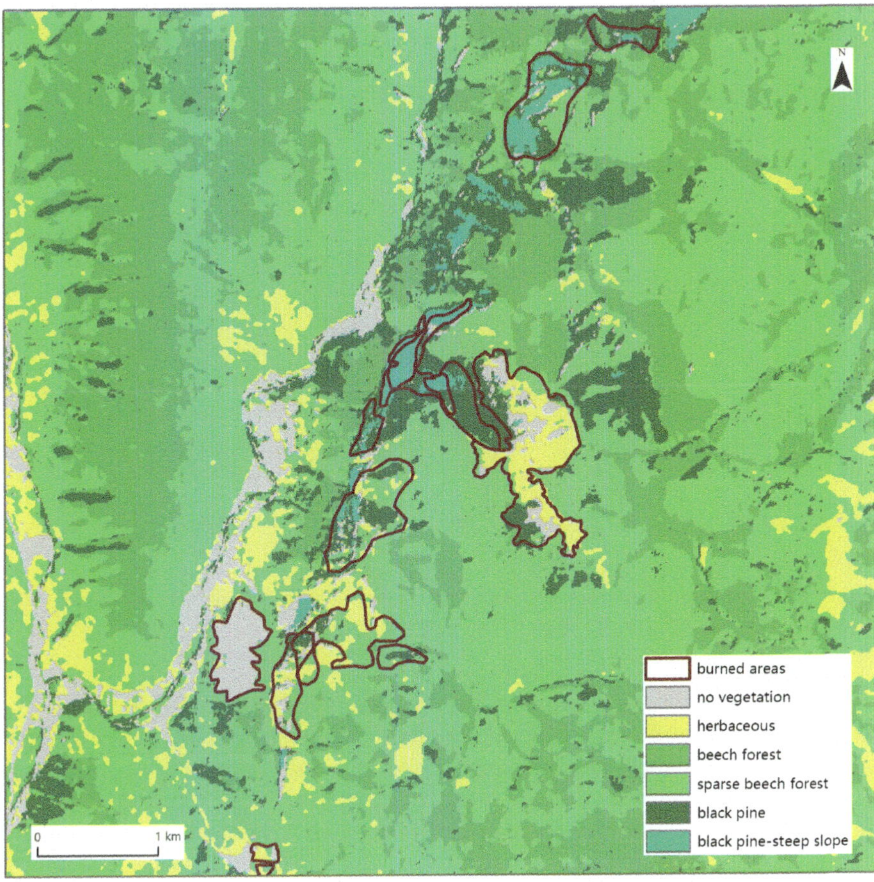

Fig. 6 Thematic classification of Sentinel-2 MSI derived data (10 m resolution) and the spatial configuration of burned areas (2000–2013) in Domogled-Băile Herculane area

2012; Matricardi et al. 2010), boreal forests from Canada (Serbin et al. 2013) and Mediterranean vegetation from Spain (Maffei and Menenti 2014). Most of them tested different indices on MODIS imagery at coarser spatial resolution and very few on medium resolution imagery like Landsat. Caspard et al. (2015) produced data for the vegetation recovery after forest fires in Reunion Island by integrating higher resolution imagery, like SPOT, Pleiades and WorldView-2. All these contributions tested a limited number of indices like LAI, SAVI and NDVI. This case study applies a larger collection of indices for a detailed analysis of the forest fire susceptibility in a protected area with forest stands at risk (Domogled-Valea Cernei National Park).

The selected indices are evaluated with statistical parameters and validated with the canopy spatial coverage data. We propose a collection of complementary radiometric and biophysical indices and characteristic thresholds to be used for the temperate region forests, and mainly for mountain and hilly regions of Central and Eastern Europe in order to separate the forest fire susceptibility different canopies.

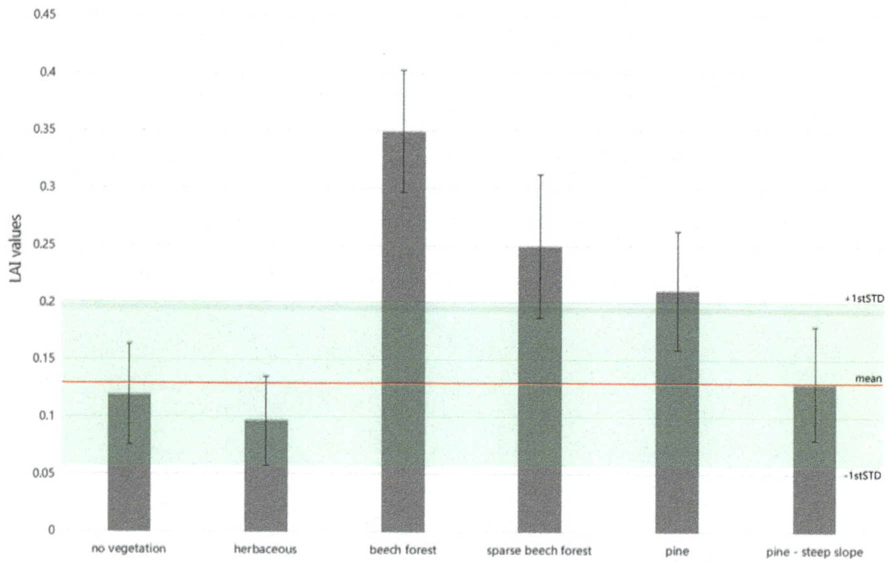

Fig. 7 Statistical correspondence and validation between selected canopies and LAI values, produced from Sentinel-2 MSI imagery from August, 31 2015

The approach of using two satellite imagery types with complementary characteristics, at two different spatial scales offers the advantage to explore two series of spectral reflectance measurements collected right after forest fire events. Sentinel-2 MSI images have an advantage in comparison with Pleiades 1B, because of the superior spectral resolution in the infrared domain. In the same time, they have a good temporal resolution of five days, which is useful for these phenomena, especially when they affect larger surfaces of hectares. Pleiades 1B are superior in spatial resolution, and this is an advantage for isolated features recognition like fire ignition point, but their spectral resolution is rather limited, while the temporal one is strictly controlled between archive data or pre-ordered imagery.

SAVI returns good results, being a measure of the reflectance in red of the pixels showing sparse vegetation areas. This index includes in the NDVI equation the L factor which indicates the variation of the soil pixels in the vegetation plots. In temperate regions, the sparsely vegetation is an indicator of the limited conditions for growing. For example, the Banat black pine is a species which can grow in these limited conditions, resulting a high exposure to fire events.

The indices successfully used in the Mediterranean region for this type of analysis (i.e. NDBR—Normalized Difference Burning Ratio, indices that evaluate the water content in the soil, MSI—Moisture Stress Index, NPCRI—Normalized Pigment Chlorophyll Ratio Index) cannot be applied in the temperate areas, where our study area is located. For example, high values of vegetation density in the Mediterranean region are associated with a high burning potential, opposite to the situations observed in the temperate region.

Still, there are some limitations in the use of the indices related to the vegetation specific features such as chlorophyll, carbon or water content. For example, the vegetation index fAPAR, didn't have a good result because it evaluates the carbon content and nutrients cycle, which does not help in differentiating the burning potential. Despite this, in temperate regions, the vegetation indices are more suitable for the identification/mapping of the forest fire susceptibility.

Acknowledgements The research was done in the framework of SIAFIM project (Satellite Image Analysis for Fire Monitoring), 2012–2015, financed by ROSA-Romanian Space Agency and ESA-European Space Agency.

References

Agapiou A, Alexakis DD, Sarris A, Hadjimitsis DG (2014) Evaluating the potentials of Sentinel-2 for archaeological perspective. Remote Sens-Basel 6(3):2176–2194

Anderson LO (2012) Biome-scale forest properties in Amazonia based on field and satellite observations. Remote Sens-Basel 4(5):1245–1271

Angayarkkani K, Radhakrishnan N (2009) Efficient forest fire detection system: a spatial data mining and image processing based approach. Int J Comput Sci Netw Secur 9(3):100–107

Arpaci A, Malowerschnig B, Sass O, Vacik H (2014) Using multi variate data mining techniques for estimating fire susceptibility of Tyrolean forests. Appl Geogr 53:258–270

Baret F, Fourty T (1997) Radiometric estimates of nitrogen status of leaves and canopies. In: Diagnosis of the nitrogen status in crops. Springer, Berlin, pp 201–227

Bowman DM, Balch JK, Artaxo P, Bond WJ, Carlson JM, Cochrane MA, D'Antonio CM, DeFries RS, Doyle JC, Harrison SP (2009) Fire in the earth system. Science 324(5926):481–484

Caspard M, Yésou H, Selle A, Tinel C, Tessier P, Durand A, Clandillon S, de Fraipont P (2015) Forest recolonization monitoring based on HR and VHR imagery: the case of the Maido forest fire exploiting Pléiades and spot Kalideos database. Rev Fr Photogram Télédétection 209:149

Cheng T, Wang J (2008) Integrated Spatio-temporal data mining for forest fire prediction. Trans GIS 12(5):591–611

Chuvieco E (2000) Remote sensing of forest fires—current limitations and future prospects. Observing Land Space: Sci Customers Technol 4:47–51

Chuvieco E, Aguado I, Jurdao S, Pettinari ML, Yebra M, Salas J, Hantson S, de la Riva J, Ibarra P, Rodrigues M (2014) Integrating geospatial information into fire risk assessment. Int J Wildland Fire 23(5):606–619

Chuvieco E, Aguado I, Yebra M, Nieto H, Salas J, Martín MP, Vilar L, Martínez J, Martín S, Ibarra P (2010) Development of a framework for fire risk assessment using remote sensing and geographic information system technologies. Ecol Model 221(1):46–58

Chuvieco E, Cocero D, Riano D, Martin P, Martınez-Vega J, de la Riva J, Perez F (2004) Combining NDVI and surface temperature for the estimation of live fuel moisture content in forest fire danger rating. Remote Sens Environ 92(3):322–331

Clima României (2008) Clima României. Editura Academiei Române, Bucureşti

Cortez P, Morais AdJR (2007) A data mining approach to predict forest fires using meteorological data

Dale VH, Joyce LA, McNulty S, Neilson RP, Ayres MP, Flannigan MD, Hanson PJ, Irland LC, Lugo AE, Peterson CJ (2001) Climate change and forest disturbances: climate change can affect forests by altering the frequency, intensity, duration, and timing of fire, drought, introduced species, insect and pathogen outbreaks, hurricanes, windstorms, ice storms, or landslides. Bioscience 51(9):723–734

Datcu M, Daschiel H, Pelizzari A, Quartulli M, Galoppo A, Colapicchioni A, Pastori M, Seidel K, Marchetti PG, d'Elia S (2003) Information mining in remote sensing image archives: system concepts. IEEE Trans Geosci Remote Sens 41(12):2923–2936

Datcu M, Seidel K (2000) Image information mining: exploration of image content in large archives. In: Aerospace Conference Proceedings, 2000 IEEE, pp 253–264

Delegido J, Verrelst J, Alonso L, Moreno J (2011) Evaluation of sentinel-2 red-edge bands for empirical estimation of green LAI and Chlorophyll Content. Sens Basel 11(7):7063–7081

Drusch M, Del Bello U, Carlier S, Colin O, Fernandez V, Gascon F, Hoersch B, Isola C, Laberinti P, Martimort P, Meygret A, Spoto F, Sy O, Marchese F, Bargellini P (2012) Sentinel-2: ESA's optical high-resolution mission for GMES operational services. Remote Sens Environ 120:25–36

Fernandez A, Illera P, Casanova JL (1997) Automatic mapping of surfaces affected by forest fires in Spain using AVHRR NDVI composite image data. Remote Sens Environ 60(2):153–162

Flannigan MD, Stocks BJ, Wotton BM (2000) Climate change and forest fires. Sci Total Environ 262(3):221–229

Frampton WJ, Dash J, Watmough G, Milton EJ (2013) Evaluating the capabilities of sentinel-2 for quantitative estimation of biophysical variables in vegetation. Isprs J Photogramm 82:83–92

Gao B-C (1996) NDWI—a normalized difference water index for remote sensing of vegetation liquid water from space. Remote Sens Environ 58(3):257–266

Gitelson A, Merzlyak MN (1994) Spectral reflectance changes associated with autumn senescence of *Aesculus hippocastanum* L. and *Acer platanoides* L. leaves. Spectral features and relation to chlorophyll estimation. J Plant Physiol 143(3):286–292

Gitelson AA, Kaufman YJ, Merzlyak MN (1996) Use of a green channel in remote sensing of global vegetation from EOS-MODIS. Remote Sens Environ 58(3):289–298

Gitelson AA, Merzlyak MN (1998) Remote sensing of chlorophyll concentration in higher plant leaves. Adv Space Res 22(5):689–692

Han JG, Ryu KH, Chi KH, Yeon YK (2003) Statistics based predictive geo-spatial data mining: Forest fire hazardous area mapping application. Web Technol Appl 2642:370–381

Hsu W, Lee ML, Zhang J (2002) Image mining: trends and developments. J Intell Inf Syst 19 (1):7–23

Huang Y-L, Devan MN, U'Ren JM, Furr SH, Arnold AE (2016) Pervasive effects of wildfire on foliar endophyte communities in montane forest trees. Microb Ecol 71(2):452–468

Huete AR (1988) A soil-adjusted vegetation index (SAVI). Remote Sens Environ 25(3):295–309

Jacquemoud S, Baret F (1990) PROSPECT: a model of leaf optical properties spectra. Remote Sens Environ 34(2):75–91

Joint Research Center I, Land Management and Natural Hazard Unit (2014) Forest Fire in Europe, Middle East and North Africa 2013. European Commission. https://doi.org/10.2788/99870

Maffei C, Menenti M (2014) A MODIS-based perpendicular moisture index to retrieve leaf moisture content of forest canopies. Int J Remote Sens 35(5):1829–1845. https://doi.org/10.1080/01431161.2013.879348

Main-Knorn M, Pflug B, Debaecker V, Louis J (2015) Calibration and validation plan for the L2A processor and products of the sentinel-2 mission. Int Arch Photogrammetry Remote Sens Spat Inf Sci 40(7):1249

Majasalmi T, Rautiainen M (2016) The potential of Sentinel-2 data for estimating biophysical variables in a boreal forest: a simulation study. Remote Sens Lett 7(5):427–436

Matricardi EAT, Skole DL, Pedlowski MA, Chomentowski W, Fernandes LC (2010) Assessment of tropical forest degradation by selective logging and fire using Landsat imagery. Remote Sens Environ 114(5):1117–1129. https://doi.org/10.1016/j.rse.2010.01.001

Maxant J, Proy C, Fontannaz D, Clandillon S, Allenbach B, Yesou H, Battiston S, Uribe C, De Fraipont P (2013) Contribution of Pleiades-HR imagery for disaster damage mapping: initial feedback over Asia, Africa, Europe or the Caribbean. In: Proceedings of 33th EARSeL Symposium Towards Horizon

Mazzoni D, Logan JA, Diner D, Kahn R, Tong LL, Li QB (2007) A data-mining approach to associating MISR smoke plume heights with MODIS fire measurements. Remote Sens Environ 107(1–2):138–148
Mihai B, Săvulescu I (2014) Mapping forest fire susceptibility in temperate mountain areas with submediteranean influences with expert knowledge. A case study from Domogled ridge—Mehedinţi Mountains, Southern Carpathians. Paper presented at the S4C (Science for the Carpathians). Forum Carpaticum 2014: Local Responses to Global Challenges, Lviv, Ukraine
Mithal V, Garg A, Boriah S, Steinbach M, Kumar V, Potter C, Klooster S, Castilla-Rubio JC (2011) Monitoring global forest cover using data mining. ACM Trans Intell Syst Tec 2(4)
Pătroescu M, Chincea I, Rozylowicz L, Sorescu C, Goia I, Groza G, Frăţilă E, Iojă C, Bădescu B, Crişan A, Crăciun N (2007) Forests with Banat black pine (Pinus nigra subsp. banatica) NATURA 2000 site. Editura BRUMAR Timişoara
Pourghasemi HR (2016) GIS-based forest fire susceptibility mapping in Iran: a comparison between evidential belief function and binary logistic regression models. Scand J Forest Res 31 (1):80–98
Pourtaghi ZS, Pourghasemi HR, Aretano R, Semeraro T (2016) Investigation of general indicators influencing on forest fire and its susceptibility modeling using different data mining techniques. Ecol Indic 64:72–84
Quintano C, Fernandez-Manso A, Stein A, Bijker W (2011) Estimation of area burned by forest fires in Mediterranean countries: a remote sensing data mining perspective. Forest Ecol Manag 262(8):1597–1607
Richter K, Hank TB, Vuolo F, Mauser W, D'Urso G (2012) Optimal exploitation of the Sentinel-2 spectral capabilities for crop leaf area index mapping. Remote Sens-Basel 4(3):561–582
Rouse JW, Haas RH, Schell JA, Deering DW (1973) Monitoring vegetation systems in the Great Plains with ERTS. Paper presented at the Third ERTS Symposium
Serbin SP, Ahl DE, Gower ST (2013) Spatial and temporal validation of the MODIS LAI and FPAR products across a boreal forest wildfire chronosequence. Remote Sens Environ 133:71–84. https://doi.org/10.1016/j.rse.2013.01.022
Stojanova D, Panov P, Kobler A, Džeroski S, Taškova K (2006) Learning to predict forest fires with different data mining techniques. In: Conference on data mining and data warehouses (SiKDD 2006), Ljubljana, Slovenia, pp 255–258
Tay SC, Hsu W, Lim KH, Yap LC (2003) Spatial data mining: clustering of hot spots and pattern recognition. In: Geoscience and Remote Sensing Symposium, 2003. IGARSS'03. Proceedings. 2003 IEEE International. IEEE, pp 3685–3687
Török-Oance M, Török-Oance R (2002) Consideraţii asupra propagării şi efectelor incendiilor în regiunile montane. Studiu de caz: incendiul din Masivul Domogled (August 2000). Studii şi cercetări de Geologie 47:221–232
Umamaheshwaran R, Bijker W, Stein A (2007) Image mining for modeling of forest fires from Meteosat images. IEEE Trans Geosci Remote Sens 45(1):246–253
Wang D, Wang J, Liang S (2010) Retrieving crop leaf area index by assimilation of MODIS data into a crop growth model. Sci China Earth Sci 53(5):721–730
Whittaker RH, Niering WA (1975) Vegetation of the Santa Catalina Mountains, Arizona. V. Biomass, production, and diversity along the elevation gradient. Ecology 56(4):771–790

Monitoring and Management of Land Subsidence Induced by Over-exploitation of Groundwater

Maryam Dehghani and Mohammad Reza Nikoo

Abstract Most plains in Iran are subject to land subsidence due to over-exploitation of groundwater mainly for agricultural purposes. Synthetic Aperture Radar (SAR) interferometry has shown its ability to provide precise measurements of the ground surface displacement at high spatial and temporal resolution. In SAR interferometry, the processed interferograms are combined together via interferogram stacking or time series analysis. Stacking is a temporal averaging of the interferograms which results in mean displacement velocity. However, time series analysis of a significant number of interferograms enables us to study the short-term as well as ling-term behavior of the subsidence. In this research, three different case studies were accomplished for subsidence monitoring. The subsidence in the Varamin plain was studied using 13 ENVISAR ASAR images spanning between 2003/08/03 and 2005/11/20. The maximum subsidence rate extracted from Small Baseline Subset (SBAS) time series was estimated as 0.4 m/year. The second case study was to monitor the subsidence in the Neyshabour plain. In this area, the interferogram stacking using 9 ENVISAT ASAR images spanning between 2004/01/10 and 2005/06/18 was applied. The maximum subsidence rate was estimated as 0.16 m/year. Groundwater level measurements made at piezometric wells were applied to compare to the interferometry results. The piezometric wells mostly show the increase in water level depth caused by over-exploitation of groundwater. The groundwater information jointly with stratigraphic profiles highly correlate with subsidence in the area. In the last case study, the Persistent Scatterer Interferometry (PSI) which is a proper method of time series analysis in areas with high decorrelation effects, was used in the Shahriar plain. A hybrid method of conventional and PSI was proposed in order to address the problem of monitoring the high-rate deformation. There are 22 ENVISAT ASAR images available in the study area spanning between 2003 and 2008. The maximum subsidence rate was estimated as 0.25 m/year. The time series analysis results were then compared to the groundwater level information at piezometric wells. Due to the low correlation between water

M. Dehghani · M. R. Nikoo (✉)
Department of Civil and Environmental Engineering,
School of Engineering, Shiraz University, Shiraz, Iran
e-mail: nikoo@shirazu.ac.ir

H. R. Pourghasemi and M. Rossi (eds.), *Natural Hazards GIS-based Spatial Modeling Using Data Mining Techniques*, Advances in Natural and Technological Hazards Research 48, https://doi.org/10.1007/978-3-319-73383-8_12

level decline and subsidence rate at some piezometric wells, it can be concluded that other geology and hydrogeological factors play important role in controlling the subsidence occurrence. To show this, two data mining methods including Multi-Layer Perceptron (MLP) neural network as well as Support Vector Regression (SVR) were applied to model the subsidence in Shahriar plain using 6 different geology and hydrogeology factors as input and the subsidence rate extracted from interferometry as output of the model. These models can be further applied to estimate the subsidence rate in pixels in which the interferometry technique cannot measure the deformation due to some reasons including insufficient correlation.

Keywords Subsidence · Interferometry · Persistent scatterer
Groundwater information · Data mining

1 Introduction

Deficiency in precipitation over an extended period makes people exploit groundwater mostly used for agricultural activities. Over-exploitation of groundwater causes increasing of the effective stress within an aquifer system which consists of compressible fine-grained sediments. Hence, the fine-grained interbeds within aquifer system are generally compacted due to the changes in stress. This leads to land subsidence which results in damage to structures and buildings.

Land subsidence caused by exploitation of groundwater is considered as a main concern in different parts of the world (e.g. Poland and Davis 1969; Tolman and Poland 1940; Galloway et al. 1998, 1999). In Iran which suffers from lack of rainfall, land subsidence mainly occurs in the cultivated areas. In some cases, the subsidence extends to the residential areas causing environmental consequences including damage to buildings, pipelines, roadways, well casings and surface runoff. Moreover, earth fissures may be appeared on the ground surface subject to land subsidence. Sinkholes as devastating consequences are common phenomena in areas subject to land subsidence. Sediment compaction within the aquifer system as a result of groundwater withdrawal makes such big spaces.

Geological Survey of Iran (GSI) jointly with other organizations such as National Cartographic Center (NCC), initiated a comprehensive program in 2004 to monitor the subsidence in most parts of Iran. The first and foremost step is to measure the ground deformation caused by subsidence. Precise leveling and Global Positioning System (GPS) observations as two reliable sources of information have been vastly used for subsidence monitoring; however, these techniques are able to measure the amount of displacement only at the leveling and GPS stations. Hence, the spatial pattern of the subsidence can be barely identified using the leveling and GPS measurements.

Interferometric Synthetic Aperture Radar (InSAR) as a space-based method can measure the ground surface displacement at large coverage and high spatial resolution (Galloway et al. 1998; Lundgren et al. 2001; Dehghani et al. 2013). Using two SAR satellite images taken at different times, this technique is able to measure the land surface deformation occurred between two acquisitions. InSAR has been utilized in different studies to monitor the land subsidence (Dehghani et al. 2010, 2013). Similarly, InSAR has been widely applied to study the subsidence behavior in several parts of Iran since 2006. In most studies carried out in Iran, besides the subsidence spatial pattern, its temporal behavior has been studied as well using interferometry time series analysis. Various methods of time series analysis were employed based on the study areas characteristic and the data availability. Conventional InSAR fails to measure the deformation when there is large amount of vegetation and/or the temporal and spatial sampling of the data is poor. In this case, a newly-developed approach, namely Persistent Scatterer Interferometry (PSI) based on InSAR, has been used. The main goal of this article is to present the history of using InSAR in different parts of Iran as well as the results achieved. The next section is devoted to introduce the InSAR followed by time series analysis techniques. The results of applying InSAR to different areas will be presented in Sect. 3. Section 4, finally, presents important conclusions on monitoring subsidence.

2 SAR Interferometry Method

SAR Interferometry makes use of phase measurements of two Single Look Complex (SLC) SAR images taken at different times to produce the deformation map. The phase difference, namely as interferometric phase composes of different components including deformation ($\phi_{D,x,i}$), earth curvature ($\phi_{Curv,x,i}$), topographic effect ($\phi_{topo,x,i}$), atmospheric signal ($\phi_{Atm,x,i}$), and noise phenomenon due to decorrelation ($\phi_{N,x,i}$) as shown in Eq. 1:

$$\psi_{x,i} = W\{\phi_{D,x,i} + \phi_{Curv,x,i} + \Delta\phi_{topo,x,i} + \Delta\phi_{Atm,x,i} + \phi_{N,x,i}\} \tag{1}$$

It should be noted that the interferometric phase is wrapped which means that its value is in the $[-\pi,\pi)$ interval. W{.} presents the wrapping operator.

In order to retrieve the component due to the deformation phase $\phi_{D,x,i}$, several steps has to be accomplished according to Fig. 1. The method starts with two SLC SAR images: master and slave images. A critical procedure in InSAR is image coregistration which is the pixel-to-pixel alignment of two SAR images (Li and Bethel 2008). The phase difference between master and slave images, which is called interferogram, is then computed.

The phase component due to earth curvature is estimated using the spatial baseline value and subtracted from the interferogram. A residual phase, called residual orbital error, however is remained due to imprecise baseline value which can be further easily estimated and removed from the interferometric phase. Topographic contribution to the phase is removed by means of an available Digital Elevation Model (DEM). In this differential interferogram, each phase cycle is

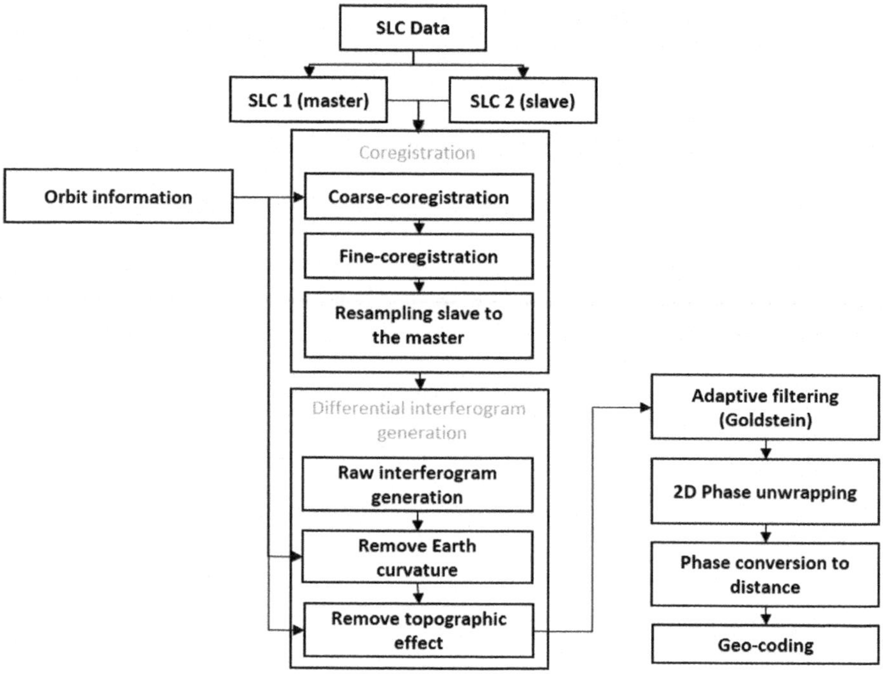

Fig. 1 Block diagram of conventional interferometry

called a fringe originated from ground displacement of half the radar wavelength. Once the differential interferogram is generated, it is low-pass filtered using an adaptive power-spectrum filter to suppress the phase noise mainly induced by the decorrelation of master and slave images. It should be noted that the phase component due to atmosphere is further reduced through the time series analysis. Phase unwrapping as a procedure of adding up the consecutive fringes is then performed in order to produce a continuous deformation field. Finally the displacement is calculated by using a multiplication factor of $\frac{\lambda}{4\pi}$. The displacement interferogram is finally resampled from the slant-range radar geometry into the geographic coordinate system. This interferogram illustrates the ground surface displacement between master and slave acquisitions; however, it is possible to infer the spatial and temporal behavior of the subsidence by using considerable number of differential interferograms through the time series analysis (Lanari et al. 2004). Time series analysis enables us to study the short-term and long-term behavior of a continuously-occurring phenomenon such as subsidence. The spatial and temporal separation between the master and slave, namely spatial and temporal baselines, should be small enough to essentially decrease the decorrelation phenomenon. The coherent interferograms processed are then applied in the time series analysis to generate the subsidence time series as well as the mean subsidence velocity map. In the next section, two different approaches of time series analysis will be introduced.

2.1 Conventional Time Series Analysis

InSAR time series analysis is used in order to study the temporal behavior of land subsidence. If we generate as many independent interferograms as acquisition dates, we are able to obtain the subsidence time series in a least square inversion process (Dehghani et al. 2010). Time series analysis method in which the interferograms with small temporal and spatial baselines are processed is named as Small Baseline Subset (SBAS). Several algorithms for SBAS have been presented in different studies (Berardino et al. 2002, Dehghani et al. 2009a, b, 2010). The steps of the simplest one is depicted in Fig. 2.

In the first processing stage, a plane is fitted to the phase values outside the deformation area and then subtracted from the interferograms to remove the residual orbital effect (e.g. Funning et al. 2005). The flattened interferograms are then inverted using a least squares approach in order to estimate the deformation at

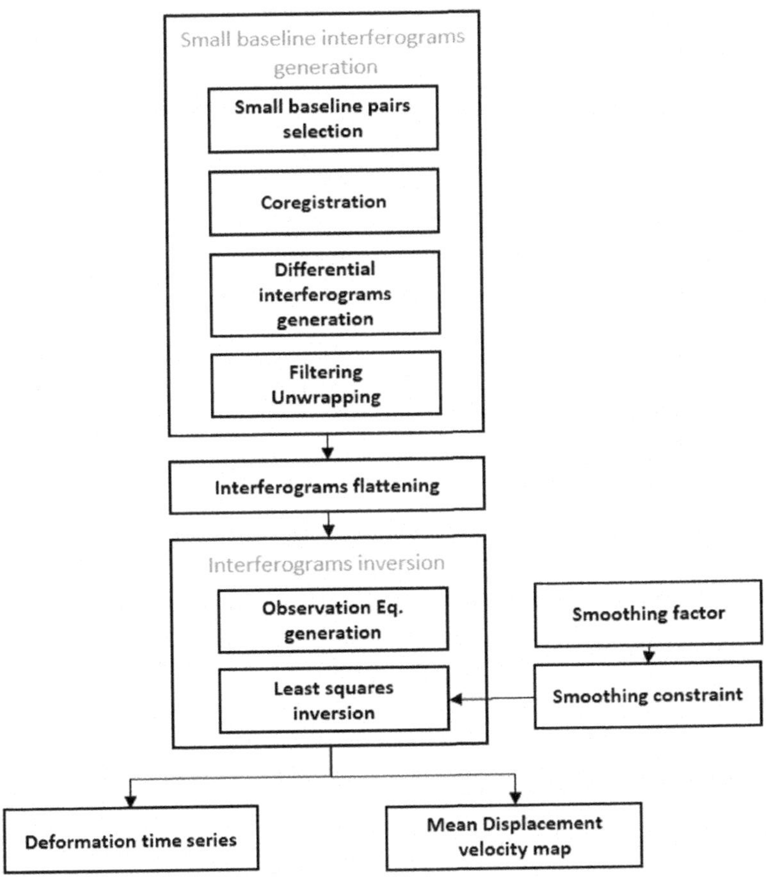

Fig. 2 Flowchart of the conventional time series analysis

each acquisition. A smoothing constraint is added into the least square inversion to decrease different errors such as noise, atmospheric effects and unwrapping errors (e.g. Schmidt and Burgmann 2003; Lundgren et al. 2001). The smoothing factor used in the smoothing constraint is determined optimally by considering the Root Mean Square Error (RMSE) of the least squares inversion problems (Dehghani et al. 2009a, b).

Time series analysis results involves two different products including deformation time series and mean displacement velocity map. The former is used to monitor the temporal behavior of the deformation while the latter indicates the main features of the deformation. The presented method has been applied to monitor the subsidence in many parts of Iran a couple of which will be presented in Sect. 3.

When the number of interferograms is less than the image acquisitions, it is not possible to apply the Least squares inversion due to the rank deficiency of the design matrix. In other words, the system of equations is underdetermined as there are fewer equations than unknowns. In this case, a method called "interferograms stacking" which temporally averages the coherent interferograms is applied. Accordingly, long-term behavior, i.e. deformation rate $\bar{\phi}_x$, of the subsidence signal is obtained as follows:

$$\bar{\phi}_x = \frac{\sum_{k=1}^{N} \phi_x^k}{\sum_{k=1}^{N} t_x^k} \tag{2}$$

where ϕ_x^k is the phase of the xth pixel in the kth coherent interferogram with the temporal baseline of t^k. Interferogram stacking as well as the SBAS time series analysis approach requires the coherent interferograms which can spatially be unwrapped. If the study area is covered by a large amount of vegetation or the spatial and temporal sampling of the available datasets is poor, the conventional interferometry cannot be used appropriately to measure the deformation due to the loss of decorrelation.

2.2 Persistent Scatterer Interferometry

In the late 1990s, it was discovered that some targets exist on the ground, named as Persistent Scatterer (PS) whose backscattering characteristics is somehow constant in time. The deformation can be easily estimated over these PS points over long time period using Persistent Scatterer Interferometry (PSI).

There are various algorithms of PSI being different in two main issues: (i) PS pixels identification, and (ii) phase unwrapping. Among all PSI algorithms, Stanford Method for PS (StaMPS) method developed by Hooper et al. (2007) is the most appropriate one for deformation monitoring in areas lacking man-made features such as cultivated lands subject to subsidence. In this technique, the amplitude dispersion index is firstly utilized to select the PS candidates (Ferretti et al. 2001).

The PS pixels are finally identified according to the phase analysis (Hooper et al. 2007). The phase unwrapping in StaMPS includes two main steps: (i) temporal unwrapping which is the unwrapping of phase difference between nearby pixels over time using Nyquist sampling criterion, (ii) spatial unwrapping in which the pixels are spatially unwrapped using a cost function resulted from the first step (Hooper 2010).

StaMPS can be well applied when the deformation rate is low; however, in the reverse case, i.e. high deformation rate, the Nyquist sampling criterion stating that the phase difference between two nearby pixels in time should be less than half a cycle is not met. In this case, the deformation rate is underestimated. In 2013 a method which is a combination of conventional and persistent scatterer interferometry, i.e. StaMPS, was developed to address this problem (Dehghani et al. 2013). The main idea of the method is to decrease the deformation rate by subtracting a mean rate extracted from a couple of coherent interferograms processed by conventional interferometry so as to reduce the likelihood of aliasing. In the presented method, the PS pixels are identified using the amplitude and phase analysis proposed by StaMPS. The unwrapping step is applied on the phase from which the low pass component of the deformation is subtracted. Figure 3 illustrates the flowchart of the proposed method.

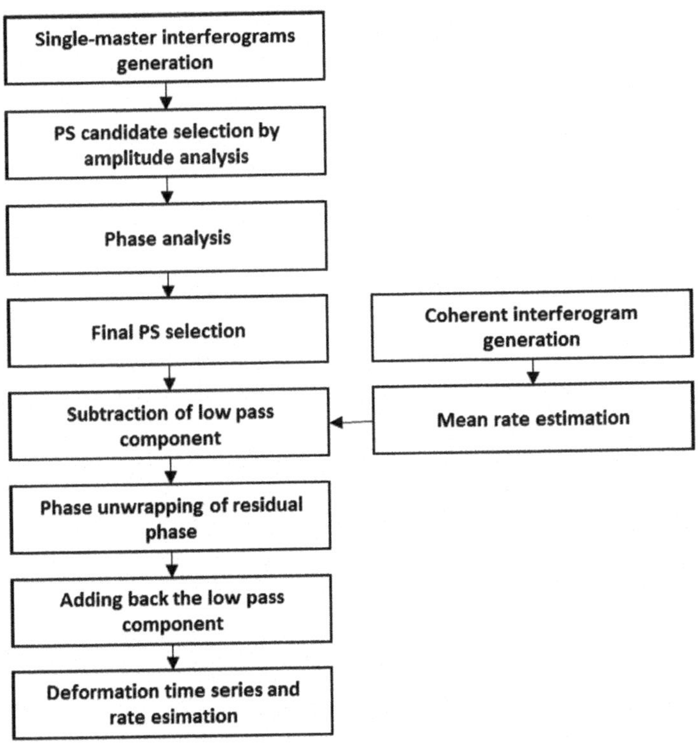

Fig. 3 flowchart of the hybrid method

The proposed algorithm based on StaMPS is originally developed for subsidence monitoring due to the number of reasons: (1) subsidence phenomenon is mostly occurred in the cultivated lands lacking in man-made features. Hence, StaMPS can be efficiently used to identify PS points; (2) in most cases, the subsidence rate is so high to meet the Nyquist sampling criterion. As a result the subsidence rate is underestimate; however, the proposed method can therefore estimate the subsidence rate correctly.

Conventional interferometry as well as the hybrid method are applied on different areas located in Iran to monitor the subsidence occurred by over-exploitation. In most case studies, the results obtained by interferometry are compared with groundwater information. In the next section, the results of some are presented.

3 Interferometry Monitoring of Subsidence in Iran

In most parts of Iran, lack of precipitation makes people extract water from aquifer systems. This results in groundwater level decline followed by subsidence. Since 2006, SAR interferometry approach as one of the essential steps in monitoring subsidence have been employed in various areas including Tehran-Shahriar, Ghazvin, Karaj, Varamin, Hashtgerd, Mashhad, Neyshabour, Mahyar, Sirjan, Rafsanjan, Kerman, Ardebil, Hamedan, Fasa, Darab plains, etc. We selected three case studies on which three different methods of interferogram stacking, SBAS and hybrid have been applied as presented in this section.

3.1 Case Study 1: Varamin Plain

Varamin plain located in south of Tehran province is subject to subsidence due to the exploitation of groundwater. Figure 4 depicts sinkhole as a consequence of the subsidence in Varamin plain.

In order to study the short term and long-term charactersitic of the subsidence, SBAS approach was applied. 13 ENVISAR ASAR SLC images spanning between 2003/08/03 and 2005/11/20 were used to process 22 differential small baseline interferograms. Acquisition geometry of the available radar data as well as the processed interferograms are presented in Fig. 5.

Time series analysis method presented in Sect. 2.1 was employed in order to generate deformation time series. To indicate the main objects of the deformation, mean subsidence velocity map was calculated by utilization of the time series analysis results (Fig. 6).

According to Fig. 6, the maximum subsidence rate along the line-of-sight (LOS) which belongs to the residential areas is 40 cm/yr. which is a cause for concern. Some closed patches probably due to unwrapping error in some single interferograms are presented as uplift which should be ignored in the interpretation process.

Fig. 4 Sinkhole as a consequence of subsidence in Varamin plain

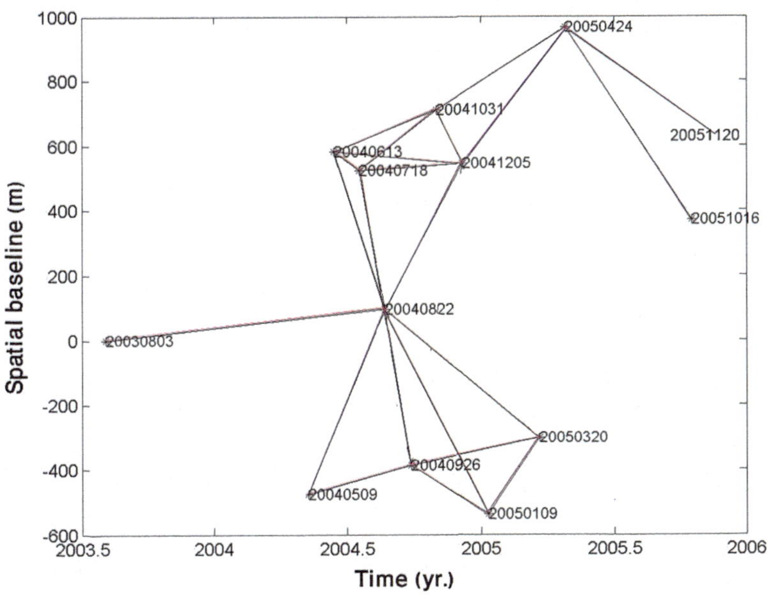

Fig. 5 Acquisition geometry of radar data: temporal baselines against spatial baselines. Solid lines indicate the processed interferograms

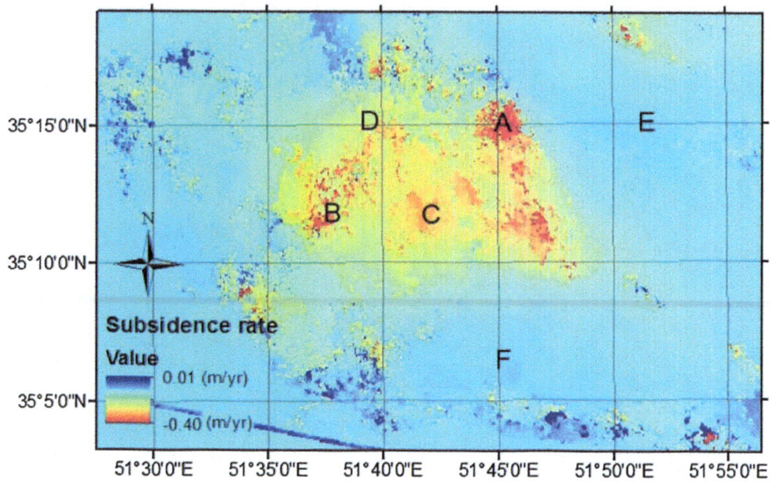

Fig. 6 Mean subsidence velocity map of Varamin subsidence. Maximum subsidence rate is estimated as 40 cm/yr

The mean subsidence velocity map presents the long-term average of the surface deformation. However, the time series analysis allows for identifying the short-term behavior of the subsidence such as seasonal fluctuations (e.g. Lanari et al. 2004). Hence, in order to study the temporal evolution of the deformation, deformation time series at a selection of points are presented. These points are located in various parts of the study area: (i) within the area of the maximum subsidence rate, (ii) along the margin of the subsidence and (iii) outside the subsidence area. Deformation time series as the chronological sequence of the deformation are demonstrated in Fig. 7.

Deformation time series at the selected points are characterized by nearly constant long-term rate on which the seasonal fluctuations due to discharge and recharge of the aquifer system are superimposed. For instance, the deformation sequences of points A and B show the decelerated subsidence in the recovery season, i.e. from September 2004 to March 2005. The reason for this change is groundwater aquifer recharge occurred in winter season.

According to the time series plots of points C and D, the seasonal effects are less pronounced probably due to insufficient recharge of the aquifer system. Moreover, in points E and F, no subsidence signal is observed since these points are located outside the subsidence area. An absence of bias in the time series plots of points E and F indicates that the residual orbital and atmospheric errors have been efficiently removed through the time series analysis process.

As a conclusion, the study area is subsiding with a rather constant rate indicating that the compaction of the aquifer system occurs inelastically associated with over-exploitation of groundwater. Other important factor affecting the subsidence rate is the soil types which constitutes the aquifer system. The subsidence is insignificant when the aquifer system composes of gravel and sand which are

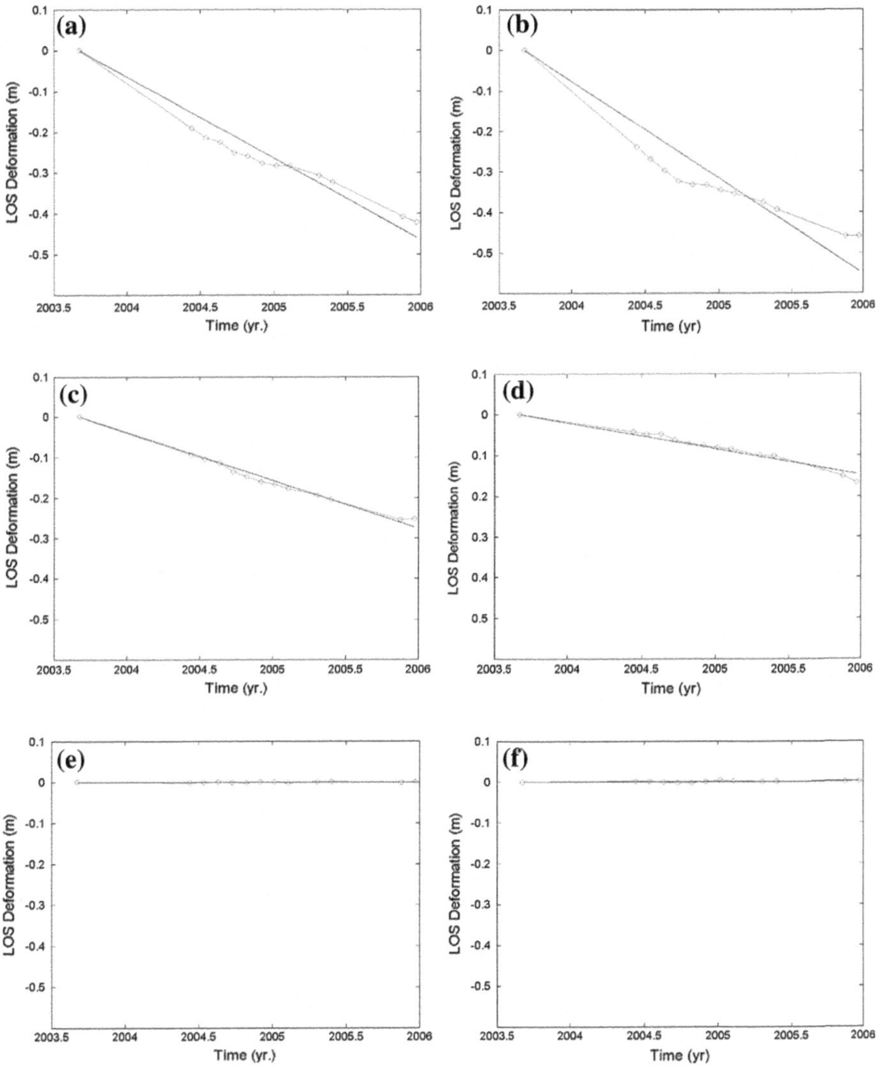

Fig. 7 Deformation time series at different points: **a** A, **b** B, **c** C, **d** D, **e** E and **f** F whose locations are shown in Fig. 6. Solid lines in the plots indicate the regression line fitted to the plots

non-compressible sediments (Dehghani et al. 2010). Integration of the Interferometry results with other sources of information such as geology and hydrogeological parameters in order to model the subsidence is considered as further studies in Varamin plain.

3.2 Case Study 2: Neyshabour Plain

During the last two decades, Neyshabour plain, located in northeast of Iran has been subject to land subsidence associated with the compaction of the aquifer system. Groundwater level has been monitored monthly by the Water Management Organization through piezometer installations. The aquifer hydraulic heads have been significantly declined due to the over-exploitation of groundwater to provide water for agricultural purposes. The aquifer system composed of highly compressible fine-grained sediments has experienced remarkable compaction leading to high-rate land subsidence. Expansion of subsidence to Neyshabour city as a historical city in Iran would be a cause for concern. Subsidence monitoring as the first step to mitigate its negative effects is required.

Interferometry as the most efficient method to monitor the subsidence in Neyshabour has been employed. 9 ENVISAT ASAR images spanning between 2004/01/10 and 2005/06/18 were utilized to process 8 coherent interferograms. Rank deficiency in Least Squares solution associated with insufficient number of interferograms, made us use interferogram stacking approach instead of time series analysis. Acquisition geometry of available data in the area is illustrated in Fig. 8.

The mean subsidence velocity map as the main product of stacking enables us to identify the long-term behavior of the subsidence. Spatial pattern of the subsidence shows a complex form surrounding Neyshabour city. The maximum deformation rate along the line-of-sight (LOS) is estimated as 16 cm/yr. as observed in Fig. 9.

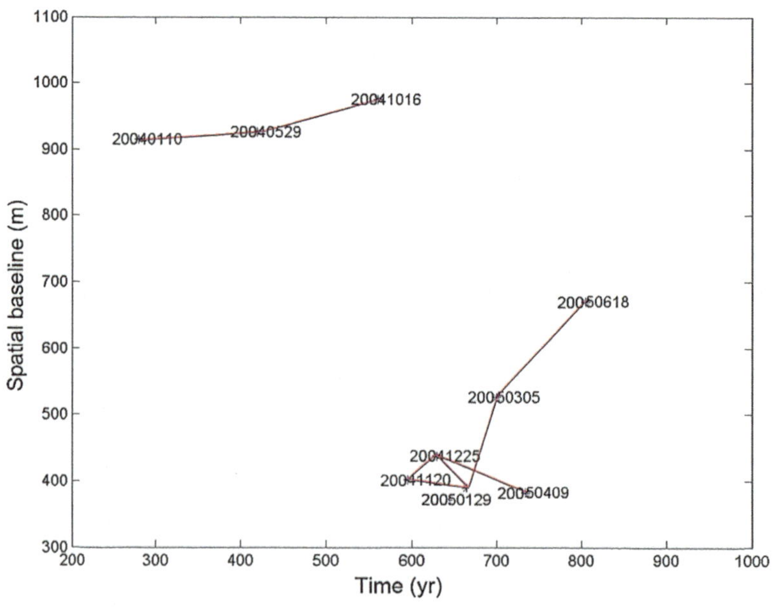

Fig. 8 Acquisition geometry of available radar data in Neyshabour area. Solid lines indicate the processed interferograms

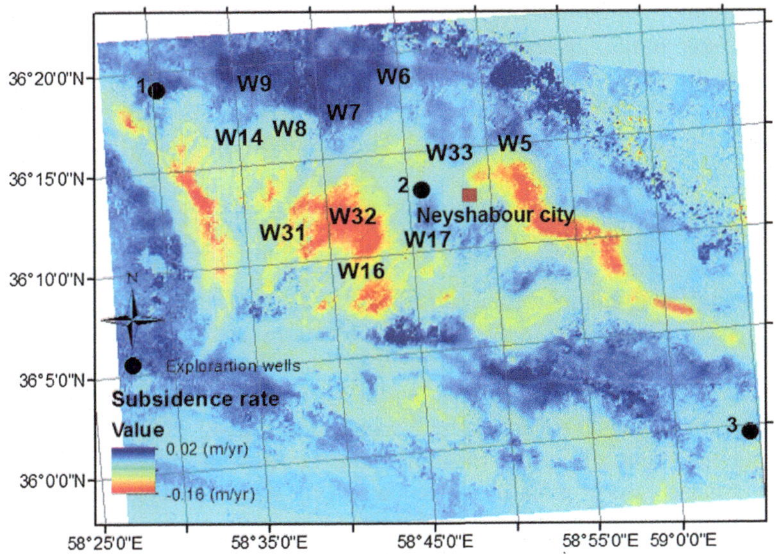

Fig. 9 Mean displacement velocity map of Neyshabour plain

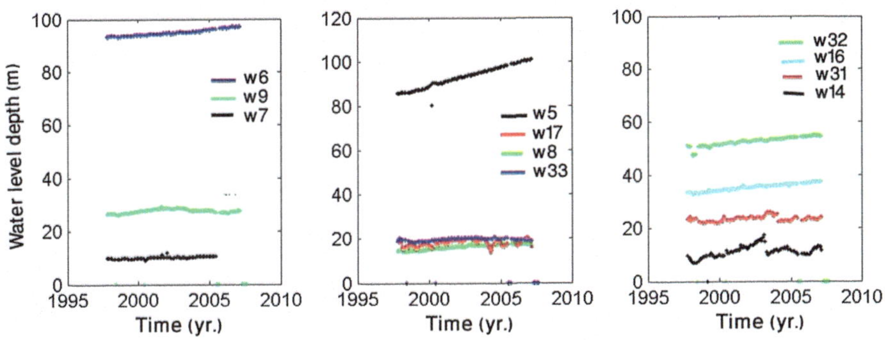

Fig. 10 Water level depth variations of piezometric well located: **a** outside (W6, W9 and W7), **b** in the margins (W5, W17, W8 and W33) and **c** in the middle of the subsidence area (W32, W16, W31 and W14)

Groundwater level measurements made at piezometric wells were applied to compare to the interferometry results. Three groups of piezometric wells located outside, in the middle and along the margin of the subsidence area have been selected for investigation. The temporal water level fluctuations at these wells are illustrated in Fig. 10. As observed in Fig. 10, the piezometric wells mostly show the increase in water level depth caused by over-exploitation of groundwater.

The water level depth of piezometric wells which are located outside the subsidence area does not show a significant increase except for W6. The reason is probably due to the soil type constituting the aquifer system at this location.

Coarse-grained non-compressible sediments does not allow for remarkable compaction of the aquifer system.

Among the piezometric wells which are located along the margin of the subsidence (Fig. 10b), the maximum water level decline of more than 15 m during 10 years belongs to W5, showing the high subsidence rate as well. A small water level decline in other wells (Fig. 10b, c) produces the high subsidence rate. Therefore, it is observed that subsidence is a function of both the hydraulic head decrease and other important factors including the soil type of sediments composing the aquifer system.

For better interpretation of the subsidence in the Neyshabour plain, stratigraphic profiles at three exploration wells whose locations are depicted in Fig. 9 were investigated (see Fig. 11). It should be noted that these exploration wells are the only ones available in the study area. All of the three exploration wells are located in the margin of the subsidence area. According to Fig. 11, fine-grained sediments including clay and silt is the most frequent constituent composing the aquifer system. An increase of effective stress caused by water level withdrawal leads to considerable compaction.

Another issue to be considered is a time delay between the water level decline and compaction of the aquifer. This happens because of the low vertical hydraulic conductivity of compressible sediments. In addition, other geology and hydrogeological parameters including specific storage coefficient, permeability and

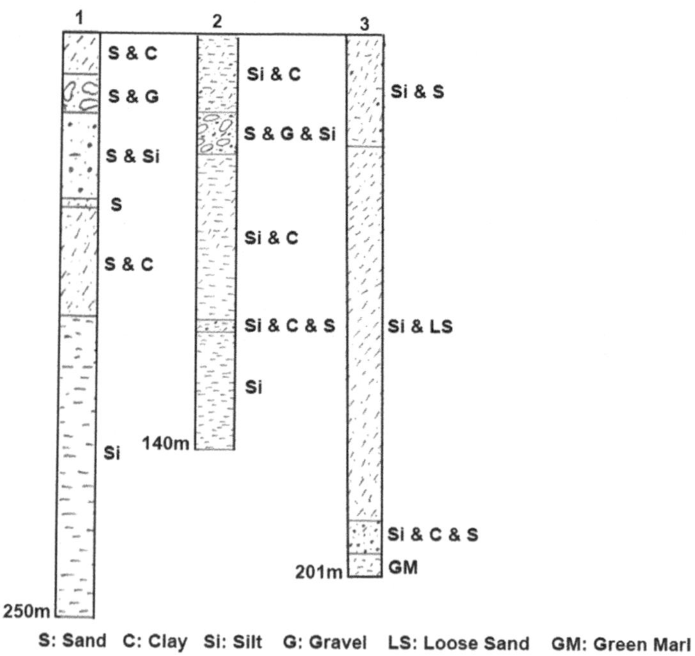

Fig. 11 Stratigraphic profiles at the exploration wells within the study area

thickness and depth of the underlying interbeds will determine the correlation between water level decline and subsidence happening in an area.

3.3 Case Study 3: Shahriar Plain

The southwestern part of the Tehran basin, Shahriar plain, has experienced significant land subsidence caused mainly by groundwater extraction from pumping wells. An annual water level decline of 40 cm makes the effective stress of the aquifer system increase resulting into compaction. In addition to the damage to buildings and structures, large fractures hav been produced on the ground as consequences of subsidence. The subsidence in the area was first detected by leveling measurements made by NCC (Arabi et al. 2005). However, leveling measurements allow for identification of subsidence rate at leveling stations. Therefore, the spatial pattern of the subsidence cannot be recognized.

Among the various techniques available, radar interferometry provides precise deformation measurements at high spatial resolution. In two other case studies of Varamin and Neyshabour, though the area is covered by the agricultural fields, the decorrelation effects were insignificant. Hence, the conventional interferometry could be easily applied to measure the deformation. However, in Shahriar plain, the conventional interferometry fails to capture the subsidence signal due to the poor spatial and temporal sampling of the data available. Large spatial and temporal baselines cause high decorrelation making unwrapping results inaccurate.

Persistent scatterer interferometry as a proper method to address the decorrelation problem has been employed in Shahriar plain. As discussed in Sect. 2.2, StaMPS algorithm is the most appropriate method to select the PS pixels in the cultivated lands. However, when the deformation rate is too high compared to the temporal sampling of the data, it underestimates the displacement due to the Nyquist sampling criterion applied in the temporal unwrapping step. The combined method of conventional and StaMPS was developed in order to mitigate the likelihood of aliasing.

There are 22 ENVISAT ASAR images available in the study area spanning between 2003 and 2008. A radar image acquired in 24 December 2004 was selected as a single master to maximize the stack coherence. Figure 12 shows the acquisition geometry of the radar data.

After single-master interferograms generation, PS pixels are selected based on the amplitude and phase stability analysis of StaMPS. 9 pairs of images with the small spatial and temporal baselines were selected to process coherent interferograms. A deformation model representing the linear component of the subsidence was estimated based on simple stacking presented in Sect. 2.1. Roughly estimated deformation model was subtracted modulo-2π from each wrapped interferogram. StaMPS unwrapping process was then applied on the residual interferometric phase. If the main part of the deformation consists of the linear component, which is the case in the study area, the residual phase can be correctly unwrapped based on

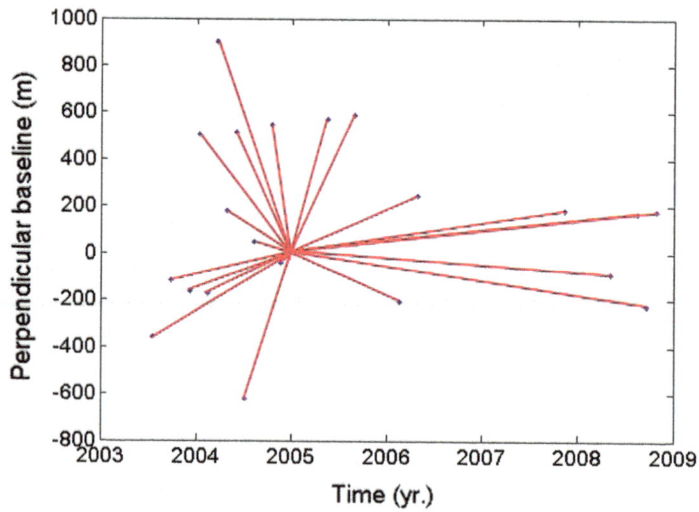

Fig. 12 Radar data available in the study area, temporal versus spatial baseline

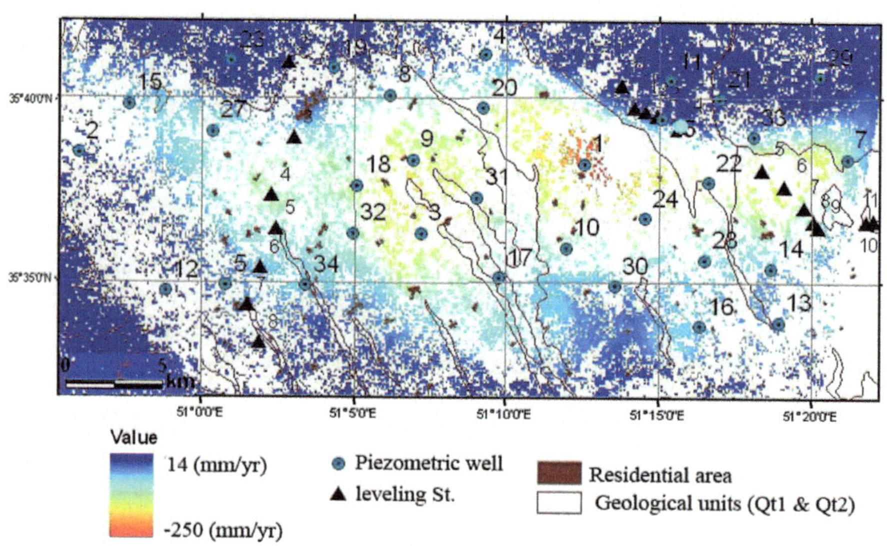

Fig. 13 Mean displacement velocity map of Shahriar plain

Nyquist sampling criterion. The linear deformation term was finally added back to the unwrapped residual phase.

The mean subsidence velocity map was calculated using the time series analysis results (Fig. 13). Maximum subsidence rate along the line-of-sight (LOS) was estimated as 25 cm/yr.

Groundwater level information shows a decline of 9 m in 20 years. Moreover, the ability of the aquifer system to yield water has dramatically decreased due to insignificant recharge (Shemshaki et al. 2005). Subsidence time series in Shahriar plain has been investigated at points including the piezometric wells. The selected wells are located in various parts of the study area: (i) outside, (ii) in the margin and (iii) in the middle of the subsidence area. The deformation evolutions at these points were compared to the water level fluctuations (Fig. 14).

Piezometric wells of W19 and W23 are located outside the deformation and their deformation time series as well as water level fluctuations are illustrated in Fig. 14a, e. There is no deformation observed in these wells though the groundwater level has significantly declined. One of the important factors affecting the compaction is the existence of find-grained compressible sediments. Therefore, for better interpretations, the geological boreholes logs for the piezometric wells illustrating different soil types at depth were employed. The geological borehole logs at these wells mainly consist of coarse-grained non-compressible sediments. Hence, despite the water level declines, no significant compaction occurs at both piezometric wells.

The second group of piezometric wells, i.e. W17 and W16, was located in the margin of the subsidence. The rate of water level decline over 10 years is 5 and 30 m in W17 and W16, respectively (Fig. 14b, f). The subsidence rate at W17 is higher than that at W17 as well. More gravel and sand as coarse-grained sediments is observed in W17 than in W16. Therefore, high water level decline produces little compaction.

W3 and W9 are two piezometric wells which are located in the middle of the subsidence. The subsidence rate at W3 and W9 is nearly the same (Fig. 14c, g). In W9 the water level dropped periodically while in W3 it has dramatically declined since 2002. According to the geological borehole logs, the sediments at W3 is mainly composed of fine-grained sediments while there is a small percentage of coarse-grained sediment exists in W9.

Due to the low correlation between water level decline and subsidence rate at some piezometric wells, it can concluded that other geology and hydrogeological factors play important role in controlling the subsidence occurrence. To show this, two data mining methods including Multi-Layer Perceptron (MLP) neural network as well as Support Vector Regression (SVR) were applied to model the subsidence in Shahriar plain using different geology and hydrogeological information.

3.3.1 Subsidence Modelling Based on Data Mining Methods

An expert system using two data mining models, namely Support Vector Regression (SVR) and Multi-Layer Perceptron (MLP) neural network was developed for predicting the subsidence rate (SR). The developed MLP and SVR models took into account 6 factors (input variables) including frequency of fine-grained sediments, thickness of fine-grained sediments, groundwater depth, amount of water level decline, transmissivity and storage coefficient on subsidence rate (output variable). Using the results previously obtained, 14,661 different input-output patterns were extracted. 70 and 30% of these data (10,263 and 4398 data sets,

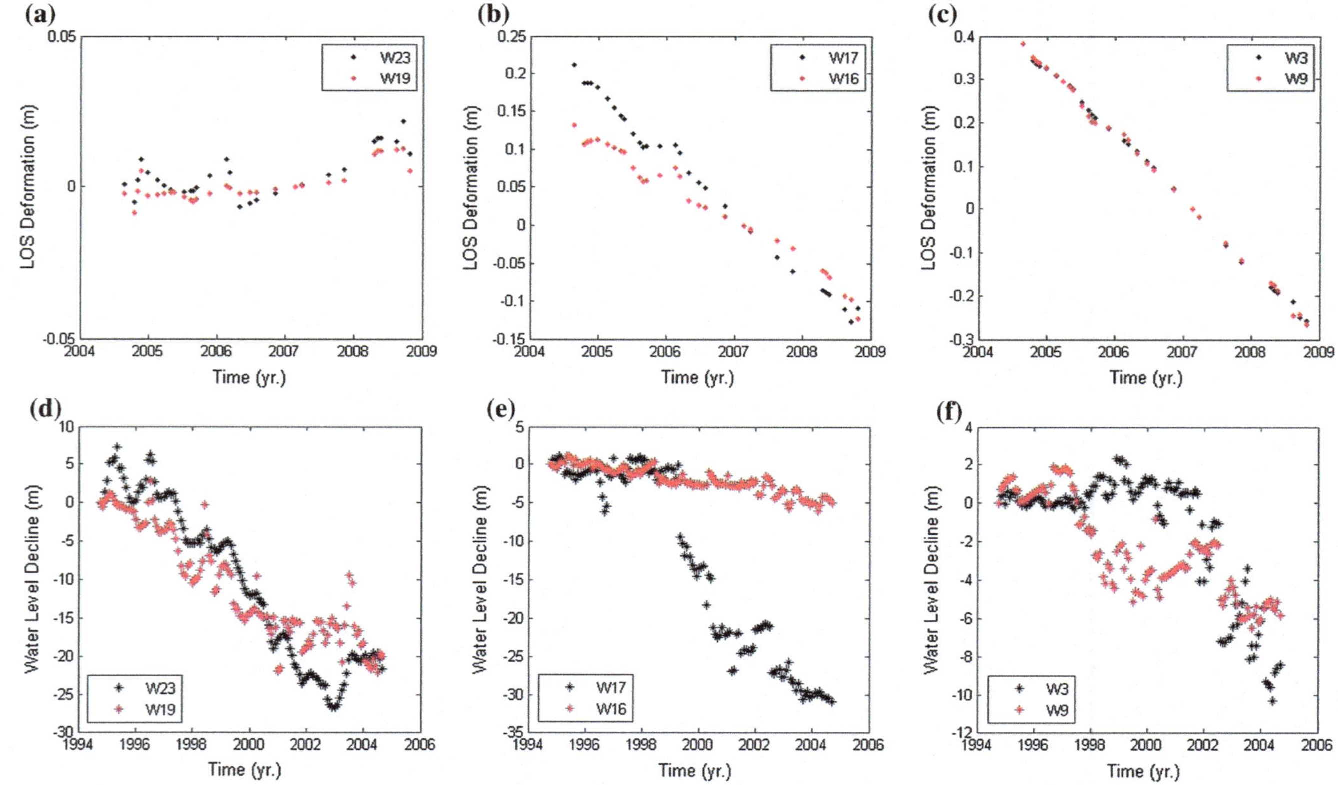

Fig. 14 Subsidence time series at the location of piezometric wells which are **a** outside (W23 and W19), **c** in the margin (W17 and W16) and **e** in the middle of the subsidence area (W3 and W9). **b**, **d** and **f** Water level fluctuations of the same wells

respectively) were utilized for training and testing (validating) the data mining models, respectively.

To verify the performance of the SVR and MLP data mining models for subsidence rate (SR) estimation, six different statistical error indices, namely Scatter Index (SI), Root Mean Square Error (RMSE), Correlation Coefficient (CC), Nash–Sutcliffe (NS) and Root Mean Relative Error (RMRE) are used in the validation process stage of the models:

$$Bias = \frac{1}{n}\sum_{i=1}^{n}(sr_i - sr_i^*) \tag{3}$$

$$SI = \frac{\sqrt{\sum_{i=1}^{n}((sr_i^* - SR^*) - (sr_i - SR))^2}}{SR} \tag{4}$$

$$RMSE = \sqrt{\frac{1}{n}\sum_{i=1}^{n}(sr_i - sr_i^*)^2} \tag{5}$$

$$CC = \frac{\sum_{i=1}^{n} sr_i \times sr_i^*}{\sqrt{\sum_{i=1}^{n} sr_i^2 \sum_{i=1}^{n} sr_i^{*2}}} \tag{6}$$

$$NS = 1 - \frac{\sum_{i=1}^{n}(sr_i - sr_i^*)^2}{\sum_{i=1}^{n}(sr_i - SR_i)^2} \tag{7}$$

$$RMRE = \sqrt{\frac{1}{n}\sum_{i=1}^{n}\left|\frac{sr_i - sr_i^*}{sr_i^*}\right|} \tag{8}$$

where sr_i^* and sr_i are respectively ith estimated and real subsidence rate (SR) and n is the number of data set. Also SR^* and SR denote the average estimated and observed SR, respectively.

The Nash–Sutcliffe (NS) coefficient is determined for both data mining models to ensure that the overfitting is not occurred (Table 1). The values of Nash–Sutcliffe coefficient are calculated for both the training and validating stages. The closer the Nash–Sutcliffe coefficient to one, the better the accuracy of the developed data mining models. According to the Table 1, the Nash–Sutcliffe coefficient is 0.991 and 0.993 for MLP and SVR models in training stage, respectively. The SVR model with more Nash–Sutcliffe coefficients in both training and validation stages indicates that the SVR is more accurate for subsidence rate prediction.

The values of six different statistical error indices (Eq. 3–8), calculated for the SVR and MLP models are presented in Table 2. The comparison of the statistical error indices makes the judgments to be made on each model's performance easier. Based on the obtained results, in terms of the Bias error, the MLP model is better than SVR in both training and validation stage. But SVR model outperforms MLP

Table 1 Nash–Sutcliffe coefficient for SVR and MLP models in subsidence rate prediction

Stage	Data mining models	
	SVR	MLP
Train	0.993	0.991
Validation	0.991	0.989

Table 2 Comparison of the SVR and MLP models performance in subsidence rate prediction using the training and validation data set

Statistical error indices		Stage			
		Train		Test	
	Data mining models	SVR	MLP	SVR	MLP
Bias	Value	0.17	0.06	0.54	0.26
	Rank	2	1	2	1
SI	Value	1.01	0.09	0.09	0.10
	Rank	1	2	1	2
RMSE	Value	4.58	5.09	5.24	5.68
	Rank	1	2	1	2
CC	Value	0.9983	0.9979	0.9979	0.9975
	Rank	1	2	1	2
NS	Value	0.993	0.991	0.991	0.989
	Rank	1	2	1	2
Summation of rankings		6	9	6	9

in other statistical error indices. Regarding the SI, RMSE, CC and NS error indices the SVR method outperforms the MLP model. Therefore, it can be concluded that SVR model is more precise than MLP model.

Scatter plots of estimated versus real SR for the SVR and MLP models (Figs. 15 and 16) confirm the fact that the SVR model outperforms the MLP model in predicting subsidence rate in both training and validation stages.

To make this comparison more visually, bar charts to mark the performance of the SVR and MLP data mining models with respect to the different statistical error indices is presented in Fig. 17.

The samples are categorized in three different classes based on the subsidence rate; Class 1, 2 and 3 include samples with the $SR < 80$, $80 < SR < 160$ and $SR > 160$ *mm/yr*, respectively for further assessments. To compare more precisely the SVR with MLP data mining models Fig. 18 shows six bar charts representing statistical error indices including Bias, SI, RMSE, CC, NS and RMRE. As an example, Fig. 18c compares the RMSE for both the SVR and MLP models in all the subsidence rate categories and the heights of bars are lower than that for the SVR model in all classes when compared to the MLP model, and thus the former has a greater accuracy in its estimation of subsidence rate. Figure 18d, e also clearly shows that the SVR model performs better than MLP model in terms of CC and NS,

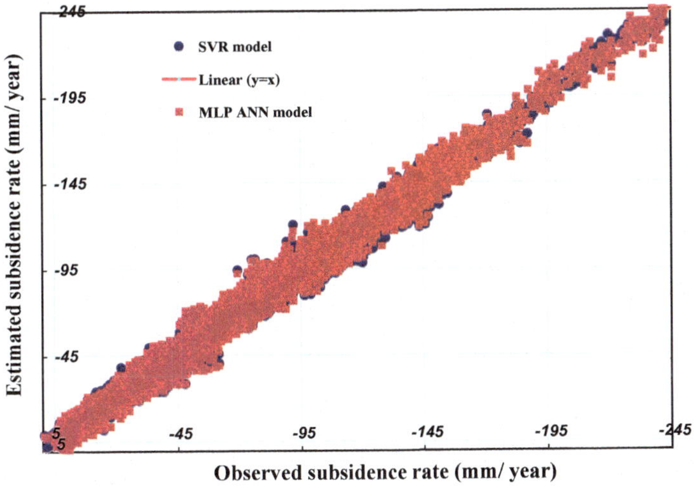

Fig. 15 Scatter plots comparing the performances of SVR and MLP data mining models for estimation SR in training stage

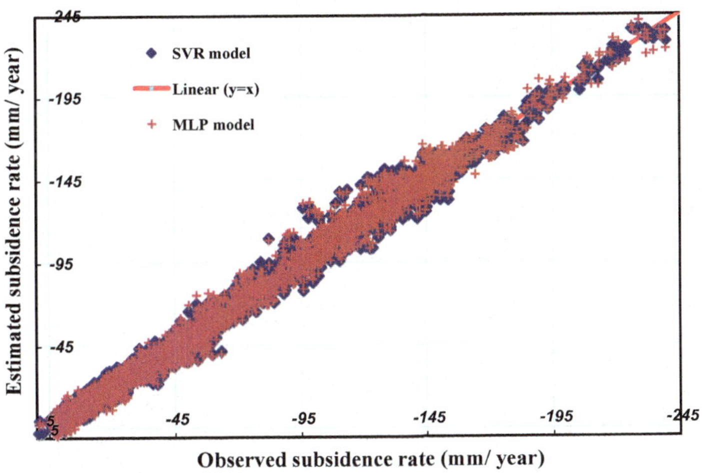

Fig. 16 Scatter plots comparing the performances of SVR and MLP data mining models for estimation SR in validation stage

respectively. In summary, the accuracy of subsidence rate estimation of the SVR model is greater than MLP when all the error indices are taken into account.

For making a final decision, the values of statistical error indices calculated for the SVR and MLP models for three different categories of SR are presented in Table 3. In addition, in the row below the obtained statistical errors, the ranks are determined, where the first rank belongs to the one with the best performance. The

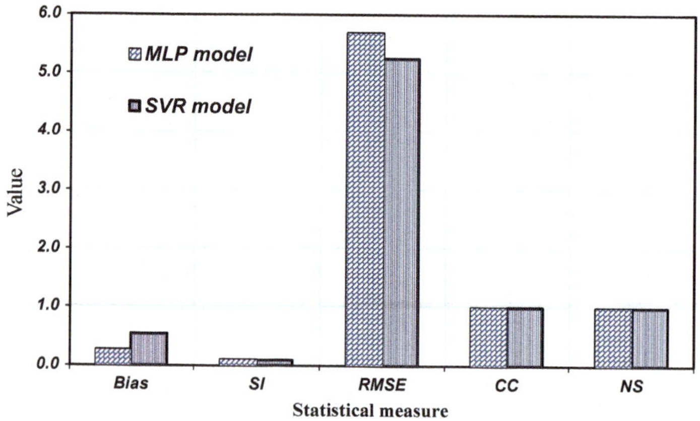

Fig. 17 Comparison of the results of the SVR and MLP models using the validation data set

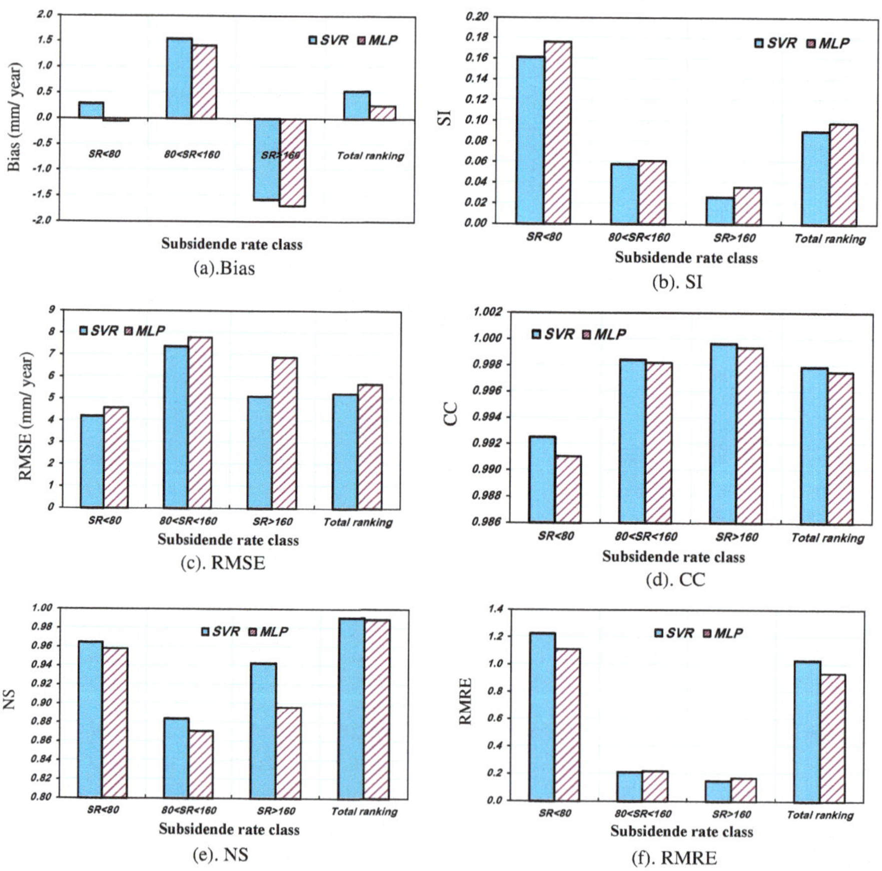

Fig. 18 Comparison of the performance of the SVR and MLP data mining models in subsidence rate estimation using the validation data set

Table 3 Performance comparison of the SVR and MLP data mining models for subsidence rate ratio (SR) estimation in validation stage

Statistical error indices	Subsidence rate class		Data mining models	
			SVR	MLP
Bias (mm/year)	SR < 80	Value	0.29	−0.05
		Rank	2	1
	80 < SR < 160	Value	1.54	1.42
		Rank	2	1
	SR > 160	Value	−1.57	−1.69
		Rank	1	2
	Total[a]	Value	0.54	0.26
		Rank	2	1
SI[b]	SR < 80	Value	0.16	0.18
		Rank	1	2
	80 < SR < 160	Value	0.058	0.061
		Rank	1	2
	SR > 160	Value	0.03	0.04
		Rank	1	2
	Total[a]	Value	0.09	0.10
		Rank	1	2
RMSE (mm/year)	SR < 80	Value	4.2	4.6
		Rank	1	2
	80 < SR < 160	Value	7.4	7.8
		Rank	1	2
	SR > 160	Value	5.1	6.9
		Rank	1	2
	Total[a]	Value	5.2	5.7
		Rank	1	2
CC[b]	SR < 80	Value	0.993	0.991
		Rank	1	2
	80 < SR < 160	Value	0.9984	0.9982
		Rank	1	2
	SR > 160	Value	1.000	0.999
		Rank	1	2
	Total[a]	Value	0.998	0.997
		Rank	1	2
NS[b]	SR < 80	Value	0.965	0.958
		Rank	1	2
	80 < SR < 160	Value	0.884	0.871
		Rank	1	2
	SR > 160	Value	0.943	0.897
		Rank	1	2

(continued)

Table 3 (continued)

Statistical error indices	Subsidence rate class		Data mining models	
			SVR	MLP
	Total[a]		0.991	0.989
		Rank	1	2
RMRE[b]	SR < 80	Value	1.22	1.11
		Rank	2	1
	80 < SR < 160	Value	0.21	0.22
		Rank	1	2
	SR > 160	Value	0.15	0.17
		Rank	1	2
	Total[a]	Value	1.03	0.94
		Rank	2	1
Summation of rankings	SR < 80		8	10
	80 < SR < 160		7	11
	SR > 160		6	12
	Total		8	10

[a]All data in the validation data set are utilized
[b]These error indices are dimensionless

comparison of the statistical error indices in the class of total makes the judgments to be made on each model's performance easier. To this regard, one can say that in terms of the Bias and RMRE, the MLP is better than SVR model. To draw on a fair judgment, other statistical error indices should be analyzed as well. Regarding the SI, RMSE, CC, and NS statistical error indices the SVR data mining model outperforms the MLP model.

To make a general assessment, the ranks are summed for each category of subsidence rate as well as the total class where the lower the rank, the better performance for the model is expected. For all the statistical error indices (Summation of rankings), the SVR model (with a total score of 8) has a superior performance compared to MLP model (with a total score of 10) in all categories of subsidence rate (last row in Table 3).

High performance of data mining methods in modelling the Shahriar subsidence shows the significant dependence of the subsidence on geology and hydrogeology characteristics of the aquifer system. The available hydrogeology information of the Tehran basin includes transmissivity, storage coefficient, frequency and thickness of fine-grained sediments, groundwater depth, and amount of water level decline of the Tehran aquifer system. Hydrogeology properties of the aquifer system were used as input variables of the model while the subsidence rate is taken as the model output. The subsidence rate could be precisely modelled using the developed data mining models. These models can be further applied to estimate the subsidence rate in pixels in which the interferometry technique cannot measure the deformation due to some reasons including insufficient correlation. Subsidence rate prediction in non-PS pixels is considered as future work.

4 Conclusion

SAR interferometry technique has shown its ability to study the land subsidence as a result of over-exploitation of groundwater at high spatial resolution. The accuracy of Interferometry has evaluated in several studies using GPS and leveling measurements (e.g. Dehghani et al. 2009a, 2013). Interferometry is able to measure the surface deformation with the accuracy of better than sub-centimeter. Two different interferometry approaches have been presented in this chapter: conventional and persistent scatterer interferometry. Conventional interferometry can efficiently be used in case where the spatial and temporal decorrelation is neglected. Time series analysis using the coherent interferogram was applied to study the temporal evolution of the subsidence in Varamin plain. Long-term as well as short-term behavior was monitored. The mean subsidence velocity map was also generated utilizing the time series analysis results. When the number of coherent interferograms are insufficient to produce a network, interferogram stacking can be employed to calculate the mean displacement velocity map. However, when the area is covered by remarkable amount of vegetation, the conventional interferometry is not able to measure the deformation due to decorrelation effect. Persistent scatterer interferometry as a proper method to address the decorrelation problem can be applied to monitor the subsidence. If the subsidence rate is so high to violate the Nyquist sampling criterion, the subsidence rate will be underestimated. In this study, a combined method of conventional and persistent scatterer interferometry was proposed to measure the high deformation rate in Shahriar plain.

The results obtained from all the interferometry approaches can be easily compared with the groundwater level information measured at the piezometric wells. This was done in Shahriar and Neyshabour plains. High correlation between subsidence signal and water level information was observed in Neyshabour plain. However, other important factors rather than water level fluctuations affect the subsidence occurrence in Shahriar plain. One of these factors is the geological type constituting the aquifer system. The existence of fine-grained deposits in the aquifer system causes high compaction when the groundwater level drops. Other geotechnical information are required to model the compaction of the aquifer system. Interferometry results as a valuable information source can be efficiently applied to determine the hydrogeological parameters of the aquifer system.

In order to show the dependencies of the subsidence phenomenon in Shahriar plain, two data mining models, i.e. MLP and SVR, were applied in order to model the subsidence rate based on the geology and hydrogeology properties of the aquifer system. The modeling results demonstrate the high performance of data mining models; however, SVR can outperform MLP for estimation of subsidence rate.

Acknowledgements We are grateful to European Space Agency (ESA) for providing ENVISAT ASAR data. We also would like to acknowledge the Geological Survey of Iran and Water Management Organization for providing the hydrogeological information.

References

Arabi S, Montazerian AR, Maleki E, Talebi A (2005) Study of land subsidence in south-west of Tehran. J Surv 69:14–24

Berardino P, Fornaro G, Lanari R, Sansosti E (2002) A new algorithm for surface deformation monitoring based on small baseline differential SAR interferograms. IEEE Trans Geosci Rem Sens 40:2375–2383. https://doi.org/10.1109/TGRS.2002.803792

Dehghani M, ValadanZoej MJ, Entezam I, Mansourian A, Saatchi S (2009a) InSAR monitoring of progressive land subsidence in Neyshabour, northeast Iran. Geophys J Int 178:47–56. https://doi.org/10.1111/j.1365-246X.2009.04135.x

Dehghani M, ValadanZoej MJ, Saatchi S, Biggs J, Parsons B, Wright T (2009b) Radar interferometry time series analysis of mashhad subsidence. J Indian Soc Remote Sens 37:147–156

Dehghani M, Valadanzoej MJ, Entezam I, Saatchi S, Shemshaki A (2010) Interferometric measurements of ground surface subsidence induced by overexploitation of groundwater. J Appl Remote Sens. 4, 041864. https://doi.org/10.1117/1.3527999

Dehghani M, Valadan Zoej MJ, Hooper A, Hanssen RF, Entezam I, Saatchi S (2013) Hybrid conventional and Persistent Scatterer SAR interferometry for land subsidence monitoring in the Tehran Basin. Iran ISPRS J Photogramm Remote Sens 79:157–170. https://doi.org/10.1016/j.isprsjprs.2013.02.012

Ferretti A, Prati C, Rocca F (2001) Permanent scatterers in SAR interferometry. IEEE Trans Geosci Remote Sens 39:8–20. https://doi.org/10.1109/36.898661

Funning GJ, Parsons B, Wright TJ, Jackson JA (2005) Surface displacements and source parameters of the 2003 Bam (Iran) earthquake from Envisat advanced synthetic aperture radar imagery. J Geophys Res 110:B09406. https://doi.org/10.1029/2004JB003338

Galloway DL, Hudnut KW, Ingebritsen SE, Phillips SP, Peltzer G, Rogez F, Rosen PA (1998) Detection of aquifer system compaction and land subsidence using interferometric synthetic aperture radar, Antelope valley, Mojave Desert, California. Water Resour Res 34:2573–2585. https://doi.org/10.1029/98WR01285

Galloway DL, Jones DR, Ingebritsen SE (1999) Land subsidence in the United States. US Geological Survey Circular 1182:175

Hooper A (2010) A statistical-cost approach to unwrapping the phase of InSAR time series. Fringe 2009 Workshop. http://home.utad.pt/~jjsousa/PARTILHA/Fringes2009/papers/p1_26hoop.pdf

Hooper A, Segall P, Zebker A (2007) Persistent scatterer interferometric synthetic aperture radar for crustal deformation analysis, with application to Volcan Alcedo. J geophys Res, Galapagos. https://doi.org/10.1029/2006JB004763

Lanari R, Lundgren P, Manzo M and Casu F (2004) Satellite radar interferometry time series analysis of surface deformation for Los Angeles, California, Geophys Res Lett 31. https://doi.org/10.1029/2004gl021294

Li Z, Bethel J (2008) Image coregistration in SAR interferometry. www.isprs.org/proceedings/XXXVII/congress/1_pdf/72.pdf

Lundgren P, Usai S, Sansosti E, Lanari R, Tesauro M, Fornaro G, Berardino P (2001) Modeling surface deformation observed with SAR interferometry at Campi Flegrei Caldera. J geophys Res 106:19 355–19 367. https://doi.org/10.1029/2001jb000194

Poland JF, Davis GH (1969) Land subsidence due to withdrawal of fluids. Rev Eng Geol 2:187–269

Schmidt DA, Burgmann R (2003) Time-dependent land uplift and subsidence in the Santa Clara valley, California, from a large interferometric synthetic aperture radar data set. J geophys Res. https://doi.org/10.1029/2002jb002267

Shemshaki A, Blourchi MJ, Ansari F (2005) Earth subsidence review at Tehran plain-Shahriar first report. http://gsi.ir/General/Lang_en/Page_27/GroupId_01-01/TypeId_All/Start_20/Action_ListView/WebsiteId_13/3.html. Accessed Aug 2005

Tolman CF, Poland JF (1940) Ground-water, salt-water infiltration, and ground-surface recession in Santa Clara Valley, Santa Clara County, California. Am Geophys Union Trans 21:23–24. https://doi.org/10.1029/TR021i001p00023

Printed by Printforce, the Netherlands